Progress in Probability and Statistics
Vol. 8

Edited by
Peter Huber
Murray Rosenblatt

Birkhäuser
Boston · Basel · Stuttgart

Philippe Bougerol
Jean Lacroix

Products of Random Matrices with Applications to Schrödinger Operators

1985

Birkhäuser
Boston · Basel · Stuttgart

Editors

Philippe Bougerol
Université Paris 7
UER de Mathématiques
2 Place Jussieu
F–75005 Paris/France

Jean Lacroix
Université de Paris XIII
Département de Mathématiques
F–93430 Villetaneuse/France

Library of Congress Cataloging in Publication Data

Bougerol, Philippe, 1951–
Products of random matrices with applications to Schrödinger operators.

(Progress in probability and statistics ; v. 8)
Bibliography: p.
1. Stochastic matrices. 2. Schrödinger operator.
I. Lacroix, Jean, 1944– . II. Title.
III. Series.
QA184.B68 1985 512'.5 85–15695

ISBN 978-1-4684-9174-6 ISBN 978-1-4684-9172-2 (eBook)
DOI 10.1007/978-1-4684-9172-2

CIP-Kurztitelaufnahme der Deutschen Bibliothek

Bougerol, Philippe:
Products of random matrices with applications to Schrödinger operators / Philippe Bougerol ; Jean Lacroix. – Boston ; Basel ; Stuttgart : Birkhäuser, 1985.
(Progress in probability and statistics ; Vol. 8)

NE: Lacroix, Jean:; GT

CONTENTS

PART B : "RANDOM SCHRÖDINGER OPERATORS"

PREFACE

This book presents two closely related series of lectures.

Part A, due to P. Bougerol, is an introduction to the works of Furstenberg, Guivarc'h, Le Page and Raugi on products of random matrices. Only invertible independent identically distributed random matrices satisfying an irreducibility condition are considered. The purpose is to prove in detail the analogues of the classical limit theorem (e.g. law of large numbers, central limit theorem). This part is based on a course given at the University of Paris 7 in 1983.

Part B, due to J. Lacroix, deals with the spectral theory of random Schrödinger operators, where the products of random matrices play a crucial role. It presents a rigorous and unified treatment of the main known results in the one-dimensional discrete case. Since we are aware that some readers are mainly interested in Schrödinger operators, any notion or result needed from part A is clearly restated (but of course not proved again !).

The book is self-contained and should be accessible to readers with minimal background. Since we feel that these topics deserve a large audience we have "tried" to write it English. It is not sure that we have succeeded and we beg the indulgence of the native speaker.

It is our pleasure to thank Yves Guivarc'h, François Ledrappier, Emile Le Page, Albert Raugi, José de Sam Lazaro and Bernard Souillard for enlightening conversations concerning the material presented here and for their encouragements.

Finally we thank Joëlle Navarron and Catherine Simon for their typing.

Paris, May 1985.

PART A

LIMIT THEOREMS FOR PRODUCTS OF RANDOM MATRICES

INTRODUCTION

This part is devoted to limit theorems for products of i.i.d. invertible random matrices. The subject matter, initiated by Bellman, was fully developed by Furstenberg, Guivarc'h, Kesten, Le Page and Raugi. This text is intented to serve as an introduction to their work.

We have chosen to keep the level as elementary as possible. This had sometimes led us to write lengthy proofs when shorter ones are available and to omit some important topics. On the other hand the text is self-contained and should be accessible to readers familiar with probability theory as usually developed at graduate level. In particular, no prior knowledge of group theory is assumed.

Let us roughly describe our general line of approach.

We consider a sequence $Y_1, Y_2, \dots$ of invertible random matrices of order d which are independent and identically distributed. We set $S_o = Id$, $S_1 = Y_1, \dots, S_n = Y_n \dots Y_2 Y_1$.

Our purpose is to prove that under an irreducibility condition, the sequence

$$\operatorname{Log} \|S_n x\| \quad ; \quad n = 1, 2, \dots$$

satisfies, for any $x \neq 0$ in $\mathbb{R}^d$, analogues of the classical limit theorems for sums of i.i.d. random variables (e.g. law of large numbers, central limit theorem, ...).

The strategy we shall adopt is based on the following observation.

For any $x \neq 0$ in $\mathbb{R}^d$, let $\bar{x}$ denote its direction (i.e. the class of x in the projective space $P(\mathbb{R}^d)$). Consider the function $F : Gl(d, \mathbb{R}) \times P(\mathbb{R}^d) \longrightarrow \mathbb{R}$ defined by

$$F(Y,\overline{x}) = \text{Log} \frac{\|Yx\|}{\|x\|} .$$

Then

$$\begin{aligned}
\text{Log} \frac{\|S_n x\|}{\|x\|} &= \sum_{i=o}^{n-1} \text{Log} \frac{\|S_{i+1} x\|}{\|S_i x\|} \\
&= \sum_{i=o}^{n-1} \text{Log} \left\| Y_{i+1} \frac{S_i x}{\|S_i x\|} \right\| \\
&= \sum_{i=o}^{n-1} F(Y_{i+1}, \overline{S_i x}) .
\end{aligned}$$

Therefore $\text{Log} \|S_n x\|$ can be written as a simple functional of the Markov chain $(Y_{i+1}, \overline{S_i x})$ on $Gl(d,\mathbb{R}) \times P(\mathbb{R}^d)$. Taking into account our irreducibility assumption, the first conclusion we can draw is the law of large numbers. Namely, there exists a real γ_1 , called the upper Lyapunov exponent, such that for any $x \neq 0$

$$\lim_{n\to\infty} \frac{1}{n} \text{Log} \|S_n x\| = \gamma_1 \qquad , \quad \text{a.s.} \quad .$$

The other results require a detailed study of the Markov chain $(Y_{i+1}, \overline{S_i x})$. Its essential properties will be derived from the fact that, loosely speaking, the random matrices S_n have asymptocally a contracting action on the directions. To be more precise, let δ be the natural angular distance on $P(\mathbb{R}^d)$

$$\delta(\overline{x},\overline{y}) = |\sin\{\text{angle}(\overline{x},\overline{y})\}| .$$

We shall prove, following Guivarc'h and Raugi [34], that under fairly weak assumptions,

$$\limsup_{n\to\infty} \frac{1}{n} \text{Log } \delta(\overline{S_n x}, \overline{S_n y}) < 0 \qquad \text{a.s.} \qquad (1)$$

for any fixed $\overline{x}$ and $\overline{y}$ in $P(\mathbb{R}^d)$.

This property is of primary importance. For instance it yields easily that the Markov chain $(Y_{i+1}, \overline{S_i x})$ has a unique invariant probability measure. More crucially we shall see, following Le Page [49] , that (1) is the cornerstone on which rest the proofs of the central limit theorem, the estimate for large deviations and many further results. We might say that, in some sense, this inequality replaces Doeblin's condition under which the precise limit theorems for functionalsof Markov chains are usually available.

Let $\gamma_1 \geqq \gamma_2 \geqq \cdots \geqq \gamma_d$ be the Lyapunov exponents associated with the sequence (Y_n). Since

$$\limsup_{n\to\infty} \frac{1}{n} \operatorname{Log} \delta(\overline{S_n x}, \overline{S_n y}) \leqq \gamma_2 - \gamma_1$$

our first task is to show that $\gamma_1 > \gamma_2$.

The main purpose of the first three chapters is to derive the inequality $\gamma_1 > \gamma_2$ from a careful analysis of the qualitative behaviour of the Markov chain $(\overline{S_n x})$.

In Chapter I we introduce the basic properties of the upper Lyapunov exponent γ_1.

In Chapter II we restrict ourselves to 2×2 matrices. For pedagogical purpose we first develop in this simple setting the general argument leading to a proof of this inequality. Most of the results which are needed in part B can already be found here.

Chapter III treats the case of matrices of arbitrary order.

In Chapter IV we digress from the main line. We apply the preceding results to the study of all the Lyapunov exponents. We also briefly present the link between these exponents and the boundary theory of Furstenberg.

In Chapter V we derive the main limit theorem on $\operatorname{Log} \|S_n x\|$ from the inequality $\gamma_1 > \gamma_2$. Further results, such as limit theorems for the coefficients of the matrices S_n, are proved in Chapter VI.

All the main results proved in this text come from Furstenberg [21], Guivarc'h and Raugi [34] and Le Page [49]. We hope that our exposition entices the reader to go back to these profound original works.

Despite its importance, we have chosen not to consider Osseledec's theorem. The reason is twofold. Firstly we have tried to keep the prerequisites to a minimum and to give a self-contained account of the subject. Secondly Ledrappier has already given in [46] a beautiful treatment of the applications of this theorem to products of i.i.d. random matrices. We felt there was nothing to gain by a reproduction of this material. In our opinion, the reader who wants to have a full picture of random products has to read Ledrappier's monograph. It will then be an easy and useful exercise to him to check how some of our proofs can be shortened by making use of his results.

We have not considered positive matrices (except in some exercises). We understand that Joel Cohen is writing a book on this subject.

For the sake of simplicity of notation we have restricted ourselves to matrices with real entries. But all the statements are also true in the complex case (the proofs carry over immediately to this case by replacing everywhere $\mathbb{R}$ by $\mathbb{C}$) .

Some chapters contain "complements" sections. They develop some additional material which is not used elsewhere and the reader can skip them. Some of their proofs are only outlined and may be of a more advanced level.

CHAPTER I

THE UPPER LYAPUNOV EXPONENT

In this chapter we define the upper Lyapunov exponent γ which gives the exponential rate of growth of the norm of products of independent identically distributed (i.i.d.) random matrices. In order to prove the analogue of the law of large numbers we develop some basic results on G-spaces which will often be used in the sequel.

I.1. Notation

We shall write $M(d,\mathbb{R})$ for the set of $d \times d$ matrices with real entries. $Gl(d,\mathbb{R})$ is the set of invertible elements of $M(d,\mathbb{R})$, $Sl(d,\mathbb{R})$ the set of matrices with determinant one. If M is in $M(d,\mathbb{R})$ and x in $\mathbb{R}^d$, Mx will be the vector equal to the product of the matrix M by the column vector x. The transpose of M is M^*.

In general the results we present do not depend on a particular norm chosen on $\mathbb{R}^d$ or $M(d,\mathbb{R})$. For convenience we shall use, unless the contrary is explicitely stated, for $x \in \mathbb{R}^d$ and $M \in M(d,\mathbb{R})$,

$$\|x\| = \{\langle x,x\rangle\}^{1/2} = \{\sum_{i=1}^{d} x_i^2\}^{1/2}$$

$$\|M\| = \operatorname{Sup}\{\|Mx\| \; ; \; x \in \mathbb{R}^d, \|x\| = 1\} .$$

Similar definitions hold for matrices with complex entries.

It will be understood that the random variables are defined on a probability space $(\Omega, \mathcal{A}, \mathbb{P})$, $\mathbb{E}(X)$ will be the expectation of a random variable X, $\mathbb{E}(X) = \int X \, d\mathbb{P}$, and if $\mathcal{F}$ is a sub σ-algebra of $\mathcal{A}$, $\mathbb{E}^{\mathcal{F}}(X)$ will be the conditional expectation with respect to $\mathcal{F}$.

A topological semigroup is a topological set T on which an associative product is defined, the map $(M_1, M_2) \longrightarrow M_1 M_2$ from T^2 to T being continuous. For instance $M(d, \mathbb{R})$ is a topological semigroup.

A topological group is a group G with a topology for which the mappings $(g,h) \longrightarrow gh$ and $g \longrightarrow g^{-1}$ are continuous.

I.2. The upper Lyapunov exponent

Consider a sequence $\{Y_n, n \geq 1\}$ of i.i.d. random matrices with common distribution μ. Let $S_n = Y_n \ldots Y_1$. For the chosen norm,

$$\|S_n\| \leq \|Y_n\| \; \|Y_{n-1}\| \ldots \|Y_1\| .$$

Therefore if $\mathbb{E}(\mathrm{Log}^+ \|Y_1\|)$ is finite (we write f^+ for $\mathrm{Sup}(f,0)$) $\mathrm{Log}^+ \|S_n\|$ is integrable. If $n,p \geq 1$ then

$$\begin{aligned} \mathbb{E}(\mathrm{Log}\, \|S_{n+p}\|) &\leq \mathbb{E}(\mathrm{Log}\, \|Y_{n+p} \ldots Y_{n+1}\|) + \mathbb{E}(\mathrm{Log}\, \|Y_n \ldots Y_1\|) \\ &\leq \mathbb{E}(\mathrm{Log}\, \|S_p\|) + \mathbb{E}(\mathrm{Log}\, \|S_n\|) . \end{aligned}$$

Thus the sequence $\{\mathbb{E}(\mathrm{Log}\, \|S_n\|), \; n \geq 1\}$ is subadditive and $\frac{1}{n} \mathbb{E}(\mathrm{Log}\, \|S_n\|)$ converges to $\inf_{m \geq 1} \frac{1}{m} \mathbb{E}(\mathrm{Log}\, \|S_m\|)$ in $\mathbb{R} \cup \{-\infty\}$, (see exercise 5.2).

DEFINITION 2.1. *If* $\mathbb{E}(\mathrm{Log}^+ \|Y_1\|) < \infty$, *the upper Lyapunov exponent associated with* μ *is the element* γ *of* $\mathbb{R} \cup \{-\infty\}$ *defined by*

$$\gamma = \lim_{n \to \infty} \frac{1}{n} \mathbb{E}(\mathrm{Log}\, \|Y_n \ldots Y_1\|) .$$

Since all norms on the finite dimensional vector space $M(d, \mathbb{R})$ are equivalent, γ is independent of the chosen norm. We shall prove (when Y_1 is invertible) the stronger result that almost surely (a.s.)

$$\gamma = \lim_{n \to \infty} \frac{1}{n} \mathrm{Log}\, \|Y_n \ldots Y_1\| .$$

Although this turns out to be an easy consequence of Kingman's subadditive ergodic theorem (see e.g. Ledrappier [46]) we shall follow the original proof of Furstenberg and Kesten [25]. The reason is twofold. On one hand it is a good illustration of the techniques we shall use. On the other hand it gives us some information that will be needed later. Let us first describe two examples.

Example 2.2. Suppose that each matrix Y_n is a diagonal matrix $\operatorname{diag}(a_1(n),\dots,a_d(n))$. This means that the entries $(Y_n)_{i,j}$ satisfy

$$(Y_n)_{i,j} = \begin{cases} 0 & \text{if} \quad i \neq j \\ a_i(n) & \text{if} \quad i = j \,. \end{cases}$$

It is clear that $\gamma = \operatorname{Sup}_i \mathbb{E}(\operatorname{Log}|a_i(1)|)$ and

$$\gamma = \lim \frac{1}{n} \operatorname{Log} \|Y_n \dots Y_1\|$$

by the usual law of large numbers.

Example 2.3. The preceding example is degenerated. Far more interesting is the following due to Cohen and Newman [15].

Let $Y_1, Y_2, \dots$ be a sequence of i.i.d. random elements in $Gl(d,\mathbb{R})$ such that the distribution of $\|Y_1 \frac{x}{\|x\|}\|$ does not depend upon $x \neq 0$ in $\mathbb{R}^d$. For each vector $x \neq 0$, the real random variables

$$U_1 = \operatorname{Log} \|Y_1 \frac{x}{\|x\|}\|, \dots, U_n = \operatorname{Log} \|Y_n \frac{Y_{n-1}\dots Y_1 x}{\|Y_{n-1}\dots Y_1 x\|}\|, \dots$$

are independent and identically distributed : for, if $\mathcal{F}_n$ is the σ-algebra generated by $U_1,\dots,U_n$, for any Borel bounded function f on $\mathbb{R}$

$$\mathbb{E}^{\mathcal{F}_n}\{f(U_{n+1})\} = \mathbb{E}\{f(\operatorname{Log}\|Y_{n+1}y\|)\} = \mathbb{E}\{f(U_1)\}$$

where $y = \dfrac{Y_n \dots Y_1 x}{\|Y_n \dots Y_1 x\|}$. Remark that

$$\operatorname{Log} \|Y_n \dots Y_1 x\| = U_n + U_{n-1} + \dots + U_1 + \operatorname{Log} \|x\| \,.$$

Therefore we may apply the classical limit theorems for sums of i.i.d. random variables in order to obtain results on the sequence

$\{\mathrm{Log}\,\|Y_n\cdots Y_1x\|\}$. If $\mathbb{E}(\mathrm{Log}^+\|Y_1\|) < \infty$, $\mathbb{E}(U_1^+) < \infty$ and, by the usual law of large numbers

$$\lim_{n\to\infty} \frac{1}{n} \mathrm{Log}\,\|Y_n\cdots Y_1 x\| = \mathbb{E}(U_1) \qquad \text{a.s.}\ .$$

Note that $\mathbb{E}(U_1)$ is independent of x. Thus using

$$\|Y_n\cdots Y_1\| \leq d \sup_{1\leq i\leq d} \|Y_n\cdots Y_1 e_i\| \leq d\,\|Y_n\cdots Y_1\|$$

where $\{e_i,\ 1 \leq i \leq d\}$ is the canonical basis of $\mathbb{R}^d$, we obtain

$$\lim_{n\to\infty} \frac{1}{n} \mathrm{Log}\,\|Y_n\cdots Y_1\| = \mathbb{E}(U_1) = \mathbb{E}\left(\mathrm{Log}\,\left\|Y_1\frac{x}{\|x\|}\right\|\right)$$

This yields $\gamma = \mathbb{E}(U_1)$. In general there will not exist such a simple expression of γ in terms of Y_1.

I.3. Cocycles

The proof of Furstenberg and Kesten's result requires a simple lemma on cocycles. Since it will be often needed in this book we present it in a general setting.

DEFINITION 3.1. *Let* G *be a topological semigroup. We say that* G *acts on the topological space* B *if one can associate continuously to each* (g,x) *in* $G \times B$ *an element* $g\cdot x$ *of* B *in such a way that*

$$(g_1g_2)\cdot x = g_1\cdot(g_2\cdot x), \qquad \textit{for } g_1,g_2 \in G,\ x \in B.$$

If G *is a group with unit* e *and if moreover* $e\cdot x = x$ *for any* $x \in B$, *we say that* B *is a* G*-space.*

DEFINITION 3.2. *Let* G *be a topological semigroup acting on a space* B. *A continuous map* $\sigma : G \times B \longrightarrow \mathbb{R}$ *is said to be an additive cocycle if*

$$\sigma(g_1g_2,x) = \sigma(g_1,g_2\cdot x) + \sigma(g_2,x)$$

for any $g_1,g_2 \in G$, $x \in B$.

When it does not depend upon x, σ is just an homomorphism from G to $\mathbb{R}$. For $G = Gl(d,\mathbb{R})$ and S the unit sphere $\{x \in \mathbb{R}^d ;\ \|x\| = 1\}$, S is a G-space if we set

$$M\cdot x = \frac{Mx}{\|Mx\|}, \quad M \in Gl(d,\mathbb{R}), \quad x \in S,$$

and $\sigma(M,x) = \mathrm{Log}\,\|Mx\|$ is an additive cocycle we shall frequently meet. If G is the set of positive matrices in $M(d,\mathbb{R})$ (i.e. matrices M whose entries satisfy $M_{i,j} > 0$), G is a topological semigroup acting on S^+, the set of x in S with nonnegative coordinates. With the same formula G acts on S^+ and σ defines an additive cocycle on $G \times S^+$.

<u>DEFINITION 3.3.</u> *Let* G *be a topological semigroup acting on* B. *If* μ *(resp.* ν*) is a probabitity measure on* G *(resp. on* B*) we denote by* $\mu \overset{\bullet}{*} \nu$ *the distribution on* B *which satisfies*

$$\int_B f(x)d(\mu \overset{\bullet}{*} \nu)(x) = \int_G \int_B f(g\cdot x)d\mu(g)d\nu(x) \qquad (1)$$

for all bounded Borel function f *on* B.

We say that ν *is* μ*-invariant if* $\mu \overset{\bullet}{*} \nu = \nu$.

If μ and λ are two probability measures on G, the ordinary convolution product $\mu * \lambda$ is defined by

$$\int fd(\mu * \lambda) = \iint f(gh)d\mu(g)d\lambda(h)$$

for f on G. Note that $(\mu * \lambda) \overset{\bullet}{*} \nu = \mu \overset{\bullet}{*} (\lambda \overset{\bullet}{*} \nu)$ if ν is a distribution on B.

<u>PROPOSITION 3.4.</u> *Let* G *be a topological semigroup acting on the space* B *and* σ *an additive cocycle on* $G \times B$. *Consider a sequence* $\{Y_n, n \geq 1\}$ *of independent random elements of* G *with common distribution* μ.

If ν *is a* μ*-invariant distribution on* B *such that* $\iint |\sigma(g,x)|d\mu(g)d\nu(x)$ *is finite, then*

$$\lim_{n\to\infty} \frac{1}{n}\,\sigma(Y_n(\omega) \ldots Y_1(\omega), x)$$

exists for $\mathbb{P}\otimes\nu$*-almost all* (ω,x) *and in* $L^1(\mathbb{P}\otimes\nu)$.

<u>Proof</u> : Without loss of generality we may suppose that $\Omega = \{\omega = (Y_n)_{n\geq 1} ; Y_n \in G\}$, A being the σ-algebra generated by the coordinate maps Y_n and $\mathbb{P}$ the probability measure for which the Y_i's are independent with distribution μ.

Let $E = \Omega \times B$ and $Q = \mathbb{P} \otimes \nu$. Define $\theta : E \longrightarrow E$ by

$$\theta(((Y_n)_{n\geq 1},x)) = ((Y_{n+1})_{n\geq 1}, Y_1\cdot x) .$$

If A_o is a Borel subset in B and $A_1, A_2, \ldots, A_n$ Borel subsets in G,

$$Q\{(\omega,x) \in \Omega \times B \; ; \; \theta((\omega,x)) \in (A_1 \times A_2 \times \ldots \times A_n \times \ldots) \times A_o\}$$

$$= (\mathbb{P} \otimes \nu) \{Y_2 \in A_1, \ldots, Y_{n+1} \in A_n, \ldots, Y_1\cdot x \in A_o\}$$

$$= (\mathbb{P} \otimes \nu) \{Y_1 \in A_1, \ldots, Y_n \in A_n, \ldots, x \in A_o\}$$

$$= Q\{(A_1 \times A_2 \times \ldots \times A_n \times \ldots) \times A_o\}$$

This proves that θ preserves Q .

For each integer p ,

$$\theta^p(((Y_n)_{n\geq 1},x)) = ((Y_{n+p})_{n\geq 1}, (Y_p \ldots Y_1)\cdot x)$$

so that if we set $F(\omega,x) = \sigma(Y_1,x)$ for $\omega = (Y_n)_{n\geq 1}$, $x \in B$, then

$$\begin{aligned}\sigma(Y_n \ldots Y_1,x) &= \sigma(Y_n,(Y_{n-1}\ldots Y_1)\cdot x) + \sigma(Y_{n-1}\ldots Y_1,x)\\ &= \sum_{p=1}^{n} \sigma(Y_p, (Y_{p-1}\ldots Y_1)\cdot x)\\ &= \sum_{p=1}^{n} F(\theta^p((\omega,x))) .\end{aligned}$$

Therefore Birkhoff's ergodic theorem (see e.g. Breiman [10] , Theorem 6.2.1) implies that

$$\frac{1}{n}\,\sigma(Y_n \ldots Y_1,x) = \frac{1}{n} \sum_{p=1}^{n} F(\theta^p(\omega,x)))$$

converges $\mathbb{P} \otimes \nu$ a.s. and in L^1 .

If B is separable and compact there exist μ-invariant distributions on B : indeed if μ^n denotes the nth convolution power of μ ($\mu^1 = \mu$, $\mu^2 = \mu * \mu$, ...) , we have

LEMMA 3.5. *Let* B *be a compact separable* G*-space and* μ *be a distribution on* G *. For any distribution* m *on* B *, each limit point of* $\{\frac{1}{n} \sum_{i=1}^{n} \mu^i \overset{\bullet}{*} m ,\; n \geq 1\}$ *is a* μ*-invariant distribution on* B .

Proof : Let $\nu_n = \frac{1}{n} \sum_{i=1}^{n} \mu^i \overset{\bullet}{*} m$. Since B is separable and compact there exists a weakly convergent subsequence of (ν_n) and its limit, say ν , is a probability measure. For any n ,

$$\mu \overset{\bullet}{*} \nu_n = \frac{1}{n} \sum_{i=1}^{n} \mu^{i+1} \overset{\bullet}{*} m$$

$$= \frac{1}{n} \sum_{i=1}^{n} \mu^{i} \overset{\bullet}{*} m + \frac{1}{n} \{\mu^{n+1} \overset{\bullet}{*} m - \mu \overset{\bullet}{*} m\}$$

$$= \nu_n + \frac{1}{n} \{\mu^{n+1} \overset{\bullet}{*} m - \mu \overset{\bullet}{*} m\}$$

so that, letting $n \longrightarrow \infty$ along a subsequence, $\mu \overset{\bullet}{*} \nu = \nu$.

I.4. The Theorem of Furstenberg and Kesten

We can now prove the following result of [25] .

THEOREM 4.1. *Let* $Y_1, Y_2, \ldots$ *be i.i.d. matrices in* $Gl(d,\mathbb{R})$. *If* $\mathbb{E}(\mathrm{Log}^+ \|Y_1\|) < \infty$, *then with probability one*

$$\lim_{n\to\infty} \frac{1}{n} \mathrm{Log} \, \|Y_n \ldots Y_1\| = \gamma .$$

Proof : Fix an integer m and for every integer n , write $n = pm+q$, $p \geq 0$, $0 \leq q < m$. We have

$$\frac{1}{n} \mathrm{Log} \, \|Y_n \ldots Y_1\| \leq \frac{1}{n} \mathrm{Log} \, \|Y_n \ldots Y_{pm+1}\| + \frac{1}{n} \mathrm{Log} \, \|Y_{pm} \ldots Y_1\|$$

$$\leq \frac{1}{n} \sum_{i=pm+1}^{n} \mathrm{Log} \, \|Y_i\| + \frac{1}{n} \sum_{j=o}^{p-1} \mathrm{Log} \, \|Y_{(j+1)m} \ldots Y_{jm+1}\|$$

Therefore by the usual law of large numbers

$$\overline{\lim_{n\to\infty}} \frac{1}{n} \mathrm{Log} \, \|Y_n \ldots Y_1\| \leq \frac{1}{m} \mathbb{E}(\mathrm{Log} \|Y_m \ldots Y_1\|) \qquad \text{a.s.}$$

and

$$\overline{\lim_{n\to\infty}} \frac{1}{n} \mathrm{Log} \, \|Y_n \ldots Y_1\| \leq \lim_{m\to\infty} \frac{1}{m} \mathbb{E}(\mathrm{Log} \|Y_m \ldots Y_1\|) = \gamma . \qquad (2)$$

This proves the theorem if $\gamma = -\infty$.

Suppose now that γ is finite. Let B be the compact space of all $d \times d$ matrices whose norm is equal to one. If for Y in $Gl(d,\mathbb{R})$ and M in B we set

$$Y \cdot M = \frac{YM}{\|YM\|}$$

B becomes a $Gl(d,\mathbb{R})$-space. Let m be the probability measure on B

equal to the Dirac mass at the identity matrix Id and μ the distribution of Y_1. For any integer n consider the distribution ν_n on B defined by

$$\nu_n = \frac{1}{n} \sum_{i=1}^{n} \mu^i \dot{*} m .$$

If $\nu_{n(i)}$ is a convergent subsequence of $(\nu_n)_{n \geq 1}$ its limit, say ν, is a μ-invariant distribution (see lemma 3.5).

Define $\sigma : Gl(d,\mathbb{R}) \times B \longrightarrow \mathbb{R}$ by

$$\sigma(Y,M) = \text{Log} \|YM\| , \quad Y \in Gl(d,\mathbb{R}) , \quad M \in B .$$

For any Y_1, Y_2 in $Gl(d,\mathbb{R})$ and M in B

$$\begin{aligned} \sigma(Y_1 Y_2, M) &= \text{Log} \|Y_1 Y_2 M\| \\ &= \text{Log} \left\|Y_1 \frac{Y_2 M}{\|Y_2 M\|}\right\| + \text{Log} \|Y_2 M\| \\ &= \sigma(Y_1, Y_2 \cdot M) + \sigma(Y_2, M) \end{aligned}$$

so that σ is an additive cocycle.

We have $\sigma(Y,M) \leq \text{Log} \|Y\| \, \|M\| \leq \text{Log} \|Y\|$

and, for $\sigma^+ = \text{Sup}(\sigma, 0)$,

$$\sigma^+(Y,M) \leq \text{Log}^+ \|Y\| .$$

We have assumed that $\mathbb{E}(\text{Log}^+\|Y_1\|) < \infty$, $\int \sigma^+(Y,M) d\mu(Y)$ is thus a bounded continuous function on B and

$$\lim_{i \to \infty} \iint \sigma^+(Y,M) d\mu(Y) d\nu_{n(i)}(M) = \iint \sigma^+(Y,M) d\mu(Y) d\nu(M) .$$

On the other hand, for $\sigma^- = \text{Sup}(-\sigma, 0)$ and any $k > 0$,

$$\lim_{i \to \infty} \iint \inf\{k, \sigma^-(Y,M)\} d\mu(Y) d\nu_{n(i)}(M) = \iint \inf\{k, \sigma^-(Y,M)\} d\mu(Y) d\nu(M)$$

thus

$$\varliminf_{i \to \infty} \iint \sigma^-(Y,M) d\mu(Y) d\nu_{n(i)}(M) \geq \iint \sigma^-(Y,M) d\mu(Y) d\nu(M)$$

and

$$\varlimsup_{i \to \infty} \iint \sigma(Y,M) d\mu(Y) d\nu_{n(i)}(M) \leq \iint \sigma(Y,M) d\mu(Y) d\nu(M) . \tag{3}$$

But

$$\iint \sigma(Y,M) d\mu(Y) d\nu_n(M) = \frac{1}{n} \sum_{i=1}^{n} \iint \sigma(Y,M) d\mu(Y) d(\mu^i \dot{*} m)(M)$$

$$= \frac{1}{n} \sum_{i=1}^{n} \mathbb{E}\,(\sigma(Y_{i+1},(Y_i \ldots Y_1)\cdot \mathrm{Id}))$$

$$= \frac{1}{n} \sum_{i=1}^{n} \mathbb{E}\,(\sigma(Y_n \ldots Y_1, \mathrm{Id}))$$

$$= \frac{1}{n}\, \mathbb{E}(\mathrm{Log}\, \|Y_n \ldots Y_1\|)\,.$$

Therefore

$$\lim_{i\to\infty} \iint \sigma(Y,M)d\mu(Y)d\nu_{n(i)}(M) = \gamma$$

and, by (3) ,

$$\iint \sigma(Y,M)d\mu(Y)d\nu(M) \geqq \gamma \quad . \tag{4}$$

This proves that σ is actually in $L^1(\mu \otimes \nu)$. By Proposition 3.5 for some $f : \Omega \times B \longrightarrow \mathbb{R}$, as $n \longrightarrow \infty$,

$$\frac{1}{n} \mathrm{Log}\, \|Y_n(\omega)\ldots Y_1(\omega)M\| = \frac{1}{n}\sigma(Y_n(\omega)\ldots Y_1(\omega),M) \longrightarrow f(\omega,M)$$

with $\mathbb{P} \otimes \nu$ probability one. With (3) we obtain

$$f(\omega,M) \leqq \liminf_{n\to\infty} \frac{1}{n} \mathrm{Log}\, \|Y_n(\omega)\ldots Y_1(\omega)\| \leqq \gamma$$

and, using (4) ,

$$\gamma \leqq \iint f(\omega,M)d\,\mathbb{P}(\omega)d\nu(M)\,.$$

So $f(\omega,M) = \gamma$ and almost surely

$$\gamma \leqq \liminf_{n\to\infty} \frac{1}{n} \mathrm{Log}\, \|Y_n \ldots Y_1\| \quad .$$

This, together with (2), proves the theorem.

I.5. Exercises

Exercise 5.1. Using the notations of I.1, prove that

(a) $\|M\|^2 = \|M^*\|^2$ = the largest eigenvalue of M^*M.

(b) If $|\det M| = 1$ then $\|M\| \geqq 1$, and $\|M\| = 1$ if and only M is orthogonal (i.e. $MM^* = \mathrm{Id.}$).

Exercise 5.2. Let $\{a_n, n \geqq 1\}$ be a sequence of real numbers such that for any integers $m,p \geqq 1$,

$$a_{m+p} \leqq a_m + a_p \,.$$

Writing $n = mp + q$, $p \geq 0$, $0 \leq q < m$, prove that for any $m \geq 1$

$$\overline{\lim_{n\to\infty}} \frac{a_n}{n} \leq \frac{a_m}{m} .$$

Deduce that

$$\lim_{n\to\infty} \frac{a_n}{n} = \inf_n \frac{a_m}{m} .$$

Exercise 5.3. [26], [35]. Let d_1, d_2 be two positive integers. Consider a sequence $\{Y_n, n \geq 1\}$ of i.i.d. random matrices which can be written as

$$Y_n = \begin{bmatrix} A_n & B_n \\ 0 & C_n \end{bmatrix}$$

where A_n (resp. C_n) is in $Gl(d_1, \mathbb{R})$ (resp. $Gl(d_2, \mathbb{R})$) and B_n is a random $d_1 \times d_2$ matrix. We suppose $\mathbb{E}(\mathrm{Log}^+ \|Y_1\|) < \infty$ and denote by γ, γ_1, γ_2 the upper Lyapunov exponents associated with (Y_n), (A_n), (C_n).

a. Prove that $\gamma_1 \leq \gamma$, $\gamma_2 \leq \gamma$.

b. Write $Y_n \ldots Y_1 = \begin{bmatrix} A'_n & B'_n \\ 0 & C'_n \end{bmatrix}$ and $Y_{2n} \ldots Y_{n+1} = \begin{bmatrix} A''_n & B''_n \\ 0 & C''_n \end{bmatrix}$

Prove that a.s. for each $\tau > 1$ and n large enough,

$$\|A'_n\| \leq \exp n\tau\gamma_1 , \quad \|A''_n\| \leq \exp n\tau\gamma_1 , \quad \|C'_n\| \leq \exp n\tau\gamma_2 ,$$
$$\|C'_n\| \leq \exp n\tau\gamma_2 , \quad \|B'_n\| \leq \exp n\tau\gamma , \quad \|B''_n\| \leq \exp n\tau\gamma .$$

Deduce from this that $\|Y_{2n} \ldots Y_1\| \leq 8 \exp\{\tau \,\mathrm{Sup}(\gamma+\gamma_1 , \gamma+\gamma_2)\}$.
Prove that $\gamma = \mathrm{Sup}(\gamma_1, \gamma_2)$.

Exercise 5.4. Let S be the unit sphere in $\mathbb{R}^d$ and m the normalized Lebesgue measure on S (dm gives the element of surface area and $m(S) = 1$).

a. Let Y be a matrix with $|\det Y| = 1$. Show that if Ym is the image of m under the map $\tau : S \longrightarrow S$ defined by $\tau(x) = \dfrac{Yx}{\|Yx\|}$, then

$$\frac{d(Ym)}{dm}(x) = \|Y^{-1}x\|^{-d}$$

(Hint : the measure ρ defined for each Borel set A of $\mathbb{R}^d$ by $\rho(A) = \int_0^\infty \int_S 1_A(rx) r^{d-1}\, dr\, dm(x)$ is a multiple of Lebesgue measure.

Hence it is invariant under multiplication by Y).

b. Suppose that Y is a random matrix with $|\det Y| = 1$ such that the distribution of $\|Yx\|$ does not depend on x in S (see Example 2.3). For $\gamma = \mathbb{E}(\mathrm{Log}\,\|Yx\|)$ verify that

$$\gamma = -\frac{1}{d}\iint \mathrm{Log}\,\frac{d(Y^{-1}m)}{dm}(x)\,d\mathbb{P}(Y)\,dm(x)\,.$$

Using Jensen inequality show that if $\mathbb{P}$ (Y is orthogonal) < 1, then $\gamma > 0$.

Exercise 5.5. Suppose that $\mathbb{E}(\mathrm{Log}^+\|Y_1\|)$ is finite. Show that $(\mathbb{E}\,(\mathrm{Log}^+\|Y_1^{-1}\|) < \infty) \Longrightarrow (\gamma > -\infty) \Longrightarrow (\mathbb{E}\,(\mathrm{Log}\,\|Y_1\|) > -\infty)$ but that the reverse implications are false.

CHAPTER II

MATRICES OF ORDER TWO

We present here the main arguments we shall use for studying products of random matrices of arbitrary order, but in the case of 2×2 matrices. Our main interest is not the theorems in themselves and actually shorter proofs are available when dealing with 2×2 matrices. We rather intend to explicit in this simple situation the general approach, valid for matrices of any order. This chapter is thus introductive. Although all the results are particular cases of more general statements to be proved later, we give them in full detail for the convenience of the reader who is only interested in matrices of order 2.

II.1. The Set-up

Consider a sequence $\{Y_n, n \geq 1\}$ of i.i.d. invertible random matrices and set $S_n = Y_n \dots Y_2 Y_1$. As explained in the introduction our first task is to show that under broad conditions, for any non zero vectors x and y, the random vectors $S_n x$ and $S_n y$ tend to "line up" in the same direction exponentially fast.

We say that two non zero vectors x, y in $\mathbb{R}^d$ have the same direction if for some $\lambda \in \mathbb{R}$, $x = \lambda y$. This defines an equivalence relation Γ on $\mathbb{R}^d - \{0\}$. Formally,

DEFINITION 1.1. *The set of directions in* $\mathbb{R}^d$ *is the projective space* $P(\mathbb{R}^d)$ *defined as the quotient space* $\mathbb{R}^d - \{0\}/\Gamma$.
For $x \in \mathbb{R}^d - \{0\}$, $\bar{x}$ *denotes its direction (i.e. its class in* $P(\mathbb{R}^d)$*).*
For M *in* $Gl(d,\mathbb{R})$ *we set* $M \cdot \bar{x} = \overline{Mx}$.

If x and y are unit vectors in $\mathbb{R}^d$ we define

$$\delta(\bar{x},\bar{y}) = \{1 - \langle x,y\rangle^2\}^{1/2} \quad .$$

It is the modulus of the sine of the angle between $\bar{x}$ and $\bar{y}$. From now on we suppose that $d = 2$. In this case, if $x = \binom{x_1}{x_2}$ and $y = \binom{y_1}{y_2}$

$$\delta(\bar{x},\bar{y}) = \frac{|x_1 y_2 - x_2 y_1|}{\|x\| \ \|y\|} \quad . \tag{1}$$

It is easily seen that this defines a distance on $P(\mathbb{R}^2)$ and we want to prove that $\delta(S_n \cdot \bar{x}, S_n \cdot \bar{y})$ converges to 0 exponentially fast as $n \longrightarrow \infty$. Note that if we replace each Y_i by $|\det Y_i|^{-1/2} Y_i$, $S_n \cdot \bar{x}$ is unchanged, we can thus suppose without loss of generality that $|\det Y_i| = 1$. In that case, by (1) ,

$$\delta(S_n \bar{x}, S_n \bar{y}) = \frac{\|x\| \ \|y\|}{\|S_n x\| \ \|S_n y\|} \delta(\bar{x},\bar{y}) \tag{2}$$

so we are looking for conditions ensuring that $\|S_n x\| \longrightarrow \infty$ exponentially fast, i.e. essentially that $\gamma > 0$, γ being the upper Lyapunov exponent.

This is the content of the now so-called "Furstenberg's theorem". The original proof of Furstenberg in [21] has important consequences and we strongly recommend the luminous account of it in Ledrappier [46]. But for $d > 2$ it doesn't help to prove that $\delta(S_n \cdot \bar{x}, S_n \cdot \bar{y})$ converges to 0. So we rather follow Guivarc'h and Raugi [34]. For $d = 2$ this route is longer than [21] but it is worth seeing it first in this simple situation. If we spend some time here we shall see that the general case $(d > 2)$ will be just a careful copy of this one.

Let us outline the general idea of the proof in [34]. We need the following.

NOTATION 1.2. *If* M *is in* $Gl(d,\mathbb{R})$ *and* m *is a distribution on* $P(\mathbb{R}^d)$ *we denote by* Mm *the probability measure on* $P(\mathbb{R}^d)$ *defined by*

$$\int f \, d(Mm) = \int f(M\cdot\overline{x}) \, dm(\overline{x}) ,$$

for any bounded Borel function f *on* $P(\mathbb{R}^d)$.

If $\gamma > 0$ geometric considerations (cf. section 7) show that if m is a continuous distribution on $P(\mathbb{R}^2)$, i.e. $m(\{\overline{x}\}) = 0$ for any $\overline{x}$ in $P(\mathbb{R}^2)$, then for almost all ω, the sequence $\{S_n^*(\omega)m,\ n \geq 1\}$ converges weakly to a Dirac measure $\delta_{\overline{Z}(\omega)}$ (recall that S_n^* is the transpose of S_n). Conversely, if this holds for a continuous distribution and if the distribution of $\overline{Z}$ is continuous, a simple deterministic result (see Proposition 3.1) shows that, for any $x \neq 0$ $\|S_n x\| \longrightarrow \infty$ with probability one, which in turn implies $\gamma > 0$.

So we are led to see if, for some continuous distributions m, $S_n^* m$ converges to a Dirac measure. The trick is that if we choose m as an invariant distribution under the common law of the Y_i^*'s, then $S_n^* m$ always converges weakly. If it happens that there is such a m which is continuous we will prove using the path indicated above that either $\gamma > 0$ or all the Y_n's are isometries. When m is not continuous then the sequence (Y_n) has particularly simple properties and the Lyapunov exponent, in this case, can be explicitely computed.

II.2. Two basic lemmas

We use the notations of I.3 . The following lemma due to Guivarc'h and Raugi [34] improves a result of Furstenberg [21].

LEMMA 2.1. *Let* G *be a topological semigroup acting on a* 2nd *countable locally compact space* B . *Consider a sequence* $\{X_n,\ n \geq 1\}$ *of independent random elements of* G *with a common distribution* μ *defined on* $(\Omega, A, \mathbb{P})$. *If* ν *is a* μ*-invariant distribution on* B *then for almost all* ω *there exists a probability measure* ν_ω *on* B *such that the sequence*

$$\{X_1(\omega)X_2(\omega) \ldots X_n(\omega)g\nu ,\quad n \geq 1\}$$

converges weakly to ν_ω *as* $n \longrightarrow \infty$ *for almost all* $g \in G$ *with respect to* $\lambda = \Sigma_{n=0}^{\infty} 2^{-n-1}\mu^n$.

Moreover for each bounded Borel function f *on* B

$$\int f \, d\nu = \mathbb{E}\{\int f \, d\nu_\omega\} .$$

Proof : Let f be a continuous bounded Borel real function on B and $F : G \longrightarrow \mathbb{R}$ be defined by

$$F(g) = \int_B f(g\cdot x) \, d\nu(x) , \qquad g \in G .$$

Set $M_n = X_1X_2\ldots X_n$. If F_n is the σ-algebra generated by $X_1,\ldots,X_n$, then

$$\begin{aligned}\mathbb{E}^{F_n}\{F(M_{n+1})\} &= \int f(M_ng) \, d\mu(g) \\ &= \iint f(M_ng\cdot x) \, d\mu(g) \, d\nu(x) \\ &= \int f(M_n\cdot x) \, d(\mu \overset{\bullet}{*} \nu)(x) \\ &= \int f(M_n\cdot x) \, d\nu(x) \\ &= F(M_n)\end{aligned}$$

(we have used the independence of the X_i's and the fact that ν is μ-invariant). This shows that $\{F(M_n), n \geq 1\}$ is a martingale. Being bounded, $F(M_n)$ converges a.s. to some Γf and

$$\mathbb{E}\{\Gamma f\} = \mathbb{E}\{F(M_1)\} = \int f \, d\nu .$$

We have for any positive integers k, r,

$$\begin{aligned}\mathbb{E}\{|F(M_{k+r})-F(M_k)|^2\} &= \mathbb{E}\{F(M_{k+r})^2\} + \mathbb{E}\{F(M_k)^2\} - 2\mathbb{E}\{F(M_{k+r})F(M_k)\} \\ &= \mathbb{E}\{F(M_{k+r})^2\} - \mathbb{E}\{F(M_k)^2\} .\end{aligned}$$

Hence, by cancellation, for any p ,

$$\begin{aligned}\sum_{k=1}^{p} \mathbb{E}\{|F(M_{k+r})-F(M_k)|^2\} &\leq \sum_{k=1}^{r} \mathbb{E}\{F(M_k)^2\} + \sum_{k=1}^{r} \mathbb{E}\{F(M_{k+r})^2\} \\ &\leq 2r \sup_x |F(x)|^2 .\end{aligned}$$

This yields

$$\sum_{k=1}^{\infty} \mathbb{E}\{\int |F(M_kg) - F(M_k)|^2 \, d\lambda(g)\} =$$

$$\sum_{k=1}^{\infty}\sum_{r=0}^{\infty} 2^{-r-1}\ \mathbb{E}\{\int |F(M_k g) - F(M_k)|^2\ d\mu^r(g)\} =$$
$$\sum_{k=1}^{\infty}\sum_{r=0}^{\infty} 2^{-r-1}\ \mathbb{E}\{|F(M_{k+r}) - F(M_k)|^2\} \leq \sum_{r=0}^{\infty} r\ 2^{-r} \sup_x |F(x)|^2 .$$

So $\sum_{k=1}^{\infty} |F(M_k g) - F(M_k)|^2$ is finite and $F(M_k g)$ converges to Γf, $\mathbb{P} \otimes \lambda$ almost surely.

Choose a dense sequence $\{f_q, q \geqq 1\}$ in the space $C_o(B)$ of continuous functions on B which vanish at infinity. There exists a measurable subset A of $\Omega \times B$ with $(\mathbb{P} \otimes \lambda)(A) = 1$ such that for $(\omega,g) \in A$,

$$\int f_q(M_n(\omega)g\cdot x)\ d\nu(x) \longrightarrow \Gamma f_q(\omega) .$$

If $\nu_{\omega,g}$ is a limit point of $\{M_n(\omega)g\nu,\ n \geqq 1\}$ for the weak topology,

$$\int f_q\ d\nu_{\omega,g} = \Gamma f_q(\omega) , \qquad \forall q \geqq 1 .$$

This shows that $\nu_{\omega,g}$ does not depend on g and is the limit of $M_n(\omega)g\nu$. We therefore denote it by ν_ω. Since

$$\int f_q\ d\nu = \mathbb{E}\{\Gamma f_q\}$$

for any bounded Borel f

$$\int f\ d\nu = \mathbb{E}\{\int f\ d\nu_\omega\}$$

and ν_ω must be a.s. a probability measure.

For proving that Lyapunov exponents (or difference of Lyapunov exponents) are positive we shall need the following.

LEMMA 2.2. *Let* B *be a* G*-space*, $(Y_n)_{n\geq 1}$ *a sequence of independent random elements of* G *with distribution* μ *and* σ *an additive cocycle on* $G \times B$. *Suppose that* ν *is a* μ*-invariant distribution on* B *such that*

(i) $$\iint \sigma^+(g,x)\ d\mu(g)\ d\nu(x) < \infty .$$

(ii) For $\mathbb{P} \otimes \nu$*-almost all* (ω,x), $\lim_{n\to\infty} \sigma(Y_n(\omega)\dots Y_1(\omega),x) = +\infty$.

Then σ *is in* $L^1(\mathbb{P} \otimes \nu)$ *and* $\iint \sigma(g,x)\ d\mu(g)\ d\nu(x) > 0$.

Using the arguments developed in the proof of Proposition I.3.4 this follows at once from the following lemma. Its proof is borrowed from Dekking [16].

LEMMA 2.3. *Let* (E, F, λ) *be a probability space and* $\theta : E \longrightarrow E$ *a measurable transformation which preserves* λ *. If* $f : E \longrightarrow \mathbb{R}$ *is so that* $\int f^+ \, d\lambda < \infty$ *and* $\lim_{n\to\infty} \sum_{i=1}^{n} f \circ \theta^i = \infty$ *almost everywhere then* $f \in L^1(d\lambda)$ *and* $\int f \, d\lambda > 0$.

Proof : By the Ergodic Theorem (see Exercise 2.6)

$$\lim_{n\to\infty} \frac{1}{n} \sum_{i=1}^{n} f \circ \theta^i = \mathbb{E}^{\mathcal{J}}(f) \qquad \text{a.s.}$$

where $\mathcal{J} = \{A \in F \; ; \; \lambda(A \,\Delta\, \theta^{-1}A) = 0\}$. Since $\sum_{i=1}^{n} f \circ \theta^i \longrightarrow +\infty$, $\mathbb{E}^{\mathcal{J}}(f) \geq 0$ so f is actually integrable.

If $\int f \, d\lambda = 0$ then

$$\frac{1}{n} \sum_{i=1}^{n} f \circ \theta^i \longrightarrow 0 \qquad \text{a.s.} \quad . \tag{3}$$

For each $\varepsilon > 0$ let $I_\varepsilon(t) = [t-\varepsilon, t+\varepsilon]$, $S_n(x) = \sum_{i=1}^{n} (f \circ \theta^i)(x)$ and, for m being the Lebesgue measure on $\mathbb{R}$,

$$R_n^\varepsilon(x) = m\{ \bigcup_{i=1}^{n} I_\varepsilon(S_i(x))\} \; .$$

By (3), for λ-almost all x and any $\delta > 0$ there exists $n_o = n_o(x,\delta)$ such that, for $k > n_o$, $|S_k(x)| \leq k\delta$. Hence, for all n , $R_n^\varepsilon(x) \leq R_{n_o}^\varepsilon(x) + 2n\delta$ and $\overline{\lim_{n\to\infty}} \frac{1}{n} R_n^\varepsilon(x) \leq 2\delta$ for all $\delta > 0$. Therefore $\frac{1}{n} R_n^\varepsilon(x) \longrightarrow 0$ a.s. and

$$\lim_{n\to\infty} \frac{1}{n} \mathbb{E}(R_n^\varepsilon) = 0 \quad . \tag{4}$$

Note that, since $S_n \circ \theta = S_{n+1} - S_1$

$$\begin{aligned} R_{n+1}^\varepsilon - R_n^\varepsilon \circ \theta &= m\{\bigcup_{i=1}^{n+1} I_\varepsilon(S_i)\} - m\{\bigcup_{i=1}^{n} I_\varepsilon(S_{i+1}-S_i)\} \\ &= m\{\bigcup_{i=1}^{n+1} I_\varepsilon(S_i)\} - m\{\bigcup_{i=2}^{n+1} I_\varepsilon(S_i)\} \; . \end{aligned}$$

It is plain that

$$R_{n+1}^\varepsilon - R_n^\varepsilon \circ \theta \geq 2\varepsilon \; 1_{\{|S_i - S_1| > 2\varepsilon, \; i = 2,\dots,n+1\}}$$

so $\mathbb{E}\{R^{\varepsilon}_{n+1}\} - \mathbb{R}\{R^{\varepsilon}_{n} \circ \theta\} \geq 2\varepsilon\,\lambda\{x \,;\, |S_i(x)-S_1(x)| > 2\varepsilon, \quad i = 2,\dots,n+1\}$

and, since θ preserves λ ,

$$\mathbb{E}\,\{R^{\varepsilon}_{n+1}\} - \mathbb{E}\,\{R^{\varepsilon}_{n}\} \geq 2\varepsilon \quad \lambda\{x \,;\, |S_i(x)| > 2\varepsilon, \; i = 1,\dots,n\} \;.$$

This implies that

$$\lim_{n\to\infty} \frac{1}{n}\, \mathbb{E}\,\{R^{\varepsilon}_{n}\} \geq 2\varepsilon\, \lambda\{x \,;\, |S_i(x)| > 2\varepsilon \;,\quad \forall i \geq 1\} \;.$$

By (4), for any $\varepsilon > 0$,

$$\lambda\{x \,;\, |S_i(x)| > \varepsilon \quad , \qquad \forall i \geq 1\} \;=\; 0 \;.$$

The processes $\{S_i,\ i \geq 1\}$ and $\{S_{i+p}-S_p,\ i \geq 1\}$ have the same distribution. Therefore for any integer $p > 0$ and $\varepsilon > 0$,

$$\lambda\{x \,;\, |S_{i+p}(x) - S_p(x)| > \varepsilon \;,\quad \forall i \geq 1\} = 0$$

and $$\lambda\{x \,;\, |S_{i+p}(x) - S_p(x)| > \varepsilon \;,\quad \forall i \geq 1 \;\text{ for some }\; p\} = 0 \;.$$

This contradicts the fact that $\lim S_n = \infty$ a.s. .

<u>Exercise 2.4.</u> Let T be a compact separable semigroup contained in a group and μ a probability measure on T whose support is T . Prove that if m is a distribution on T such that $\mu * m = m$, then $gm = m$ for any g in T . If H is the smallest closed group which contains T show that $gm = m$ for any g in H. Deduce from this that T is actually a group.

<u>Exercise 2.5.</u> Let $\{Y_n,\ n \geq 1\}$ be i.i.d. matrices of the form $Y_i = \begin{bmatrix} 1 & x_i \\ 0 & 1 \end{bmatrix}$. Prove that one can have

$$\lim_{n\to\infty} \|Y_n \dots Y_1\| = \infty \qquad \text{a.s.}$$

but

$$\lim_{n\to\infty} \frac{1}{n} \,\mathrm{Log}\, \|Y_n \dots Y_1\| = 0 \;.$$

<u>Exercise 2.6.</u> With the notations of Lemma 2.3 we want to prove that $\lim \frac{1}{n} \sum_{i=1}^{n} f \circ \theta^i = E^{\mathcal{J}}(f)$ under the sole condition $\int f^+ \, d\lambda < \infty$.

(i) For any $K < 0$ let $E_K = \{x \,;\, \mathbb{E}^{\mathcal{J}}(f)(x) > K\}$. Using that $E_K \in \mathcal{J}$ prove that $\lim \frac{1}{n} \sum_{i=1}^{n} f \circ \theta^i = \mathbb{E}^{\mathcal{J}}(f)$ on E_K .

(ii) Show that on $\{x \,;\, \mathbb{E}^{\mathcal{J}}(f)(x) = -\infty\}$, $\lim \frac{1}{n} \sum_{i=1}^{n} f \circ \theta^i = -\infty$.

(iii) Conclude.

II.3. Contraction properties

Consider a sequence $Y_1, Y_2, \ldots$ of i.i.d. random matrices of order 2 with $|\det Y_i| = 1$. We intend to show that under broad conditions, for any non zero x in $\mathbb{R}^2$, $\|Y_n \ldots Y_1 x\| \longrightarrow \infty$ with probability one as $n \longrightarrow \infty$. The following deterministic result makes it clear that we have to look at the action of $Y_1^* \ldots Y_n^*$ on continuous distributions on $P(\mathbb{R}^2)$.

PROPOSITION 3.1. *Let* $\{A_n, n \geq 1\}$ *be a sequence of* 2×2 *matrices with* $|\det A_n| = 1$. *Suppose that there exists a continuous distribution* m *on* $P(\mathbb{R}^2)$ *such that* $A_n m$ *converges weakly to a Dirac measure* $\delta_{\overline{z}}$. *Then*

$$\lim_{n\to\infty} \|A_n\| = \lim_{n\to\infty} \|A_n^*\| = \infty \tag{5}$$

and if z *is a unit vector in* $\mathbb{R}^2$ *with direction* $\overline{z}$, *then for any* x *in* $\mathbb{R}^2$

$$\lim_{n\to\infty} \frac{\|A_n^* x\|}{\|A_n^*\|} = |\langle x,z\rangle| \quad . \tag{6}$$

We first prove :

LEMMA 3.2. *Let* A *be a non zero (non necessarily invertible) matrix of order 2 and* m *be a continuous distribution on* $P(\mathbb{R}^2)$. *Then the equation*

$$\int f \, d(Am) = \int f(A\cdot\overline{x}) \, dm(\overline{x}) ,$$

valid for all bounded Borel functions, defines a probability mecsure Am *on* $P(\mathbb{R}^2)$.

If (A_n) *is a sequence of non zero matrices which converges to* A, $A_n m$ *converges weakly to* Am .

Proof of the lemma : For any x in $\mathbb{R}^2$ such that $Ax \neq 0$, $A\cdot\overline{x} = \overline{Ax}$ is well defined. Since $A \neq 0$ there exists at most one direction $\overline{y}$

such that if y is a vector with this direction, $Ay = 0$. Hence $A\cdot\bar{x}$ is defined for all $\bar{x}$ except $\bar{y}$ but $m(\{\bar{y}\}) = 0$ since m is continuous and therefore Am can be defined. If $A_n \longrightarrow A$ then for all $\bar{x}$ outside a countable set D, $A_n\cdot\bar{x}$ exists and converges to $A\cdot\bar{x}$. Since $m(D) = 0$, $A_n m \longrightarrow Am$.

Proof of the proposition : Suppose first that $\|A_n\|^{-1} A_n$ converges to some matrix A. By the lemma (note that $\|A\| = 1$, so $A \neq 0$)

$$Am = \delta_{\bar{z}} .$$

If $\det A \neq 0$, m is the Dirac measure at $A^{-1}\cdot\bar{z}$ and is not continuous. Hence $\det A = 0$ and

$$0 = |\det A| = \lim_{n\to\infty} \left|\frac{\det A_n}{\|A_n\|^2}\right| = \lim_{n\to\infty} \frac{1}{\|A_n\|^2}$$

proving (5).

Moreover the range of A is just a line and since $Am = \delta_{\bar{z}}$ the direction of this line is $\bar{z}$. Thus if $\|z\| = 1$ and $\{e_1, e_2\}$ is the canonical basis of $\mathbb{R}^2$, $Ae_1 = \pm \|Ae_1\| z$ and $Ae_2 = \pm \|Ae_2\| z$.
For any x,

$$\begin{aligned} \|A^*x\|^2 &= \langle A^*x , e_1\rangle^2 + \langle A^*x , e_2\rangle^2 \\ &= \langle x , Ae_1\rangle^2 + \langle x , Ae_2\rangle^2 \\ &= (\|Ae_1\|^2 + \|Ae_2\|^2) \langle x,z\rangle^2 . \end{aligned}$$

Using $\|A\| = 1$ one obtains $\|Ae_1\|^2 + \|Ae_2\|^2 = 1$ and

$$\lim_{n\to\infty} \frac{\|A_n^*x\|^2}{\|A_n^*\|^2} = \|A^*x\|^2 = \langle x,z\rangle^2 .$$

This is true for any convergent subsequence of $\|A_n\|^{-1} A_n$, so we have proved the proposition.

So we are led to show that $Y_1^* \ldots Y_n^* m$ converges to a Dirac measure at least for one continuous distribution m on $P(\mathbb{R}^2)$. Put $X_i = Y_i^*$ and let μ be the distribution of X_1. If ν is μ-invariant (i.e. $\mu \overset{\bullet}{*} \nu = \int M\nu \, d\mu(M) = \nu$) we already know that $X_1 \ldots X_n \nu$ converges by Lemma 2.1 . Actually we have

PROPOSITION 3.3. *Let* $X_1, X_2, \ldots$ *be independent* 2×2 *matrices with* $|\det X_i| = 1$ *and the same distribution* μ . *Suppose that there exists on* $P(\mathbb{R}^2)$ *a continuous* μ*-invariant distribution* ν . *Then, if the support of* μ *is not contained in a compact subgroup of* $Gl(2,\mathbb{R})$, *there exists with probability one a direction* $\overline{Z}(\omega)$ *such that* $X_1(\omega)\ldots X_n(\omega)\nu$ *converges weakly to* $\delta_{\overline{Z}(\omega)}$.

Moreover the distribution of $\overline{Z}$ *is* ν *and is the unique* μ*-invariant continuous distribution on* $P(\mathbb{R}^2)$.

Proof : Let $A_n = X_1 \ldots X_n$. From Lemma 2.1 we know that there exist probability measures ν_ω such that, for almost all ω ,

$$A_n(\omega)\nu \xrightarrow{w} \nu_\omega$$

and $A_n(\omega)M\nu \xrightarrow{w} \nu_\omega$ for μ-almost all M .

Fix such an ω . For any limit point $A(\omega)$ of $\|A_n(\omega)\|^{-1} A_n(\omega)$ we have, by Lemma 3.2

$$A(\omega)\nu = A(\omega)M\nu = \nu_\omega$$

for μ-almost all M .

Notice that $H = \{M \in Gl(2,\mathbb{R})\ ;\ M\nu = \nu\ ,\ |\det M| = 1\}$ is a compact subgroup. For otherwise there would exist M_n in H such that $\|M_n\| \longrightarrow \infty$ and $\|M_n\|^{-1} M_n$ converges to some matrix C . In that case $C\nu = \nu$ and

$$|\det C| = \lim \left|\det\left(\frac{M_n}{\|M_n\|}\right)\right| = \lim \frac{1}{\|M_n\|^2} = 0 ,$$

so C sends all $\mathbb{R}^2$ onto a line and $\nu = C\nu$ must be a Dirac measure, which is absurd.

If $A(\omega)$ is invertible then $M\nu = \nu$ for μ-almost all M and $\mu(H) = 1$ which contradicts our assumption. So $A(\omega)$ is a rank-one matrix. If $\overline{Z}(\omega)$ is the direction of its range, then since $A(\omega)\nu = \nu_\omega$, $\nu_\omega = \delta_{\overline{Z}(\omega)}$. This proves that $X_1(\omega)\ldots X_n(\omega)\nu \xrightarrow{w} \delta_{\overline{Z}(\omega)}$.

Note that $\overline{Z}$ does not depend on ν . Since by Lemma 2.1 for any Borel function f on $P(\mathbb{R}^2)$

$$\int f\, d\nu = \mathbb{E}\left\{\int f\, d\nu_\omega\right\} = \mathbb{E}\{f(\overline{Z})\}$$

the distribution of $\overline{Z}$ is ν and ν is unique.

Remark 3.4. We do not say that there exists a unique μ-invariant distribution on $P(\mathbb{R}^2)$, but only that there is at most one such distribution which is continuous.

Remark 3.5. If $\mathrm{Supp}(\mu)$ is contained in a compact group, the result does not hold. We have seen in this case that $M\nu = \nu$ for each $M \in \mathrm{Supp}(\mu)$ so $X_1 \dots X_n \nu = \nu$.

We now prove the following basic result :

THEOREM 3.6. *Let* $\{Y_n, n \geq 1\}$ *be a sequence of independent random matrices of order 2 with the same distribution* μ .
Suppose that :

(a) $|\det Y_i| = 1$ a.s. .

(b) The support of μ *is not contained in a compact subgroup of* $Gl(2,\mathbb{R})$.

(c) There exists a continuous μ^**-invariant distribution on* $P(\mathbb{R}^2)$ *where* μ^* *is the distribution of* Y_1^* .

Then (i) For each $\bar{x}, \bar{y}$ *in* $P(\mathbb{R}^2)$, *with probability one*

$$\lim_{n\to\infty} \delta(Y_n \dots Y_1 \cdot \bar{x} \,,\, Y_n \dots Y_1 \cdot \bar{y}) = 0 .$$

(ii) If $\mathbb{E}(\mathrm{Log}^+ \|Y_1\|) < \infty$ *, there exists* $\gamma > 0$ *such that for each* $x \neq 0$,

$$\lim_{n\to\infty} \frac{1}{n} \mathrm{Log}\, \|Y_n \dots Y_1 x\| = \lim_{n\to\infty} \frac{1}{n} \mathrm{Log}\, \|Y_n \dots Y_1\| = \gamma \qquad \text{a.s.}$$

and if ν *is a* μ*-invariant distribution on* $P(\mathbb{R}^2)$

$$\gamma = \iint \mathrm{Log} \frac{\|Mx\|}{\|x\|} \, d\mu(M) d\nu(\bar{x}) .$$

Proof : Set $S_n = Y_n \dots Y_1$. If m is a continuous μ^*-invariant distribution on $P(\mathbb{R}^2)$ there exists by Proposition 3.3 a measurable set Ω_o with $\mathbb{P}(\Omega_o) = 1$ and a random direction $\bar{Z}$ such that, for $\omega \in \Omega_o$,

$$\lim_{n\to\infty} S_n^*(\omega) m = \delta_{\bar{Z}(\omega)} .$$

By Proposition 3.1 if Z is a unit vector with direction $\bar{Z}$ and $x \in \mathbb{R}^2$, then for all $\omega \in \Omega_o$,

$$\lim_{n\to\infty} \|S_n(\omega)\| = \infty \tag{7}$$

and
$$\lim_{n\to\infty} \frac{\|S_n(\omega)x\|}{\|S_n(\omega)\|} = |\langle Z(\omega),x\rangle| \ . \tag{8}$$

For each fixed $x \neq 0$, if $\bar{u}$ is the direction orthogonal to x,

$\mathbb{P}(\{\omega;\ \langle Z(\omega),x\rangle = 0\}) = \mathbb{P}(\{\omega;\ \bar{Z}(\omega) = \bar{u}\}) = m(\{\bar{u}\}) = 0$

since m is continuous.

So for each fixed $x \neq 0$, by (7) and (8)

$$\lim_{n\to\infty} \|S_n(\omega)x\| = +\infty \qquad \text{a.s.} \tag{9}$$

and
$$\lim_{n\to\infty} \frac{\|S_n(\omega)x\|}{S_n(\omega)\|} \neq 0 \qquad \text{a.s.} \tag{10}$$

We thus obtain by (2), for $x,y \in \mathbb{R}^2-\{0\}$,

$$\lim_{n\to\infty} \delta(S_n(\omega)\cdot\bar{x}\ ,\ S_n(\omega)\cdot\bar{y}) = \lim_{n\to\infty} \frac{\|x\|\ \|y\|\ \delta(\bar{x},\bar{y})}{\|S_n(\omega)x\|\ \|S_n(\omega)y\|} = 0\ , \qquad \text{a.s.}\ .$$

Suppose now that $\mathbb{E}(\text{Log}\|Y_1\|) < \infty$. (Note that $\text{Log}\|Y_1\| \geq 0$ since $|\det Y_1| = 1$, see Exercise I.5.1). We define an additive cocycle σ on $Gl(2,\mathbb{R}) \times P(\mathbb{R}^2)$ by

$$\sigma(Y,\bar{x}) = \text{Log}\frac{\|Yx\|}{\|x\|}\ , \qquad Y \in Gl(2,\mathbb{R})\ , \quad x \in \mathbb{R}^2-\{0\}\ .$$

If ν is a μ-invariant distribution on $P(\mathbb{R}^2)$, then by Fubini's theorem and (9)

$$\lim_{n\to\infty} \sigma(S_n(\omega),\bar{x}) = +\infty \qquad \mathbb{P}\otimes\nu\text{-a.s.}\ .$$

Since $\int \sigma^+(Y,\bar{x})d\mu(Y)d\nu(\bar{x}) \leq \int \text{Log}\ \|Y\|\ d\mu(Y) < \infty$ we can apply Lemma 2.2 : σ is in $L^1(\mathbb{P}\otimes\nu)$ and if γ is defined as

$$\gamma = \iint \text{Log}\ \frac{\|Yx\|}{\|x\|}\ d\mu(Y)d\nu(\bar{x})$$

then $\gamma > 0$. Moreover, by Proposition I.3.4, for some $\phi : \Omega\times\mathbb{P}(\mathbb{R}^2)\to\mathbb{R}$

$$\lim \frac{1}{n}\text{Log}\ \|S_n(\omega)\ \frac{x}{\|x\|}\| = \lim \frac{1}{n}\sigma(S_n(\omega),\bar{x}) = \phi(\omega,\bar{x}) \tag{11}$$

$\mathbb{P}\otimes\nu$ - a.s. with $\iint \phi(\omega,\bar{x})d\mathbb{P}(\omega)d\nu(\bar{x}) = \gamma$.

Fix an $\bar{x}$ such that (11) holds for almost all ω. By (10)

$$\lim_{n\to\infty} \frac{1}{n} \operatorname{Log} \|S_n(\omega)\| = \lim \frac{1}{n} \operatorname{Log} \|S_n(\omega)x\| = \phi(\omega,\bar{x}) .$$

So $\phi(\omega,\bar{x})$ is the upper Lyapunov exponent γ (one can also remark directly that $\phi(\omega,\bar{x})$ depends only on the tail σ-algebra of the sequence of the independent random variables $Y_1, Y_2, \ldots$ and hence is constant).

Finally, for any fixed $y \neq 0$, by (10)

$$\lim \frac{1}{n} \operatorname{Log} \|S_n(\omega)y\| = \lim \frac{1}{n} \operatorname{Log} \|S_n(\omega)\| = \gamma \qquad \text{a.s.}$$

proving the theorem.

One must note that the almost sure convergence statement in (i), resp. (ii), is only valid for $\bar{x}$, $\bar{y}$, resp. x, fixed. For instance, as one can guess from the proof, $\lim \frac{1}{n} \operatorname{Log} \|Y_n(\omega)\ldots Y_1(\omega)x\|$ is not equal to γ if x is in the direction orthogonal to $\bar{Z}(\omega)$. The content of Osseledec's theorem which we will prove in Chapter VI is that this limit is actually $-\gamma$. Let us show this on a typical, although degenerate, example.

<u>Example 3.7.</u> Suppose that $(b_i)_{i\geq 0}$ is a sequence of i.i.d. bounded non constant random variables and set for $i \geq 1$

$$Y_i = \frac{1}{\sqrt{2}} \begin{bmatrix} 2 & b_i \\ 0 & 1 \end{bmatrix} .$$

If $a_n = \sum_{i=1}^{n} 2^{-i} b_i$, and $S_n = Y_n \ldots Y_1$ then

$$S_n = 2^{-n/2} \begin{bmatrix} 2^n & 2^n a_n \\ 0 & 1 \end{bmatrix}$$

For $a_\infty = \sum_{i=1}^{\infty} 2^{-i} b_i$, we have

$$\frac{1}{\sqrt{2}} \begin{bmatrix} 2 & 0 \\ b_o & 1 \end{bmatrix} \begin{bmatrix} 1 \\ a_\infty \end{bmatrix} = \sqrt{2} \begin{bmatrix} 1 \\ \frac{1}{2} b_o + \frac{1}{4} a_\infty \end{bmatrix}$$

Since a_∞ has the same distribution as $\frac{1}{2} b_o + \frac{1}{4} a_\infty$ this shows that the distribution of the direction of $\binom{1}{a_\infty}$ is μ^*-invariant.

Since b_i is not constant it is easy to see that this latter distribution is continuous. So we can apply the theorem. If we choose

$x = \binom{1}{0}$ we find $\gamma = \frac{1}{2}$ Log 2 . But for $x = \binom{-a_\infty}{1}$,

$$S_n x = 2^{-n/2} \begin{bmatrix} 2^n(a_n - a_o) \\ 1 \end{bmatrix}$$

and $\|S_n x\|^2 = 2^{-n}(1 + \{\sum_{i=1}^{\infty} \frac{b_{n+1}}{2^i}\}^2)$. So

$$2^{-n} \leq \|S_n x\|^2 \leq 2^{-n} \operatorname{Sup}|b_i|$$

and $\lim \frac{1}{n} \operatorname{Log} \|S_n x\| = -\frac{1}{2} \operatorname{Log} 2 = -\gamma$.

II.4. Furstenberg's theorem

We will state this important result in several equivalent forms.

THEOREM 4.1. (Furstenberg's theorem, first form)
Let μ *be a probability measure on* $Gl(2,\mathbb{R})$ *such that if* G_μ *is the smallest closed subgroup of* $Gl(2,\mathbb{R})$ *which contains the support of* μ, *the following hold :*

(i) For M *in* G_μ , $|\det M| = 1$.

(ii) G_μ *is not compact.*

(iii) There does not exist a subset L *of* $\mathbb{R}^2$ *which is a finite union of one-dimensional subspaces such that* $M(L) = L$ *for any* M *in* G_μ .

Then the conclusions of Theorem 3.6 are satisfied for matrices with distribution μ . *Moreover there exists a unique* μ*-invariant distribution* ν *on* $P(\mathbb{R}^2)$ *and* ν *is continuous.*

This is an immediate consequence of Proposition 3.3, Theorem 3.6 and the following lemma, where μ^* denotes the image of μ under the transpose map.

LEMMA 4.2. *Let* μ *be a probability distribution on* $Gl(2,\mathbb{R})$ *for which (iii) of Theorem 4.1 holds. Then each* μ*-invariant distribution and each* μ^**-invariant distribution on* $P(\mathbb{R}^2)$ *are continuous.*

Proof : Let ν be a μ-invariant distribution on $P(\mathbb{R}^2)$. For each $\alpha > 0$, $\{\bar{x} \in P(\mathbb{R}^2) ; \nu(\{\bar{x}\}) > \alpha\}$ is finite. Thus if ν is not continuous there exists $\beta > 0$ such that

- $\nu(\{\bar{x}\}) \leq \beta$ for any $\bar{x} \in P(\mathbb{R}^2)$,
- $F = \{\bar{x} \in P(\mathbb{R}^2) ; \nu(\{\bar{x}\}) = \beta\}$ is finite and non empty.

If $\bar{x}_o \in F$ and since $\mu \dot{*} \nu = \nu$

$$\beta = \nu(\{\bar{x}_o\}) = \iint 1_{\{\bar{x}_o\}}(M\cdot\bar{x})d\mu(M)d\nu(\bar{x}) = \int \nu(\{M^{-1}\cdot\bar{x}_o\})d\mu(M) \leq \beta .$$

But $\nu(\{M^{-1}\cdot\bar{x}_o\}) \leq \beta$ so $\nu(\{M^{-1}\cdot\bar{x}_o\}) = \beta$ and $M^{-1}\cdot\bar{x}_o \in F$ for μ almost all M.

This shows that G_μ is contained in the closed subgroup

$$H = \{M \in Gl(2,\mathbb{R}) ; M\cdot\bar{x} \in F \quad \text{for any} \quad \bar{x} \in F\} .$$

If $L = \{0\} \cup \{x \in \mathbb{R}^2 - \{0\} ; \bar{x} \in F\}$, L is a finite union of one-dimensional subspaces of $\mathbb{R}^2$ and $M(L) = L$ for any M in G_μ, in contradiction with the hypothesis. Therefore ν is continuous.

To prove that each μ^*-invariant distribution on $P(\mathbb{R}^2)$ is continuous we verify that G_{μ^*} satisfies (iii) too. Suppose there exist one-dimensional subspaces $V_1, V_2, \ldots, V_r$ such that for each $i \in \{1,\ldots,r\}$ and M in G_{μ^*}

$$M(V_i) = V_j \tag{12}$$

for some $j \in \{1,\ldots,r\}$. If W_i is the orthogonal of V_i, (12) gives

$$M^*(W_j) = W_i .$$

Therefore for each $M \in G_\mu = \{M^* ; M \in G_{\mu^*}\}$, $M(\bigcup_{i=1}^r W_i) = \bigcup_{i=1}^r W_i$ and (iii) doesn't hold for G_μ.

One can in fact weaken the hypothesis (iii) using :

<u>PROPOSITION 4.3.</u> *If* μ *satisfies (i) and (ii) of Theorem 4.1, (iii) is equivalent to*

(iii)' For any $\bar{x}$ *in* $P(\mathbb{R}^2)$, $\{M\cdot\bar{x} , M \in G_\mu\}$ *has more than two elements.*

<u>Proof</u> : (iii) $\Longrightarrow$ (iii)' is obvious. To prove the converse it suffices to verify that if $\bar{x}_1,\ldots,\bar{x}_r$ are distinct points of $P(\mathbb{R}^2)$ with

$$M\cdot\{\bar{x}_1,\ldots,\bar{x}_r\} = \{\bar{x}_1,\ldots,\bar{x}_r\} , \qquad \text{for each } M \in G_\mu ,$$

then $r \leqq 2$. Each M of G_μ induces a permutation $\phi(M)$ of $\{\bar{x}_1,\dots,\bar{x}_r\}$ and ϕ is a group homomorphism. The kernel H of ϕ, i.e. $H = \{M \in G_\mu \; ; \; M\cdot\bar{x}_i = \bar{x}_i \;, \; \forall i = 1,\dots,r\}$ is a closed normal subgroup of G_μ and G_μ/H is finite. By (ii) H is not finite. If $r \geqq 3$ consider three non zero vectors x_1, x_2, x_3 with directions $\bar{x}_1, \bar{x}_2, \bar{x}_3$. We can write

$$x_3 = \alpha x_1 + \beta x_2 \qquad \text{with } \alpha \neq 0 \;, \; \beta \neq 0 \;,$$

and for each M in H, for some $\lambda_i \neq 0$

$$Mx_i = \lambda_i x_i \;, \qquad i = 1, 2, 3 \;.$$

This yields

$$\lambda_3 \alpha x_1 + \lambda_3 \beta x_2 = \lambda_3 x_3 = Mx_3 = \alpha M x_1 + \beta M x_2 = \alpha\lambda_1 x_1 + \beta\lambda_2 x_2$$

and $\alpha\lambda_3 = \alpha\lambda_1$, $\beta\lambda_3 = \beta\lambda_2$. So $\lambda_1 = \lambda_2 = \lambda_3$ and $M = \lambda_1 \mathrm{Id}$. By (i), $|\det M| = 1$ and each M of H must be Id or $-$Id. But H is not finite, so $r \leqq 2$.

In some applications it is more convenient to use the following equivalent form :

<u>THEOREM 4.4.</u> (Furstenberg's theorem, second form)
Let $Y_1, Y_2, \dots$ *be a sequence of independent invertible matrices with the same distribution* μ *. Suppose there is no distribution* m *on* $P(\mathbb{R}^2)$ *such that* $Mm = m$ *for each* M *in the group generated by the support of* μ *. Then if* $|\det Y_1| = 1$ *a.s. the hypotheses of Theorem 4.1 are satisfied. In particular* $\gamma > 0$ *and there exists a unique* μ*-invariant distribution* ν .

<u>Proof</u> : To check (ii) of Theorem 4.1 one can either use Remark 4.5 or, more simply, notice that if G_μ is compact and λ is its Haar measure, then for all $\bar{x}$ in $P(\mathbb{R}^2)$, and M in G_μ,

$$M(\lambda \,\dot{*}\, \delta_{\bar{x}}) = (\delta_M * \lambda) \,\dot{*}\, \delta_{\bar{x}} = \lambda \,\dot{*}\, \delta_{\bar{x}} \;.$$

This contradicts the hypothesis.

Suppose that (iii) doesn't hold. We then could find $\{\bar{x}_1, \bar{x}_2, \dots, \bar{x}_r\}$ in $P(\mathbb{R}^2)$ such that, for any M in G_μ

$$M\cdot\{\bar{x}_1,\dots,\bar{x}_r\} = \{\bar{x}_1,\dots,\bar{x}_r\} \;.$$

But this would imply that $M(\frac{1}{r}\sum_{i=1}^{r}\delta_{\bar{x}_i}) = \frac{1}{r}\sum_{i=1}^{r}\delta_{\bar{x}_i}$, which is absurd.

II.5. Some simple examples

Given a sequence $\{Y_n, n \geq 1\}$ of independent random matrices in $Gl(2,\mathbb{R})$ with the same distribution μ satisfying Furstenberg's theorem one would like to compute the upper Lyapunov exponent γ and/or the μ-invariant distribution on $P(\mathbb{R}^2)$. Unlike in the case of the classical law of large numbers γ cannot be written as an integral with respect to μ only. Therefore this computation is most often unfeasible.

In I.2 we have described one of the only known models where γ has a simple expression. We give here two related examples due to Furstenberg, valid in any dimension and a third one which appears in the study of Schrödinger operators.

We shall see more intricate examples in Chapter VI.

5.1. Rotation invariant distributions

Let $SO(d)$ be the group of orthogonal $d \times d$ matrices with determinant one and m_K the Haar measure on $SO(d)$ which satisfies $m_K(SO(d)) = 1$. For any $\bar{x}$ in $P(\mathbb{R}^d)$, $m_d = m_K \star \delta_{\bar{x}}$ is a probability measure on $P(\mathbb{R}^d)$ which does not depend on $\bar{x}$. For any distribution ν on $P(\mathbb{R}^d)$,

$$m_K \star \nu = \int m_K \star \delta_{\bar{x}}\, d\nu(\bar{x}) = \int m_d\, d\nu(\bar{x}) = m_d \tag{13}$$

and m_d is the unique distribution on $P(\mathbb{R}^d)$ wich satisfies

$$Mm_d = m_d \qquad \text{for any } M \in SO(d) .$$

It is clear that m_d is actually the image of the probability measure on the unit sphere which gives the element of surface area under the canonical projection onto $P(\mathbb{R}^d)$.

As in Exercise I.5.4 one proves that if x is a unit vector

$$\frac{d(Ym_d)}{dm}(\bar{x}) = \|Y^{-1}x\|^{-d} \qquad \text{for each } Y \text{ s.t. } |\det Y| = 1.$$

Let $Y_1, Y_2, \ldots$ be independent matrices in $Gl(d,\mathbb{R})$ with distribution μ . Suppose that the distribution of kY_1k^{-1} is the same for all $k \in SO(d)$ (in other words $\delta_k * \mu * \delta_{k^{-1}} = \mu$, $\forall k \in SO(d)$). Then m_d is a μ-invariant distribution on $P(\mathbb{R}^d)$ (Proof : for any $k \in SO(d)$,

$$\mu \overset{\bullet}{*} m_d = \mu \overset{\bullet}{*} (\delta_k \overset{\bullet}{*} m_d) = (\delta_k * \mu * \delta_{k^{-1}} * \delta_k) \overset{\bullet}{*} m_d = (\delta_k * \mu) \overset{\bullet}{*} m_d$$

so, by (13) ,

$$\mu \overset{\bullet}{*} m_d = \int (\delta_k * \mu) \overset{\bullet}{*} m_d \, dm_K(k) = (m_K * \mu) \overset{\bullet}{*} m_d$$

$$= m_K \overset{\bullet}{*} (\mu \overset{\bullet}{*} m_d) = m_d) .$$

In this case the distribution of $\|Y_1 x\|$ does not depend on x on the unit sphere. So this example is a particular case of the one introduced in I.2.3 .

An other example where the invariant distribution can be calculated is the following. Suppose that there are two probability measures λ_1 and λ_2 on $Gl(d,\mathbb{R})$ such that $\mu = \lambda_1 * m_K * \lambda_2$, (in other words each Y_i can be written as $A_i K_i B_i$ where A_i, K_i, B_i are independent matrices and the distribution of K_i is the Haar measure on $SO(d)$). Then $\lambda_1 \overset{\bullet}{*} m_d$ is the unique μ-invariant distribution on $P(\mathbb{R}^d)$, because for any probability measure ν on $P(\mathbb{R}^d)$, by (13) ,

$$(\lambda_1 * m_K * \lambda_2) \overset{\bullet}{*} \nu = \lambda_1 \overset{\bullet}{*} (m_K \overset{\bullet}{*} (\lambda_2 \overset{\bullet}{*} \nu)) = \lambda_1 \overset{\bullet}{*} m_d .$$

So if ν is μ-invariant $\nu = \lambda_1 \overset{\bullet}{*} m_d$ and this distribution is invariant. We thus have

$$\gamma = \iint \text{Log} \frac{\|Yx\|}{\|x\|} \, d\mu(Y) \, d(\lambda_1 \overset{\bullet}{*} m_d)(\bar{x})$$

$$= \int \text{Log} \frac{\|AKB\ A'x\|}{\|A'x\|} \, d\lambda_1(A) \, d\lambda_2(A') \, dm_K(K) \, d\lambda_2(B) \, dm_d(\bar{x}) .$$

Writing $\text{Log} \dfrac{\|AKB\ A'x\|}{\|A'x\|} = \text{Log} \dfrac{\|AKB\ A'x\|}{\|B\ A'x\|} - \text{Log} \|A'x\| + \text{Log} \|B\ A'x\|$

one finds $\gamma = \displaystyle\int \text{Log} \frac{\|B\ Ax\|}{\|x\|} \, d\lambda_2(B) \, d\lambda_1(A) \, dm_d(\bar{x})$.

One can also check this directly by noticing that γ is the exponent associated with the distribution of $K_2 B_2 A_1 K_1$.

5.2. The Cauchy distribution

Define a bijective map $\pi : P(\mathbb{R}^2) \longrightarrow \mathbb{R} \cup \{\infty\}$ as follows :

if $x = \binom{x_1}{x_2}$ is non zero then $\pi(\bar{x}) = \begin{cases} \frac{x_1}{x_2} & \text{if} \quad x_2 \neq 0 \\ \infty & \text{if} \quad x_2 = 0 \end{cases}$

One can transport the action of $Gl(2,\mathbb{R})$ on $P(\mathbb{R}^2)$ by setting $g\cdot t = \pi\{g\cdot\pi^{-1}(t)\}$ for $t \in \mathbb{R} \cup \{\infty\}$. If $g = \begin{pmatrix} a & b \\ c & d \end{pmatrix}$

$$g\cdot t = \frac{at+b}{ct+d} \,. \tag{14}$$

This is just a concrete realization of $P(\mathbb{R}^2)$.

DEFINITION 5.2. *The space* $\mathbb{R} \cup \{\infty\}$ *endowed with the action of* $Gl(2,\mathbb{R})$ *defined by* (14) *will be called the projective line.*

The image under π of the probability measure m_2 defined in 5.1 is the image of the Lebesgue measure on the circle $\frac{1}{2\pi} 1_{[0,2]} d\theta$, under $\frac{1}{\tan\theta}$. Thus it is the Cauchy distribution, i.e. the probability measure on $\mathbb{R}$ defined by the density $\frac{1}{\pi(1+t^2)}$.

For $z = \alpha + i\beta$ in $\mathbb{C}$ with $\beta > 0$, write C_z for the distribution of $\beta U + \alpha$ where U is a Cauchy-distributed random variable. Using the fact that each matrix g in $Gl(2,\mathbb{R})$ can be written

$$g = \begin{bmatrix} a & b \\ 0 & a^{-1} \end{bmatrix} k \qquad \text{with} \qquad k \in O(2)$$

one sees that $g\, C_i = C_{g\cdot i}$ and more generally, for any z in $\mathbb{C}$ with $\operatorname{Im} z > 0$, if $g'\cdot i = z$

$$g\, C_z = gg'\, C_i = C_{gg'\cdot i} = C_{g\cdot z} \,.$$

Consider now the random matrix $Y = \begin{pmatrix} X & -1 \\ 1 & 0 \end{pmatrix}$ where X is a random variable with distribution C_z. Denote by μ_z its distribution. For any $x \in \mathbb{R}$, $z' \in \mathbb{C}$,

$$\begin{pmatrix} x & -1 \\ 1 & 0 \end{pmatrix} C_{z'} = \begin{pmatrix} 1 & x \\ 0 & 1 \end{pmatrix} \begin{pmatrix} 0 & -1 \\ 1 & 0 \end{pmatrix} C_{z'} = \begin{pmatrix} 1 & x \\ 0 & 1 \end{pmatrix} C_{-1/z'} \,.$$

So if U is a random variable with distribution $C_{z'}$, independent of

X , the distribution of $Y \cdot U$ is $C_z * C_{-1/z'}$ that is $C_{z-1/z'}$. We can thus write

$$\mu_z \overset{\bullet}{*} C_{z'} = C_{z-1/z'}$$

and if $z = z' + \frac{1}{z'}$, $C_{z'}$ is a μ_z-invariant distribution on $\mathbb{R} \cup \{\infty\}$. It is easily seen that μ_z satisfies Furstenberg's theorem. Therefore by Theorem 4.1, if $\gamma(z)$ is its upper Lyapunov exponent,

$$\begin{aligned}\gamma(z) &= \iint \operatorname{Log} \frac{\|M\binom{t}{1}\|}{\|\binom{t}{1}\|}\, d\mu_z(M)\, d\, C_{z'}(t)\\ &= \frac{1}{2}\iint \operatorname{Log} \frac{(tx-1)^2+t^2}{1+t^2}\, dC_z(x)\, dC_{z'}(t)\\ &= \frac{1}{2}\iint \operatorname{Log}\{(\tfrac{tx-1}{t})^2+1\} dC_z(x)\, dC_{z'}(t) +\\ &\quad + \frac{1}{2}\int \operatorname{Log} t^2\, dC_{z'}(t) - \frac{1}{2}\int \operatorname{Log}(1+t^2) dC_{z'}(t)\end{aligned}$$

But since $\mu_z \overset{\bullet}{*} C_{z'} = C_{z'}$, for any bounded Borel f on $\mathbb{R}$

$$\iint f(\tfrac{tx-1}{t}) dC_z(x)\, dC_{z'}(t) = \int f(t) dC_{z'}(t)$$

therefore $\gamma(z) = \int \operatorname{Log}|t| dC_{z'}(t)$.

By computation or using the fact that $\operatorname{Log}|z|$ is harmonic in the upper half plane and that $C_{z'}$ is the Poisson kernel one obtains

$$\gamma(z) = \operatorname{Log}|z'| .$$

II.6. Exercises

Remarks : After the exercises (6.4), (6.5) and (6.6) it should be easy for the reader to study the Lyapunov exponent of arbitrary distributions on $Gl(2,\mathbb{R})$. (6.5) shows that in Theorem 3.6 one cannot replace μ^* by μ in the hypothesis (c). Exercise (6.6) implies that it does not suffice to suppose that G_μ does not leave any one-dimensional subspace invariant to obtain Fustenberg's theorem 4.1 . One really needs (iii).

(6.1) Let $\{A_n, n \geq 1\}$ be a sequence of matrices in $Gl(2,\mathbb{R})$ such that for some continuous distribution m on $P(\mathbb{R}^2)$, $A_n m$ converges weakly to $\delta_{\bar{z}}$. Show that this is actually true for all continuous distributions.

(6.2) Verify that working with $P(\mathbb{C}^2)$ instead of $P(\mathbb{R}^2)$, all the results of this chapter are valid for matrices with complex entries (with the same proofs).

(6.3) Prove that Theorems 4.1 and 4.4 are equivalent.

(6.4) Let μ be a probability measure on $Sl(2,\mathbb{R}) = \{M \in Gl(2,\mathbb{R})\,;\ \det M = 1\}$. Denote as usual by G_μ the smallest closed subgroup of $Gl(2,\mathbb{R})$ which contains the support of μ . Prove that one of the following holds

(i) G_μ is finite.

(ii) For any $\bar{x}$ in $P(\mathbb{R}^2)$, $G_\mu \cdot \bar{x}$ has more than two points.

(iii) $Q\, G_\mu\, Q^{-1} \subset \{\begin{pmatrix} a & b \\ 0 & a^{-1} \end{pmatrix} ,\ a \neq 0 ,\ b \in \mathbb{R}\}$, for some $Q \in Gl(2,\mathbb{R})$

(iv) $Q\, G_\mu\, Q^{-1} \subset \{\begin{pmatrix} a & 0 \\ 0 & a^{-1} \end{pmatrix} ,\ a \neq 0\} \cup \{\begin{pmatrix} 0 & b^{-1} \\ -b & 0 \end{pmatrix} ,\ b \neq 0\}$, for some Q .

(6.5) Let $\{Y_n,\ n \geq 1\}$ be independent matrices with distribution μ of the form

$$Y_i(\omega) = \begin{bmatrix} a_i(\omega) & b_i(\omega) \\ 0 & 1 \end{bmatrix} , \qquad a_i(\omega) \neq 0 .$$

a. Prove that if for some $x \in \mathbb{R}$, $a_i x + b_i = x$ a.s., there exists an invertible matrix Q such that each $Q\, Y_i\, Q^{-1}$ is diagonal (the converse is clear).

b. Suppose that for any Q , $Q\, Y_i\, Q^{-1}$ is not a.s. diagonal and that $\mathbb{E}(\mathrm{Log}\|Y_1\|) < \infty$.

(i) Assume $\mathbb{E}(\mathrm{Log}|a_1|) < 0$. Prove that $u = \sum_{n=1}^{\infty} a_1 \ldots a_{n-1} b_n$ exists with probability 1 and that the distribution ν of the direction of $\binom{u}{1}$ is μ-invariant. For $e_1 = \binom{1}{0}$ prove that $\{\alpha\, \delta_{\bar{e}_1} + (1-\alpha)\nu ,\ 0 \leq \alpha \leq 1\}$ is the set of μ-invariant distributions on $P(\mathbb{R}^2)$.

(ii) Assume $\mathbb{E}(\mathrm{Log}\,|a_1|) \geq 0$. Prove that the only μ-invariant distribution on $P(\mathbb{R}^2)$ is $\delta_{\bar{e}_1}$ (Hint : if not, there would be a continuous μ-invariant distribution and Theorem 3.6 would lead to a contradiction).

(iii) Prove that, on all cases, $\lim \frac{1}{n} \text{Log} \|Y_n \dots Y_1\| = \text{Sup}(0, \mathbb{E}(\text{Log}|a_1|))$. (Hint : use Exercise I.5.3).

(6.6) Let $G = \{\begin{pmatrix} a & 0 \\ 0 & a^{-1} \end{pmatrix}, a \neq 0\} \cup \{\begin{pmatrix} 0 & b^{-1} \\ -b & 0 \end{pmatrix}, b \neq 0\}$.

a. Prove that G is a subgroup of $Gl(2,\mathbb{R})$.

b. Let $\{Y_n, n \geq 1\}$ be i.i.d. matrices in G with distribution μ. Suppose that

$$\mathbb{P}(Y_i = \pm I \text{ or } \pm\begin{pmatrix} 0 & -1 \\ 1 & 0 \end{pmatrix}) \neq 1 \quad \text{and} \quad \mathbb{P}(Y_1 \text{ is diagonal}) \neq 1 .$$

Prove that there is no continuous μ-invariant distribution on $P(\mathbb{R}^2)$ (use Proposition 3.3). Deduce that $\nu = \frac{1}{2}(\delta_{\bar{e}_1} + \delta_{\bar{e}_2})$ is the only μ-invariant distribution. Show that $\lim \frac{1}{n} \text{Log} \|Y_n \dots Y_1\| = 0$ (this latter question is more involved).

II.7. Complements

Let $G = Sl(2,\mathbb{R})$, the matrices of $Gl(2,\mathbb{R})$ with determinant 1, $H = \{z = x+iy, x \in \mathbb{R}, y > 0\}$ and $\partial H = \{z = x+iy, x \in \mathbb{R}, y = 0\} \cup \{\infty\}$. For $g = \begin{pmatrix} a & b \\ c & d \end{pmatrix}$ in G and z in $\bar{H} = H \cup \partial H$ we set

$$g \cdot z = \frac{az + b}{cz + d} \tag{15}$$

With this action H and ∂H are G-spaces. Notice that ∂H is the projective line introduced Definition 5.2.

PROPOSITION 7.1. *Let* $\{A_n, n \geq 1\}$ *be a bounded sequence in* G *such that* $\liminf_{n\to\infty} \frac{1}{n} \text{Log} \|A_1 \dots A_n\| > 0$. *Then there is* x *in* ∂H *such that*

(i) For any z *in* H, $A_1 A_2 \dots A_n z$ *converges to* x *as* $n \to \infty$.

(ii) If ν *is a continuous distribution on* ∂H, $A_1 \dots A_n \nu$ converges weakly to δ_x.

Proof : It is more convenient to work with the unit disk $D = \{z \in \mathbb{C}, |z| < 1\}$ and $\partial D = \{z \in \mathbb{C}, |z| = 1\}$. Let $\tau : H \longrightarrow D$ be defined by $\tau(z) = \frac{z-i}{z+i}$. This map extends to ∂H and $\tau(\partial H) = \partial D$.

Let $Q = \begin{pmatrix} 1 & -i \\ 1 & i \end{pmatrix}$ and $G' = \{\begin{pmatrix} a & b \\ \bar{b} & \bar{a} \end{pmatrix}$; $a, b \in \mathbb{C}$, $|a|^2 - |b|^2 = 1\}$.
Consider the map $\psi : G \longrightarrow G'$ defined by

$$\psi(g) = Q\, g\, Q^{-1} \quad , \qquad g \in G .$$

If for $g = \begin{pmatrix} a & b \\ \bar{b} & \bar{a} \end{pmatrix}$ in G' and z in $\bar{D} = D \cup \partial D$ we set

$$g\cdot z = \frac{az + b}{\bar{b}z + \bar{a}}$$

it is clear that

$$\psi(g)\cdot\tau(z) = \tau(g\cdot z) .$$

Therefore it suffices to show that if $A_i' = \psi(A_i)$,

(i)' for some $x \in \partial D$, $A_1' \ldots A_n'\cdot z \longrightarrow x$ for any $z \in D$.

(ii)' if ν is a continuous distribution on ∂D , $A_1' \ldots A_n'\nu \longrightarrow \delta_x$.

Write $A_1' \ldots A_n' = \begin{pmatrix} a_n & b_n \\ \bar{b}_n & \bar{a}_n \end{pmatrix}$ with $|a_n|^2 - |b_n|^2 = 1$. Since $\liminf \frac{1}{n} \log \|A_1 \ldots A_n\| > 0$, $\liminf \frac{1}{n} \operatorname{Log} |a_n| > 0$ and for some $\alpha > 0$ and n large enough ,

$$|a_n| > e^{n\alpha} .$$

For any u in $\bar{D}$,

$$|A_1' \ldots A_n'\cdot u - A_1' \ldots A_n'\cdot 0| = \left|\frac{a_n u + b_n}{\bar{b}_n u + \bar{a}_n} - \frac{b_n}{\bar{a}_n}\right| = \frac{|u|}{|a_n|^2 \left|\frac{\bar{b}_n}{\bar{a}_n} u + 1\right|} \tag{16}$$

and if $u \in D$,

$$|A_1' \ldots A_n'\cdot u - A_1' \ldots A_n'\cdot 0| \leq \frac{1}{|a_n|^2 \{1 - |u|\}} \tag{17}$$

Since (A_n) is bounded, there exists $\rho < 1$ such that $|A_n'\cdot 0| < \rho$.
So if $z_n = A_1' \ldots A_n'\cdot 0$, (17) yields

$$|z_{n+1} - z_n| = |A_1' \ldots A_n' A_{n+1}'\cdot 0 - A_1' \ldots A_n'\cdot 0| \leq \frac{1}{|a_n|^2 (1-\rho)} \leq \frac{e^{-2n\alpha}}{1-\rho} \quad ,$$

and (z_n) is a Cauchy sequence. If $x = \lim z_n$, x is in ∂D and by (15) for any z in D , $\lim A_1' \ldots A_n'\cdot z = x$, proving (i)' and (i).

Let ν be a continuous distribution on ∂D and m a weak limit point of the sequence $\{A_1' \ldots A_n'\nu,\ n \geq 1\}$. Since $\frac{b_n}{a_n}$ is bounded we can suppose that for some subsequence n_i and some y

$$A_1' \ldots A_{n_i}'\nu \longrightarrow m$$

and

$$a_{n_i}^{-1} b_{n_i} \longrightarrow y \qquad \text{as} \quad i \longrightarrow \infty .$$

By (16), if $u \in \partial D$ and $y\bar{u} + 1 \neq 0$, $A'_1 \dots A'_{n_i} \cdot u \longrightarrow x$.

Therefore, since $\nu(\{-\bar{y}^{-1}\}) = 0$, $A'_1 \dots A'_{n_i} \nu \longrightarrow \delta_x$ and $m = \delta_x$ which proves (ii) and (ii)'.

COROLLARY 7.1. *If* $Y_1, Y_2, \dots$ *are i.i.d. bounded matrices in* $Sl(2,\mathbb{R})$ *and if* $\liminf \frac{1}{n} \|Y_n \dots Y_1\| > 0$ *there exists a random* x *in* ∂H *such that, a.s.*,

(i) $Y_1^* \dots Y_n^* z \longrightarrow x$ *for any* x *in* H.

(ii) $Y_1^* \dots Y_n^* \nu \longrightarrow \delta_x$ *for any continuous distribution* ν *on* ∂H.

We bring to the reader's notice that since $\partial H = P(\mathbb{R}^2)$, our method of proof of Fustenberg's theorem, due to [34], consits simply in trying to find a converse of this corollary.

Consider independent matrices $Y_1, Y_2, \dots$ in $Sl(2,\mathbb{R})$ with the same distribution such that $k\,Y_1 k^{-1}$ has a distribution independent of $k \in SO(2)$. Then for any g in $Sl(2,\mathbb{R})$ the sequence $Z_o = g \cdot i$, $Z_1 = g\,Y_1 \cdot i, \dots, Z_n = g\,Y_1 \dots Y_n \cdot i, \dots$ is a Markov chain on the upper half plane. To verify this first note that if we set, for any Borel set A in H and h in $Sl(2,\mathbb{R})$

$$P(h \cdot i, A) = \int 1_A(hg \cdot i)\,d\mu(A) \tag{18}$$

then if $h' \cdot i = h \cdot i$

$$\begin{aligned} P(h' \cdot i, A) &= \int 1_A(h\{h^{-1}h'\}g\{h^{-1}h'\}^{-1} \cdot i)\,d\mu(g) \\ &= \int 1_A(hg \cdot i)\,d\mu(g) \\ &= P(h \cdot i, A) \end{aligned}$$

because $h^{-1}h' \cdot i = i$ and $h^{-1}h' \in SO(2)$. Therefore $P(h \cdot i, A)$ depends really only on $h \cdot i$.

Let $\mathcal{F}_n$ be the σ-algebra generated by $Y_1, \dots, Y_n$ and $\mathcal{G}_n$ the σ-algebra generated by $Z_o, \dots, Z_n$. One has for $h = g\,Y_1 \dots Y_n$

$$\mathbb{E}^{\mathcal{F}_n}\{1_A(Z_{n+1})\} = \mathbb{E}\{1_A(h\,Y_{n+1} \cdot i)\} = P(h \cdot i, A) = P(Z_n, A)$$

and

$$\mathbb{E}^{\mathcal{G}_n}\{1_A(Z_{n+1})\} = \mathbb{E}^{\mathcal{G}_n}\{\mathbb{E}^{\mathcal{F}_n}(1_A(Z_{n+1}))\} = P(Z_n, A).$$

So Z_n is a Markov chain with transition probability P. By Proposition 7.1 this chain converges a.s. to some point of the boundary ∂H. From 5.1 and 5.2 the distribution of this point is C_{Z_o}.

An example of such a Markov chain is obtained by choosing, for a fixed $g \in Sl(2,\mathbb{R})$, $\mu = m_K * \delta_g * m_K$ where m_K is the Haar measure on $SO(2)$. Starting from $x \in H$ it goes to a point of the hyperbolic circle $\{z' \in H \; ; \; d(z,z') = \text{Log}\,\|g\|\}$ (see Exercise 7.3), this point being chosen with uniform distribution.

Remark that if P is defined by (18), then for g,h in $Sl(2,\mathbb{R})$,

$$P(gh.i,g.A) = \int 1_{g\cdot A}(ghg'\cdot i)d\mu(g') = \int 1_A(hg'\cdot i)d\mu(g') = P(h\cdot i,A).$$

So for any z in H,

$$P(g\cdot z,\ g\cdot A) = P(z,A)\ . \tag{19}$$

Conversely consider a transition probability P on H which satisfies (19). Define a distribution μ on $Sl(2,\mathbb{R})$ by, if B is a Borel set in $Sl(2,\mathbb{R})$ and $\overline{B} = \{g\cdot i \; ; \; g \in B\}$

$$\mu(B) = P(i,\overline{B})\ .$$

If $k, k' \in SO(2)$

$$\mu(kBk') = P(i,\ k\cdot\overline{B}) = P(k^{-1}\cdot i,\ k^{-1}k\cdot\overline{B}) = P(i,\overline{B}) = \mu(B)$$

so $\varepsilon_k * \mu * \varepsilon_{k'} = \mu$ and it is easily seen that if $\{Y_n,\ n \geq 1\}$ are i.i.d. matrices with distribution μ, $Z_n = gY_1\dots Y_n\cdot i$ is a Markov chain with transition probability P. Since μ satisfies the hypothesis required in example I.2.3, if $Z_o = i$ and $\|x\| = 1$,

$$\lim_{n\to\infty} \frac{1}{n} d(z,Z_n) = \gamma = \int \text{Log}\,\|Yx\|d\mu(Y) = \mathbb{E}\{d(i,Z_1)\}$$

where d is the hyperbolic metric introduced in Exercise 7.3.

For instance let P_t be the transition semigroup with infinitesimal generator $\Delta = \frac{1}{2}\, y^2\, \{\frac{\partial^2}{\partial x^2} + \frac{\partial^2}{\partial y^2}\}$ on H, i.e. the Laplace-Beltrami operator associated with the hyperbolic metric. Since each g of $Sl(2,\mathbb{R})$ acts on H as an isometry Δ commutes with g and each P_t satisfies (19). It follows from the preceding results that if X_t is the associated Markov process starting from i, $\lim_{t\to\infty} X_t = X_\infty$ exists with probability one and is in $\partial H \cong \mathbb{R}$. But in that case this result is obvious : the operator Δ is a scalar multiple of the ordinary

Laplacian $\frac{1}{2}\{\frac{\partial^2}{\partial x^2}+\frac{\partial^2}{\partial y^2}\}$. So if B_t is the complex Brownian motion with $B_0 = i$, $X_t = B_{\tau_t}$ for some clock τ_t . So $X_\infty = B_T$ where $T = \inf\{t > 0 , B_t \in \partial H\}$.

Exercise 7.3. a. Prove that $G\cdot i = H$ and $\{g \in G ; g\cdot i = i\} = SO(2)$.

b. If $z = g\cdot i$ and $z' = g'\cdot i$ with $g,g' \in Sl(2,\mathbb{R})$ set

$$d(z',z) = \operatorname{Log} \|g^{-1}g'\| .$$

Prove that this defines a distance on H (the so-called hyperbolic distance).

c. Let $\{z_n , n \geqq 1\}$ be a sequence in H such that for some C , $d(z_n,z_{n+1}) \leqq C$ and $\liminf \frac{1}{n} d(i,z_n) \neq 0$. Prove that there exists x in ∂H with $\lim_{n\to\infty} z_n = x$ (Hint : write $z_n = A_1\ldots A_n\cdot i$ with $A_n \in G$).

Exercise 7.4. Let $Y_1,Y_2,\ldots$ be i.i.d. matrices in G with $\mathbb{E}(\operatorname{Log}\|Y_1\|) < \infty$. Prove that for any $r > 0$ and almost all ω , $\|Y_n(\omega)\| \leqq n^r$ for n large enough. Deduce the analogue of Corollary 7.2 under these hypotheses.

CHAPTER III

CONTRACTION PROPERTIES

We now consider i.i.d. invertible random matrices $Y_1, Y_2, \ldots$ of arbitrary order, say d. This chapter is devoted to the study of the basic almost sure properties of the products $S_n = Y_n \ldots Y_1$. We shall derive their salient feature, which is the contracting action of S_n on the set of directions. In particular we shall give (see 3.4, 4.3 and 6.1), following Guivarc'h and Raugi [34], a condition ensuring that

(a) For any $x \neq 0$ in $\mathbb{R}^d$, a.s.

$$\lim_{n\to\infty} \frac{1}{n} \operatorname{Log} \|S_n x\| = \gamma .$$

(b) There exists a unique invariant distribution on $P(\mathbb{R}^d)$.

(c) For any $\bar{x}, \bar{y}$ in $P(\mathbb{R}^d)$, a.s.

$$\overline{\lim_{n\to\infty}} \frac{1}{n} \operatorname{Log} \delta(S_n \cdot \bar{x}, S_n \cdot \bar{y}) < 0$$

where δ is the natural angular metric on $P(\mathbb{R}^d)$.

(d) The two upper Lyapunov exponents associated with (S_n) are distinct.

This result is basic for further developments.

A close look at the proof of these statements in Chapter II (for $d = 2$) leads us to introduce the following conditions. Let μ (resp. μ^*) be the distribution of Y_1 (resp. Y_1^*) and T_μ be the smallest closed semigroup in $Gl(d,\mathbb{R})$ which contains the support of μ. One has to require that

(i) There exists a μ^*-invariant distribution m on $P(\mathbb{R}^d)$ such that $m(\overline{V}) = 0$ for any proper projective subspace $\overline{V}$.

(ii) There is a sequence (M_n) in T_μ such that $M_n m$ converges to a Dirac measure.

In the first two sections we develop natural and tractable conditions on T_μ itself ensuring that (i) and (ii) hold. In Section 3 we generalize the method introduced in Chapter II. Namely we study the action of S_n^* on measures on $P(\mathbb{R}^d)$. We derive from this analysis a key property from which we shall deduce the main results and we prove statements (a) and (b).

In Section 4 we give, under an irreducibility assumption, a necessary and sufficient condition ensuring that $\delta(S_n \cdot \overline{x}, S_n \cdot \overline{y}) \longrightarrow 0$ a.s. as $n \longrightarrow \infty$. For later use we shall be needing the stronger statements (c) and (d). They are proved in Section 6. (The d Lyapunov exponents are introduced in Section 5).

We end this chapter with a complement section, which can be skipped at first reading, where we slightly improve Furstenberg's theorem.

For the reader's convenience we only consider matrices with real entries. But all the given results are valid with essentially the same proofs for matrices in $Gl(d,\mathbb{C})$. Finally we must say that the method we use is influenced by the paper [32] of Guivarc'h.

III.1. Contracting sets

We first define what we will call a polar decomposition of a matrix. Let $\mathrm{diag}(a_1,\ldots,a_d)$ denote the $d \times d$ matrix A whose entries $A_{i,j}$ satisfy

$$A_{i,j} = \begin{cases} a_i & \text{if} \quad i = j \\ 0 & \text{if} \quad i \neq j \,. \end{cases}$$

We write $O(d)$ for the set of orthogonal matrices, i.e.

$$O(d) = \{M \in Gl(d,\mathbb{R})\,;\ M^*M = Id\}\,.$$

DEFINITION 1.1. *Let* M *be a matrix in* $Gl(d,\mathbb{R})$. *We say that*

$$M = K\,A\,U$$

is a "polar decomposition" of M *if* K *and* U *are in* $O(d)$ *and if*

$A = \mathrm{diag}(a_1,\dots,a_d)$ *with* $a_1 \geqq a_2 \geqq \dots \geqq a_d > 0$.

LEMMA 1.2. *Any invertible matrix has a polar decomposition. Moreover* $a_1 \geqq a_2 \geqq \dots \geqq a_d > 0$ *are necessarily the square roots of the eigenvalues of* M^*M.

Proof : Let $\{e_i,\ 1 \leqq i \leqq d\}$ be the canonical basis of $\mathbb{R}^d$ and $\{f_i,\ i \leqq 1 \leqq d\}$ be an orthonormal basis such that $M^*Mf_i = a_i^2 f_i$, $i = 1,\dots,d$. Define the orthonormal matrix U by $Uf_i = e_i$, $1 \leqq i \leqq d$. We have

$$U^*A^2Uf_i = U^*A^2e_i = a_i^2U^{-1}e_i = a_i^2f_i = M^*Mf_i$$

so that $U^*A^2U = M^*M$. If we set $K = MU^{-1}A^{-1}$ then $K^*K = A^{-1}UM^*MU^{-1}A^{-1} = A^{-1}A^2A^{-1} = Id$, whence K is orthogonal. Since $M = KAU$, M has a polar decomposition. The last statement is obvious for if $M = KAU$ then $M^*M = U^{-1}A^2U$.

Recall the rank of a matrix is the dimension of its range.

DEFINITION 1.3. *Given a subset* T *of* $Gl(d,\mathbb{R})$, *we define the index of* T *as the least integer* p *such that there exists a sequence* $\{M_n,\ n \geqq 0\}$ *in* T *for which* $\|M_n\|^{-1} M_n$ *converges to a rank* p *matrix.*

We say that T *is contracting when its index is one.*

LEMMA 1.4. *Let* T *be a subset of* $Gl(d,\mathbb{R})$. *For any* M *in* T, *let* $a_1(M) \geqq a_2(M) \geqq \dots \geqq a_d(M) > 0$ *be the square roots of the eigenvalues of* M^*M. *Then, if* p *is the index of* T,

$$\sup\{a_i(M)^{-1}\, a_1(M)\ ;\ M \in T\} = \infty$$

if and only if $i > p$.

Proof : Fix an integer i, $2 \leqq i \leqq d$, and consider a sequence $\{M_n,\ n \geqq 0\}$ in T such that

$$\lim_{n\to\infty} a_i(M_n)^{-1}\, a_1(M_n) = \sup\{a_i(M)^{-1}\, a_1(M),\ M \in T\}.$$

If $M_n = K_nA_nU_n$ is a polar decomposition of M_n we may always assume, passing to a subsequence, that K_n (resp. U_n) converges to some orthogonal matrix K (resp. U) and that $\|A_n\|^{-1}A_n$ converges to some $\mathrm{diag}(\alpha_1,\alpha_2,\dots,\alpha_d)$ with $\alpha_1 \geqq \alpha_2 \geqq \dots \geqq \alpha_d \geqq 0$. But orthogonal matrices are isometries, whence $\|M_n\| = \|A_n\|$. This implies that

$$\lim_{n\to\infty} \|M_n\|^{-1} M_n = K \operatorname{diag}(\alpha_1,\ldots,\alpha_d)U .$$

By assumption the index of T is p so that $\alpha_p \neq 0$ and $\alpha_{p+1} = 0$ (since the rank of $K \operatorname{diag}(\alpha_1,\ldots,\alpha_d)U$ is $\operatorname{Max}\{i; \alpha_i \neq 0\}$). The conclusion comes thus from the relation

$$\lim_{n\to\infty} a_i(M_n)^{-1} a_1(M_n) = \frac{1}{\alpha_i} .$$

DEFINITION 1.5. *A probability measure* m *on* $P(\mathbb{R}^d)$ *is said to be proper if for any hyperplane* H *in* $\mathbb{R}^d$

$$m(\bar{x} \in P(\mathbb{R}^d);\quad x \in H - \{0\}) = 0 .$$

The following result shows that the word "contracting" refers to the action on proper measures on $P(\mathbb{R}^d)$.

PROPOSITION 1.6. *Let* T *be a contracting subset of* $Gl(d,\mathbb{R})$. *For any proper probability measure* m *on* $P(\mathbb{R}^d)$ *there exists a sequence* $\{M_n, n \geq 0\}$ *in* T *such that* $M_n m$ *converges weakly to a Dirac measure as* $n \longrightarrow \infty$.

Proof : Let (M_n) be a sequence in T such that $\|M_n\|^{-1} M_n$ converges to a rank one matrix, say M. For any $x \in \mathbb{R}^d$ such that $Mx \neq 0$, $\lim_{n\to\infty} M_n\cdot\bar{x} = M\cdot\bar{x}$. But since m is proper and since $\{x \in \mathbb{R}^d ; Mx = 0\}$ is a hyperplane, we see that

$$\lim_{n\to\infty} M_n\cdot\bar{x} = M\cdot\bar{x} \qquad \text{for m-almost all } \bar{x} .$$

It is now clear that if $\bar{z}$ is the direction of the range of M,

$$\lim_{n\to\infty} M_n\cdot\bar{x} = \bar{z} \qquad \text{for m-almost all } \bar{x}$$

so that $M_n m$ converges weakly to $\delta_{\bar{z}}$.

It is often difficult to compute the index of a given set. But we shall use the following.

PROPOSITION 1.7. *Let* T *be a semigroup of* $Gl(d,\mathbb{R})$. *If the index of* T *is equal to* d *then* $\{|\det M|^{-1/d} M ; M \in T\}$ *is contained in a compact subgroup of* $Gl(d,\mathbb{R})$.

Proof : Let $T' = \{|\det M|^{-1/d} M ; M \in T\}$. It is clear that T' is a semigroup with index equal to d. Furthermore its closure C is itself a semigroup and it is easily seen that its index is also d. Let us show that C is compact. If (M_n) is a sequence in C we can find a subsequence, denoted also by M_n for convenience, such that $\|M_n\|^{-1} M_n$ converges to some matrix M. We have

$$|\det M| = \lim |\det (\frac{M_n}{\|M_n\|})| = \lim \|M_n\|^{-d} .$$

Since the index of C is d, M is nonsingular and this relation shows that $\|M_n\|$ converges. Therefore M_n converges in $Gl(d,\mathbb{R})$ and the limit is in C since C is closed. So every sequence in C contains a converging subsequence and C is a compact semigroup contained in $Gl(d,\mathbb{R})$. By Theorem 9.16 of Hewitt and Ross [36] or by Exercise II.2.4 this implies that C is in fact a compact group.

Remark 1.8. It is useful to notice that under the hypothesis of Proposition 1.7 there exists a scalar product $[\cdot,\cdot]$ on $\mathbb{R}^d$ such that for $M \in T$, $|\det M|^{-1/d}M$ is orthogonal for this scalar product. This is due the fact that if C is a compact subgroup of $Gl(d,\mathbb{R})$ and if λ is its normalized Haar measure, then each element of C is orthogonal for the scalar product defined by

$$[x,y] = \int_C \langle Mx,My\rangle \, d\lambda(M) .$$

Exercise 1.9. Suppose that a semigroup T in $Gl(d,\mathbb{R})$ contains a matrix M with a unique eigenvalue of maximal modulus, this eigenvalue being simple. Show that T is contracting. (Solution : Corollary IV.2.2).

Exercise 1.10. Show that any real matrix of order d has a polar decomposition. Prove that any M in $Gl(d,\mathbb{C})$ can be written as $M=KAU$ where $A = \operatorname{diag}(a_1,\dots,a_d)$, $a_1 \geq a_2 \geq \dots \geq a_d > 0$ being the square roots of the eigenvalues of M^*M, and where K and U are in $U(d) = \{M \in Gl(d,\mathbb{C}) ;\ M^*M = Id\}$.

Exercise 1.11. Consider two polar decompositions $M = KAU$, $M = K'AU'$ of a matrix M in $Gl(d,\mathbb{R})$.

(i) Prove that $U^*e_1,\dots,U^*e_d$ are eigenvectors of M^*M.

(ii) Show that there exists an orthogonal matrix B such that $AB = BA$ and $K = K'B$, $U' = BU$. If $A = \operatorname{diag}(a_1,\dots,a_d)$ prove that if $a_i \neq a_j$ the entries $B_{i,j}$ and $B_{j,i}$ of B are zero. If $a_1 > a_2 > \dots > a_d$

verify that $B = \mathrm{diag}(\varepsilon_1,\ldots,\varepsilon_d)$ with $\varepsilon_i = \pm 1$.

Exercise 1.12. Let (M_n) be a sequence in $Gl(d,\mathbb{R})$. Show that the following are equivalent

(i) For some proper distribution m on $P(\mathbb{R}^d)$, $M_n m$ converges to $\delta_{\bar{z}}$.

(ii) For any proper distribution m on $P(\mathbb{R}^d)$, $M_n m$ converges to $\delta_{\bar{z}}$.

(iii) The limit points of $\|M_n\|^{-1} M_n$ are rank-one matrices with range $\mathbb{R}z$.

III.2. Strong irreducibility

DEFINITION 2.1. *Given a subset* S *of* $Gl(d,\mathbb{R})$ *we say that*

(i) S *is irreducible, if there is no proper linear subspace* V *of* $\mathbb{R}^d$ *such that* $M(V) = V$ *for any* M *in* S .

(ii) S *is strongly irreducible if there does not exist a finite family of proper linear subspaces of* $\mathbb{R}^d$, $V_1,V_2,\ldots,V_k$ *such that*

$$M(V_1 \cup V_2 \ldots \cup V_k) = V_1 \cup V_2 \ldots \cup V_k$$

for any M *in* S .

For instance $S = \{\begin{pmatrix} 0 & 1 \\ 1 & 0 \end{pmatrix}\}$ is irreducible but not strongly irreducible.

NOTATION 2.2. *Given a probability measure* μ *on* $Gl(d,\mathbb{R})$ *we write* T_μ *(resp.* G_μ*) for the smallest closed semigroup (resp. subgroup) of* $Gl(d,\mathbb{R})$ *which contains the support of* μ .

Remark that if $V_1,\ldots,V_k$ are subspaces of $\mathbb{R}^d$, then the set of matrices M such that $M(V_1 \cup \ldots \cup V_k) = V_1 \cup \ldots \cup V_k$ is a closed subgroup of $Gl(d,\mathbb{R})$. Therefore

$$\begin{aligned}\text{Supp}\,\mu \text{ strongly irreducible} &\iff T_\mu \text{ strongly irreducible}\\ &\iff G_\mu \text{ strongly irreducible}\,.\end{aligned}$$

The importance of strong irreducibility comes from the following result. Recall that a probability measure ν on $P(\mathbb{R}^d)$ is said to be μ-invariant if

$$\nu(A) = \iint 1_A(M\cdot\overline{x})d\mu(M)\ d\nu(\overline{x})$$

for any Borel set A in $P(\mathbb{R}^d)$.

PROPOSITION 2.3. *Let* μ *be a probability measure on* $Gl(d,\mathbb{R})$ *such that* G_μ *is strongly irreducible. Then any* μ*-invariant distribution* ν *on* $P(\mathbb{R}^d)$ *is proper.*

Proof : Let $\Gamma(\ell)$ be the set of ℓ-dimensional subspaces of $\mathbb{R}^d$. Consider

$$\ell_o = \inf\{\ell \leq d\ ;\ \nu(\overline{V}) \neq 0 \text{ for some } V \text{ in } \Gamma(\ell)\}$$

(where $\overline{V} = \{\overline{x} \in P(\mathbb{R}^d)\ ;\ x \in V - \{0\}\}$) ,

$$r = \sup\{\nu(\overline{V})\ ;\ V \in \Gamma(\ell_o)\}$$

and

$$\Lambda = \{V \in \Gamma(\ell_o)\ ;\ \nu(\overline{V}) = r\}\ .$$

If V_1 and V_2 are in Λ either $V_1 = V_2$ or $\dim(V_1 \cap V_2) < \ell_o$ and the definition of ℓ_o implies that $\nu(\overline{V}_1 \cap \overline{V}_2) = 0$. Therefore if $V_1, V_2, \ldots, V_k$ are distinct elements of Λ

$$\nu(\overline{V}_1 \cup \overline{V}_2 \cup \ldots \cup \overline{V}_k) = \sum_{i=1}^{k} \nu(\overline{V}_i) = kr$$

so Λ must be finite.

On the other hand, for each V in Λ

$$r = \nu(\overline{V}) = \iint 1_{\overline{V}}(M\cdot\overline{x})d\mu(M)d\nu(\overline{x}) = \int \nu(M^{-1}\cdot\overline{V})\ d\mu(M)\ .$$

But $\nu(M^{-1}\cdot\overline{V}) \leq r$, so that for μ-almost all M $\nu(M^{-1}\cdot\overline{V}) = r$ and $M^{-1}\cdot\overline{V}$ is in Λ . Write L for the set of vectors which belong to some V in Λ . From the foregoing , $\{M \in Gl(d,\mathbb{R})\ ;\ M(L) = L\}$ is a closed subgroup which contains the support of μ . It thus contains T_μ . If $\ell_o < d$, L is a finite union of proper subspaces and T_μ is not strongly irreducible. Therefore $\ell_o = d$ and ν is proper.

Exercise 2.4. Prove that a closed subgroup G of $Gl(d,\mathbb{R})$ is strongly irreducible if and only if each closed normal subgroup H of G such that G/H is finite is irreducible.

Exercise 2.5. Show that G_μ is strongly irreducible if and only if each μ-invariant distribution on $P(\mathbb{R}^d)$ is proper.

III.3. A key property

The main properties of product of random matrices will be consequences of the following.

THEOREM 3.1. *Let* $X_1, X_2, \dots$ *be independent random matrices in* $Gl(d,\mathbb{R})$ *with a common distribution* μ *, defined on* $(\Omega, A, \mathbb{P})$. *We suppose that* T_μ *is strongly irreducible and write* p *for its index.*

Then, if $M_n = X_1 X_2 \dots X_n$ *, for almost all* ω *there exists a* p*-dimensional subspace* $V(\omega)$ *of* $\mathbb{R}^d$ *such that any limit point of* $\{ \|M_n(\omega)\|^{-1} M_n(\omega),\ n \geq 1\}$ *is a rank* p *matrix with range* $V(\omega)$. *For any non zero vector* x *,*

$$\mathbb{P}\{x \text{ is orthogonal to } V(\omega)\} = 0 \ .$$

When T_μ *is contracting there exists a unique* μ*-invariant distribution* ν *on* $P(\mathbb{R}^d)$ *and* $M_n(\omega)\nu$ *converges weakly to* $\delta_{\overline{Z}(\omega)}$, where $Z(\omega)$ *is any non zero vector of* $V(\omega)$.

Proof : Consider a μ-invariant distribution ν on $P(\mathbb{R}^d)$. By Lemma II.2.1 there exists, for almost all ω , a probability measure ν_ω such that, in the weak topology ,

$$M_n(\omega) g\, \nu \longrightarrow \nu_\omega \qquad , \qquad \text{as } n \longrightarrow \infty \ ,$$

for almost all g in $Gl(d,\mathbb{R})$ w.r.t. $\lambda = \sum_{n=0}^{\infty} 2^{-n-1} \mu^n$.

Henceforth we fix such an ω . Consider a limit point $A(\omega)$ of $\{ \|M_n(\omega)\|^{-1} M_n(\omega),\ n \geq 1\}$. If $A(\omega)x \neq 0$, $M_n(\omega)\cdot\overline{x} \longrightarrow A(\omega)\cdot\overline{x}$ when n goes to infinity along a subsequence. By Proposition 2.3

$$\nu(\{\overline{x} \in P(\mathbb{R}^d) \ ; \ A(\omega) g\, x = 0\}) = 0$$

and $A(\omega) g\, \nu = \lim M_n(\omega) g\, \nu = \nu_\omega$ for λ- almost all g . (We can of course define $A(\omega)g\,\nu$ by Proposition 2.3 , see the analogous Lemma II.3.2). It is clear that $\{g \in Gl(d,\mathbb{R})\ ;\ A(\omega) g\, \nu = \nu_\omega\}$ is closed. The support of λ being $T_\mu \cup \{Id\}$, this yields

$$A(\omega) g\, \nu = \nu_\omega \qquad \text{for all } g \text{ in } T_\mu \ . \tag{1}$$

Consider now a sequence $\{g_n,\ n \geq 1\}$ in T_μ such that $\|g_n\|^{-1} g_n$ converges to a matrix h of rank p (which exists, since the index of T_μ is p). For any g in T_μ , $g g_n \in T_\mu$ so that

$$A(\omega) g\, g_n\, \nu = \nu_\omega \ .$$

If $A(\omega)gh\,x \neq 0$, $A(\omega)gg_n\cdot\bar{x}$ converges to $A(\omega)gh\cdot\bar{x}$. Therefore if

$$\nu(\{\bar{x} \in P(\mathbb{R}^d)\ ;\ A(\omega)gh\,x = 0\}) = 0 \qquad (2)$$

then $A(\omega)gg_n\,\nu$ converges weakly to $A(\omega)gh\,\nu$. Since by Proposition 2.3 ν is proper, (2) yields that

$$A(\omega)gh\,x = 0 \qquad \text{for any } x \text{ in } \mathbb{R}^d. \qquad (3)$$

If (3) were true for all g in T_μ, the subspace V of $\mathbb{R}^d$ spanned by $\{gx\ ;\ g \in T_\mu\ ,\ x \in \operatorname{Im} h\}$ would be contained in the null space of $A(\omega)$, and hence would be proper. But for any g in T_μ $gV = V$ and this would be in contradiction with the strong irreducibility of T_μ. We thus can find $g \in T_\mu$ for which (2) holds and then

$$\nu_\omega = \lim A(\omega)g_n g\,\nu = A(\omega)gh\,\nu\ . \qquad (4)$$

Let $V(\omega)$ be the linear span of $\{x \in \mathbb{R}^d - \{0\}\ ;\ \bar{x} \in \operatorname{Supp} \nu_\omega\}$. By (4), $V(\omega)$ is contained in the range of $A(\omega)gh$ and $\dim V(\omega) \leq p$ for the rank of h is p.

On the other hand the support of the distribution of M_n is $(\operatorname{Supp}\mu)^n$, hence $\mathbb{P}(M_n \in T_\mu) = 1$ and $\mathbb{P}(M_n \in T_\mu\ ,\ \forall n \geq 1) = 1$. By definition of the index, the rank of $A(\omega)$ is at least p. Since $\nu_\omega = A(\omega)\nu$, $V(\omega) = \operatorname{Im} A(\omega)$ and $\dim V(\omega) = p$.

We have thus proved that $V(\omega)$ is a p-dimensional subspace which is the range of each limit point of $\|M_n(\omega)\|^{-1} M_n(\omega)$.

For any non zero x in $\mathbb{R}^d$ set $H(x) = \{\bar{y} \in P(\mathbb{R}^d)\ ,\ \langle x,y\rangle = 0\}$. We know (see Lemma II.2.1) that $\nu = \int \nu_\omega \, d\mathbb{P}(\omega)$. Therefore

$$\begin{aligned}\mathbb{P}(x \text{ is orthogonal to } V(\omega)) &= \mathbb{P}(\operatorname{Supp}\nu_\omega \subset H(x)) \\ &= \mathbb{P}(\nu_\omega(H(x)) = 1) \\ &= \nu(H(x))\end{aligned}$$

and this is equal to zero for ν is proper.

Finally if T_μ is contracting then $p = 1$ by definition and $\bar{V}(\omega)$ is reduced to a point $\bar{Z}(\omega)$ in $P(\mathbb{R}^d)$. By construction $\bar{Z}(\omega)$ does not depend on ν. Since the distribution of $\bar{Z}(\omega)$ is $\int \delta_{\bar{Z}(\omega)}\, d\mathbb{P}(\omega) = \nu$, ν must be the unique μ-invariant distribution.

The following consequence is of prime importance. We shall make an intensive use of it.

PROPOSITION 3.2. *Let* $Y_1, Y_2, \ldots$ *be i.i.d. random elements of* $Gl(d,\mathbb{R})$

with common distribution μ *and* $S_n = Y_n \dots Y_1$. *Consider a polar decomposition*

$$S_n = K_n A_n U_n$$

with K_n *and* U_n *in* $O(d)$ *and* $A_n = \mathrm{diag}(a_1(n), \dots, a_d(n))$ *with* $a_1(n) \geq \dots \geq a_d(n) > 0$.

If T_μ *is a strongly irreducible semigroup with index* p *, then*

(a) The subspace spanned by $\{U_n^*(\omega)e_1, \dots, U_n^*(\omega)e_p\}$ *converges a.s. to a* p-*dimensional subspace* $V(\omega)$.

(b) With probability one

$$\lim_{n\to\infty} \frac{a_{p+1}(n)}{\|S_n\|} = 0 \quad \text{and} \quad \operatorname{Inf}_n \frac{a_p(n)}{\|S_n\|} > 0 .$$

(c) For any sequence $\{x_n, n \geq 1\}$ *in* $\mathbb{R}^d$ *which converges to a non zero vector ,*

$$\sup_{n\geq 1} \frac{\|S_n(\omega)\|}{\|S_n(\omega)x_n\|} < \infty \qquad a.s. .$$

We first prove

LEMMA 3.3. *If* ρ *is the distribution of* Y_1^* *the index of* T_ρ *is equal to the index of* T_μ *and* T_ρ *is strongly irreducible if* T_μ *is.*

Proof : It is plain that $T_\rho = \{M^*, M \in T_\mu\}$ so that the index of T_ρ is equal to the index of T_μ (recall that rank M = rank M^*). Suppose that there exist proper subspaces $V_1, V_2, \dots, V_\ell$ of $\mathbb{R}^d$ such that

$$M\Big(\bigcup_{i=1}^{\ell} V_i\Big) = \bigcup_{i=1}^{\ell} V_i \qquad \text{for each } M \in T_\rho .$$

Looking only at the V_i's with maximal dimension we may suppose that they all have the same dimension. Clearly

$$\{MV_1, \dots, MV_\ell\} = \{V_1, \dots, V_\ell\}$$

for any M in T_ρ . If W_i is the subspace orthogonal to V_i this implies that $\{M^*W_1, \dots, M^*W_\ell\} = \{W_1, \dots, W_\ell\}$ for any M in T_ρ . Therefore $M\Big(\bigcup_{i=1}^{\ell} W_i\Big) = \bigcup_{i=1}^{\ell} W_i$ for M in T_μ and T_μ is not strongly irreducible.

Proof of the proposition : By the lemma we can apply Theorem 3.1 to the random matrices $X_i = Y_i^*$. There exists, a.s., a p-dimensional subspace $V(\omega)$ which is the range of each limit point $M(\omega)$ of $\{\|S_n^*(\omega)\|^{-1} S_n^*(\omega),$

$n \geq 1\}$. Since $\|S_n^*\| = a_1(n)$,

$$\frac{S_n^*}{\|S_n\|} = U_n^* \operatorname{diag}\left(1, \frac{a_2(n)}{a_1(n)}, \ldots, \frac{a_d(n)}{a_1(n)}\right)K_n^* .$$

Each component of this product lies in a compact set. If $U_\infty(\omega)$, $K_\infty(\omega)$ $\alpha_2(\omega), \ldots, \alpha_d(\omega)$ are limit points of $U_n(\omega)$, $K_n(\omega)$, $\frac{a_2(n)}{a_1(n)}(\omega), \ldots, \frac{a_d(n)}{a_1(n)}(\omega)$, then

$$M(\omega) = U_\infty^*(\omega) \operatorname{diag}(1, \alpha_2(\omega), \ldots, \alpha_d(\omega))K_\infty^*(\omega)$$

is a limit point of $\|S_n^*(\omega)\|^{-1} S_n^*(\omega)$. Therefore with probability one rank $M(\omega) = p$, i.e. $\alpha_2(\omega) \geq \ldots \geq \alpha_p(\omega) > 0$ and $\alpha_{p+1}(\omega) = \ldots = \alpha_d(\omega) = 0$. The linear span of $\{U_\infty^*(\omega)e_1, \ldots, U_\infty^*(\omega)e_p\}$ is thus equal to the range of $M(\omega)$, proving (a). Assertion (b) follows readily from the relations $\alpha_p(\omega) \neq 0$ and $\alpha_{p+1}(\omega) = 0$. In order to verify (c) write

$$\frac{\|S_n x_n\|^2}{\|S_n\|^2} = \frac{\|A_n U_n x_n\|^2}{\|A_n\|^2} = \sum_{i=1}^{d} \left\{\frac{a_i(n)}{a_1(n)}\right\}^2 < x_n, U_n^* e_i >^2$$

so that $$\frac{\|S_n x_n\|^2}{\|S_n\|^2} \geq \left\{\frac{a_p(n)}{a_1(n)}\right\}^2 \sum_{i=1}^{p} < x_n , U_n^* e_i >^2 .$$

If $x_n \longrightarrow x$ and $y(\omega)$ is the orthogonal projection of x onto $V(\omega)$, then

$$\liminf_{n\to\infty} \frac{\|S_n(\omega)x_n\|}{\|S_n(\omega)\|} \geq \inf_n\left\{\frac{a_p(n)}{a_1(n)}(\omega)\right\} \|y_n(\omega)\|$$

By Theorem 3.1

$$\mathbb{P}(\|y(\omega)\| = 0) = \mathbb{P}(x \text{ is orthogonal to } V(\omega)) = 0$$

and (a) follows using (b) and Lemma 1.4 .

We now deduce from this proposition an easy, but important, corollary due to Furstenberg. Actually we shall see in Section 7 an analogous result when G_μ is merely irreducible.

<u>COROLLARY 3.4.</u> *Consider a sequence* $\{Y_n, n \geq 1\}$ *of independent random matrices in* $Gl(d,\mathbb{R})$ *with common distribution* μ . *We suppose that*

(a) $\mathbb{E}(\operatorname{Log}^+\|Y_1\|) < \infty$.

(b) G_μ *is strongly irreducible* .

Then

(i) For any sequence (x_n) *which converges to a non zero vector,*

$$\lim \frac{1}{n} \operatorname{Log} \|Y_n \dots Y_1 \, x_n\| = \gamma \qquad a.s. \quad .$$

(ii) If ν *is a* μ*-invariant distribution on* $P(\mathbb{R}^d)$, *then*

$$\gamma = \iint \operatorname{Log} \frac{\|Mx\|}{\|x\|} \, d\mu(M) \, d\nu(\bar{x}) \ .$$

(iii) If moreover $\mathbb{E}(\operatorname{Log}^+ \|Y_1^{-1}\|) < \infty$, *then* $\mathbb{E}\{\frac{1}{n} \operatorname{Log} \|Y_n \dots Y_1 x\|\}$ *converges to* γ *uniformly on* $\{x \in \mathbb{R}^d ; \ \|x\| = 1\}$.

Proof : Let $S_n = Y_n \dots Y_1$ and $a_1(n) \geq \dots \geq a_d(n) > 0$ be the square roots of the eigenvalues of $S_n^* S_n$. By Proposition 3.2 if (x_n) converges to some non zero vector there exists $C(\omega)$ such that, a.s. ,

$$0 < C(\omega) \leq \frac{\|S_n(\omega) x_n\|}{\|S_n(\omega)\|} \leq 1 \ .$$

Therefore

$$\lim \frac{1}{n} \operatorname{Log} \|S_n(\omega) x_n\| = \lim \frac{1}{n} \operatorname{Log} \|S_n\| = \gamma \qquad \text{a.s.}$$

which proves (i).

Let ν be a μ-invariant distribution on $P(\mathbb{R}^d)$. Since $\iint \operatorname{Log}^+ \frac{\|Yx\|}{\|x\|} \, d\mu(Y) d\nu(\bar{x})$ is finite, one can deduce from Proposition I.3.4 using the cocycle $\sigma(Y,\bar{x}) = \operatorname{Log} \frac{\|Yx\|}{\|x\|}$, that $\frac{1}{n} \operatorname{Log} \|Y_n \dots Y_1 x\|$ converges $\mathbb{P} \otimes \nu$ a.s. to a random variable whose expectation is $\iint \operatorname{Log} \frac{\|Yx\|}{\|x\|} \, d\mu(Y) d\nu(\bar{x})$ (see also exercise II.2.6). Since by the foregoing this limit is γ , this proves (ii).

Finally we prove (iii). Since $\{x \in \mathbb{R}^d ; \ \|x\| = 1\}$ is compact it suffices to show that if (x_n) is a converging sequence on the unit sphere then

$$\lim_{n \to \infty} \frac{1}{n} \mathbb{E}(\operatorname{Log} \|S_n x_n\|) = \gamma \ .$$

But we have, by (i) ,

$$\lim \frac{1}{n} \operatorname{Log} \|S_n x_n\| = \gamma$$

and

$$|\operatorname{Log} \|Yx\| \, | \leq \operatorname{Log}^+ \|Y\| + \operatorname{Log}^+ \|Y^{-1}\|$$

so that

$$\left|\frac{1}{n} \operatorname{Log} \|S_n x_n\|\right| \leq \frac{1}{n} \sum_{i=1}^{n} (\operatorname{Log}^+ \|Y_i\| + \operatorname{Log}^+ \|Y_i^{-1}\|) .$$

The right hand side converges in L^1 by the law of large numbers, hence $\{\frac{1}{n} \operatorname{Log} \|S_n x\| , n \geq 1\}$ is uniformly integrable and (iii) follows.

Remark 3.5. If $|\det Y_1| = 1$ a.s. the condition $\mathbb{E}(\operatorname{Log}^+ \|Y_1\|) < \infty$ implies that $\mathbb{E}(\operatorname{Log}^+ \|Y_1^{-1}\|) < \infty$ since

$$\|Y_1^{-1}\| = a_d(1)^{-1} = \prod_{i=1}^{d-1} a_i(d) \leq a_1(d)^{d-1} = \|Y_1\|^{d-1} .$$

Exercise 3.6. Let m be a proper probability measure on $P(\mathbb{R}^d)$. Show that $\{M \in Gl(d,\mathbb{R}) ; Mm = m\}$ is a closed subgroup of $Gl(d,\mathbb{R})$ with index d.

Exercise 3.7. Show that the conclusion (a) of Proposition 3.2 is not necessarily true if T_μ is merely irreducible (Hint : use the example introduced in Exercise II.6.6).

Exercise 3.8. Show that if $\mathbb{E}(\operatorname{Log}^+ \|Y_1\|) < \infty$ then the conditions $\mathbb{E}(\operatorname{Log}^+ \|Y_1^{-1}\|) < \infty$ and $\mathbb{E}(\operatorname{Log} |\det Y_1|) > -\infty$ are equivalent.

III.4. Contracting action on $P(\mathbb{R}^d)$ and convergence in direction

Recall the definition of the natural angular distance on $P(\mathbb{R}^d)$.

DEFINITION 4.1. *If* x, y *are unit vectors in* $\mathbb{R}^d$ *with direction* $\bar{x}$, $\bar{y}$ *we set*

$$\delta(\bar{x},\bar{y}) = \{1 - \langle x,y\rangle^2\}^{1/2} .$$

For any orthogonal matrix K , $\delta(K\cdot\bar{x}, K\cdot\bar{y}) = \delta(\bar{x},\bar{y})$. To check that δ is a distance it thus suffices to show that

$$\delta(\bar{y},\bar{z}) \leq \delta(\bar{y},\bar{x}) + \delta(\bar{x},\bar{z})$$

for $x = e_1$, $y = y_1e_1 + y_2e_2$, $z = z_1e_1 + z_2e_2 + z_3e_3$ being three unit vectors. In that case

$$\begin{aligned}\delta(\bar{y},\bar{z}) &= \{(y_1z_2 - y_2z_1)^2 + (y_1z_3)^2 + (y_2z_3)^2\}^{1/2} \\ &= \{(y_1z_2 - y_2z_1)^2 + z_3^2\}^{1/2}\end{aligned}$$

$$\leq \{(y_2+z_2)^2 + z_3^2\}^{1/2} \leq |y_2| + (z_2^2 + z_3^2)^{1/2}$$
$$\leq \delta(\overline{x},\overline{y}) + \delta(\overline{x},\overline{z}) .$$

LEMMA 4.2. *For any* M *in* $Gl(d,\mathbb{R})$, *if* $a_1(M)$, $a_2(M)$ *are the square roots of the two top eigenvalues of* M^*M *then*

$$\frac{\delta(M\cdot\overline{x} , M\cdot\overline{y})}{\delta(\overline{x},\overline{y})} \leq a_1(M)a_2(M) \frac{\|x\| \ \|y\|}{\|Mx\| \ \|My\|} .$$

This follows from an elementary computation (See also Lemma 5.3 and (9)).

Now we prove the main result on the behaviour of the random product S_n available without moment hypotheses. It states that S_n ultimately contracts the distance δ with probability one while the direction of S_n^*x converges in probability to a random limit independent of x (we shall see in VI that under stronger assumptions this latter convergence holds also almost surely). For instance all the column vectors of the matrix S_n tend to align in the same direction and the directions of the rows converge. This important result is due to Guivarc'h and Raugi [34]. It is remarkable that it holds under a condition which depends only on T_μ .

THEOREM 4.3. *Let* $S_n = Y_n \dots Y_1$ *where* $\{Y_n, n \geq 1\}$ *are independent matrices in* $Gl(d,\mathbb{R})$ *with distribution* μ . *Suppose that* T_μ *is strongly irreducible and contracting. Then*

(i) For any $\overline{x}, \overline{y}$ *in* $P(\mathbb{R}^d)$, *with probability one*

$$\lim_{n\to\infty} \delta(S_n\cdot\overline{x} , S_n\cdot\overline{y}) = 0 .$$

(ii) There exists a random direction $\overline{Z}$ *such that* $S_n^*\cdot\overline{x}$ *converges in probability to* $\overline{Z}$, *uniformly in* $\overline{x} \in P(\mathbb{R}^d)$.

(iii) There is a unique μ*-invariant distribution* ν *on* $P(\mathbb{R}^d)$, *and for any continuous function* f *on* $P(\mathbb{R}^d)$,

$$\sup_{\overline{x}} \left| \mathbb{E}\{f(S_n\cdot\overline{x})\} - \int f \, d\nu \right| \longrightarrow 0 \quad \text{as} \quad n \longrightarrow \infty .$$

Proof : Let $a_1(n)$, $a_2(n)$ be the square roots of the two top eigenvalues of $S_n^*S_n$. If (x_n) and (y_n) are two sequences of unit vectors which converge in direction, Lemma 4.2 gives

$$\frac{\delta(S_n\cdot\bar{x}_n\ ,\ S_n\cdot\bar{y}_n)}{\delta(\bar{x}_n,\bar{y}_n)} \leq \frac{a_1(n)\ a_2(n)}{\|S_n x_n\|\ \|S_n y_n\|} \leq \frac{a_2(n)}{\|S_n\|}\ \frac{\|S_n\|}{\|S_n x_n\|}\ \frac{\|S_n\|}{\|S_n y_n\|}\ .$$

Therefore using Proposition 3.2 we have, a.s.

$$\lim_{n\to\infty} \frac{\delta(S_n\cdot\bar{x}_n\ ,\ S_n\cdot\bar{y}_n)}{\delta(\bar{x}_n,\bar{y}_n)} = 0 \tag{5}$$

which, in particular, proves (i).

Let μ^* be the common distribution of the random matrices $X_i = Y_i^*$ and $R_n = X_n \dots X_1$. I claim that as $n \longrightarrow \infty$

$$\underset{\bar{x},\bar{y}}{\operatorname{Sup}}\ \mathbb{E}\{\delta(R_n\cdot\bar{x}\ ,\ R_n\cdot\bar{y})\} \longrightarrow 0\ . \tag{6}$$

For if (6) does not hold we could find $\varepsilon > 0$, a subsequence n_i and $(\bar{x}_{n_i})$, $(\bar{y}_{n_i})$ for which

$$\mathbb{E}\{\delta(R_{n_i}\cdot\bar{x}_{n_i}\ ,\ R_{n_i}\cdot\bar{y}_{n_i})\} > \varepsilon\ .$$

Without loss of generality we may assume that $(\bar{x}_{n_i})$ and $(\bar{y}_{n_i})$ converge. By Lemma 3.3, (5) holds for R_n and

$$\lim_{i\to\infty} \delta(R_{n_i}\cdot\bar{x}_{n_i}\ ,\ R_{n_i}\cdot\bar{y}_{n_i}) = 0$$

which leads to a contradiction.

We know (Theorem 3.1) that if m is the unique μ^*-invariant distribution on $P(\mathbb{R}^d)$, $X_1 \dots X_n m$ converges weakly to some Dirac measure $\delta_{\bar{Z}}$. We have

$$\mathbb{E}\{\delta(S_n^*\cdot\bar{x},\bar{Z})\} \leq \mathbb{E}\{\delta(S_n^*\cdot\bar{x}\ ,\ S_n^*\cdot\bar{y})\} + \mathbb{E}\{\delta(S_n^*\cdot\bar{y},\bar{Z})\}$$

whence, since S_n^* has the same law as R_n ,

$$\underset{\bar{x}\in P(\mathbb{R}^d)}{\operatorname{Sup}}\ \mathbb{E}\{\delta(S_n^*\cdot\bar{x},\bar{Z})\} \leq \underset{\bar{x},\bar{y}}{\operatorname{Sup}}\ \mathbb{E}\{\delta(R_n\cdot\bar{x},R_n\cdot\bar{y})\} + \int \mathbb{E}\{\delta(S_n^*\cdot\bar{y},\bar{Z})\}dm(\bar{y})$$

The first term of the right hand side tends to 0 by (6), the second one is equal to

$$\mathbb{E}\{\int \delta(X_1\dots X_n\cdot\bar{y},\bar{Z})dm(\bar{y})\} = \mathbb{E}\{\int \delta(\bar{u},\bar{Z})d(X_1\dots X_n m)(\bar{u})\}$$

which converges to $\mathbb{E}\{\delta(\bar{Z},\bar{Z})\} = 0$. We immediately deduce (ii).

If f is a continuous function on $P(\mathbb{R}^d)$, f is uniformly continuous. Therefore, using the foregoing ,

$$\sup_{\bar{x}} \mathbb{E}\{|f(S_n^* \cdot \bar{x}) - f(\bar{Z})|\} \longrightarrow 0 \qquad \text{as } n \longrightarrow \infty .$$

Thus

$$\sup_{\bar{x}} |\mathbb{E}(f(X_n \dots X_1 \cdot \bar{x})) - \int f \, dm| \leq \sup_{\bar{x}} \mathbb{E}\{|f(S_n^* \cdot \bar{x}) - f(\bar{Z})|\}$$

converges to zero.

Similarly, working with Y_i instead of X_i, one obtains (iii).

The assumptions of this theorem are rather weak. For instance let us show, using its notations

PROPOSITION 4.4. *If* T_μ *is only supposed to be irreducible, (i) holds if and only if* T_μ *is strongly irreducible and contracting.*

LEMMA 4.5. *If* T_μ *is not strongly irreducible we can find proper subspaces* $V_1, \dots, V_\ell$ *with the same dimension such that*

(a) If $i \neq j$, $V_i \cap V_j = \{0\}$.

(b) For any $i \in \{1, \dots, \ell\}$ *and* M *in* T_μ, $MV_i = V_j$ *for some* $j \in \{1, \dots, \ell\}$.

Proof : Let r be the least positive integer such that there exists a finite union L of subspaces with dimension not greater than r, such that $M(L) = L$ for each M in T_μ. Let V_1 be one of these subspaces with dimension r. It is clear that there exist $V_2, \dots, V_\ell$ in L such that $\{MV_1, M \in T_\mu\} = \{V_1, V_2, \dots, V_\ell\}$. We suppose the V_i distinct. Note that, if $M \in T_\mu$,

$$M(\bigcup_{i \neq j} (V_i \cap V_j)) = \bigcup_{i \neq j} (V_i \cap V_j) \quad .$$

Since $\dim V_i \cap V_j < r$ the minimality assumption on r implies (a). (b) is obvious.

Proof of the Proposition : Let m be any proper distribution on $P(\mathbb{R}^d)$, (see Definition 1.5). If (i) of Theorem 4.3 holds one can find ω, $\bar{x}$ and n_i such that $S_{n_i}(\omega) \in T_\mu$ and, as $i \longrightarrow \infty$, $S_{n_i}(\omega) \cdot \bar{x}$ converges to some $\bar{z}$ and $\delta(S_{n_i}(\omega) \cdot \bar{x}, S_{n_i}(\omega) \cdot \bar{y}) \longrightarrow 0$ for m-almost all $\bar{y}$.

This implies that $S_{n_i}(\omega)m$ converges weakly to the Dirac measure at $\bar{z}$ and $\|S_{n_i}(\omega)\|^{-1} S_{n_i}(\omega)$ converges to a matrix with range $\bar{z}$. This proves that T_μ is contracting.

Suppose that T_μ is not strongly irreducible. Let $\overline{V}_i = \{\overline{x} \in P(\mathbb{R}^d);\ x \in V_i\}$, where $V_1,\dots,V_\ell$ are the proper subspaces introduced in the lemma. Since $\overline{V}_i \cap \overline{V}_j = \emptyset$,

$$\alpha = \inf\{\delta(\overline{V}_i,\overline{V}_j)\ ,\ 1 \leq i < j \leq \ell\} > 0\ .$$

If (i) holds we can find $\overline{x} \in \overline{V}_1$, $\overline{y} \in \overline{V}_2$ and $S_n(\omega) \in T_\mu$ such that $\lim \delta(S_n(\omega)\cdot\overline{x}\ ,\ S_n(\omega)\cdot\overline{y}) = 0$. But (b) of the lemma implies that for some $i \neq j$ $S_n(\omega)\cdot\overline{x} \in \overline{V}_i$ and $S_n(\omega)\cdot\overline{y} \in \overline{V}_j$. Thus $\delta(S_n(\omega)\cdot\overline{x}\ ,\ S_n(\omega)\cdot\overline{y}) > \alpha > 0$ for each positive integer n.

Remark 4.6. It is usually difficult to know whether a given semigroup is contracting or not. The study of this problem remains to be done.

Remark 4.7. For arbitrary positive matrices, i.e. matrices whose entries are > 0, a stronger statement holds. Let $\mathbb{R}^d_+ = \{x = (x_1,\dots,x_d) \in \mathbb{R}^d\ ;\ x_i > 0\ ,\ 1 \leq i \leq d\}$ and $C = \{\overline{x} \in P(\mathbb{R}^d), x \in \mathbb{R}^d_+\}$. There exists on C a distance θ compatible with the topology, called the Hilbert projective distance, with the following property : if $M > 0$ there is $\theta(M) < 1$ such that

$$\Theta(M\cdot\overline{x}\ ,\ M\cdot\overline{y}) \leq \theta(M)\ \theta(\overline{x},\overline{y}) \qquad (7)$$

for any $\overline{x}$, $\overline{y}$ in C. Individually each matrix contracts θ.
The following exercises provide the proof of this result due to G. Birkhoff [6] (one can find another one in [64]).
If $S_n = Y_n\dots Y_1$ is a product of positive i.i.d. matrices it is almost immediate to obtain the analogue of Theorem 4.3 where (C,θ) take the place of $(P(\mathbb{R}^d),\delta)$, see Exercise 6.5 .

Exercise 4.8. (case $d = 2$). Let L denote the open half line $(0,\infty)$ and $\overline{L} = [0,\infty]$. For u, v in L set $\tau(u,v) = |\text{Log } u - \text{Log } v|$, this defines a distance. Consider a projective map p (this means that $p(x) = \frac{ax+b}{cx+d}$) such that $p(\overline{L}) \subset L$. We want to prove that if λ is the diameter of $p(L)$ for the distance τ, then

$$\tau(p(u),\ p(v)) \leq \{\tanh \frac{\lambda}{4}\}\ \tau(u,v) \qquad (8)$$

a. Using $\tau(p(u),p(v)) = \tau(\frac{1}{p(u)}, \frac{1}{p(v)})$ show that one can suppose that $p(x) = \frac{ax+b}{cx+d}$ with $a > 0$, $b > 0$, $c > 0$, $d > 0$ and $ad-bc \neq 0$.

b. Prove that $\sup_{\mathbb{R}} \frac{d}{dx} \text{Log } p(e^x) = \frac{\nu-1}{\nu+2\sqrt{\nu}+1}$ if $\nu = \frac{ad}{bc}$ and that $\lambda = \text{Log } \nu$.

c. Prove the inequality (8).

d. If $x = (x_1,x_2)$, $y = (y_1,y_2)$ are in $\mathbb{R}_+^2$, define

$$\theta(\bar{x}\ ,\ \bar{y}) = \left|\mathrm{Log}\ \frac{x_1}{x_2} - \mathrm{Log}\ \frac{y_1}{y_2}\right| .$$

Show that (7) holds for positive 2×2 matrices.

Exercise 4.9. (case $d \geq 3$). Let $H = \{x = (x_1,\ldots,x_d) \in \mathbb{R}^d\ ,\ \sum_{i=1}^{d} x_i = 1\}$ and B the boundary of $H \cap \mathbb{R}_+^d$. For x, y in $H \cap \mathbb{R}_+^d$ we consider the two points a, b on the line $\{\alpha x + (1-\alpha)y,\ \alpha \in \mathbb{R}\}$ which lie on B, and define

$$\theta(\bar{x}\ ,\ \bar{y}) = \mathrm{Log}\ [a,b\ ;\ x,y]$$

where $[\ ;\ ]$ is the cross ratio (if $x = \lambda a + \mu b$, $y = \lambda' a + \mu' b$, $[a,b\ ;\ x,y] = \frac{\lambda'\mu}{\lambda\mu'}$). For any z on $[a,b] = \{\alpha a + (1-\alpha)b,\ 0 \leq \alpha \leq 1\}$ we set

$$\Psi_{\bar{x},\bar{y}}(\bar{z}) = \mathrm{Log}\ \left[a,b\ ;\ \frac{a+b}{2}\ ,\ z\right] .$$

a. Show that $\theta(\bar{x},\bar{y}) = \mathrm{Log}\{\max_{1\leq i\leq d} \frac{x_i}{y_i}\}\{\min_{1\leq i\leq d} \frac{x_i}{y_i}\}^{-1}$.

b. Prove that θ is a distance on C and that $\Psi_{\bar{x},\bar{y}}$ is an isometry from $([a,b] \cap C, \theta)$ onto $(L\ ,\ \tau)$ (see Exercise 4.8).

c. Let M be a positive matrix. Show that the diameter λ of $M \cdot \bar{C}$ is finite. For $\bar{x}$, $\bar{y}$ in C prove that $p : \bar{L} \longrightarrow \bar{L}$ defined by

$$p(u) = \Psi_{M\cdot\bar{x},M\cdot\bar{y}}(M \cdot \Psi^{-1}_{\bar{x},\bar{y}}(u))$$

is a projective map and that $p(\bar{L}) \subset L$. Prove that (7) holds with $\theta(M) = \tanh \frac{\lambda}{4}$ (Use Exercise 4.8).

III.5. Lyapunov exponents

We shall now define the d Lyapunov exponents associated with (Y_n). They will in particular help improve Theorem 4.3. This is easily done if we work with the exterior powers $\Lambda^p \mathbb{R}^d$. For the reader's convenience we recall the definition and simple properties of these spaces.

Let E be a d-dimensional vector space with dual E^*. For any positive integer p let $\Lambda^p E$ denote the vector space of alternating p-linear forms on E^*. For $u_1, u_2, \ldots, u_p$ in E, $f_1, \ldots, f_p$ in E^* we set

$$(u_1 \wedge u_2 \wedge \ldots \wedge u_p)(f_1, \ldots, f_p) = \det\left[\{f_i(u_j)\}_{i,j}\right] .$$

It is clear that $u_1 \wedge \ldots \wedge u_p$ is an element of $\Lambda^p \mathbb{R}^d$, we shall call it a decomposable p-vector.

The following facts are readily checked (see e.g. 7.7 of Loomis, Sternberg [52]).

LEMMA 5.1. *(i) If* $\{e_1, \ldots, e_d\}$ *is a basis of* E, *then* $\{e_{i_1} \wedge e_{i_2} \wedge \ldots \wedge e_{i_p}, 1 \leq i_1 < i_2 < \ldots < i_p \leq d\}$ *is a basis of* $\Lambda^p E$.

(ii) For $u_1, u_2, \ldots, u_p$ *in* E, $u_1 \wedge u_2 \wedge \ldots \wedge u_p$ *is non zero if and only if* $u_1, u_2, \ldots, u_p$ *are linearly independent.*

(iii) Two independent p-*uples* $\{u_i, 1 \leq i \leq p\}$ and $\{v_i, 1 \leq i \leq p\}$ *have the same linear span if an only if* $u_1 \wedge \ldots \wedge u_p = \lambda(v_1 \wedge \ldots \wedge v_p)$ *for some* λ.

One deduces from this lemma that the image of the set of decomposable p-vectors in the projective space $P(\Lambda^p E)$ can be identified with the set of p-dimensional subspaces of E.

NOTATION 5.2. *For* $\langle\cdot,\cdot\rangle$ *being the usual scalar product on* $\mathbb{R}^d$ *we define a scalar product on* $\Lambda^p \mathbb{R}^d$ *by the formula*

$$\langle u_1 \wedge \ldots \wedge u_p , v_1 \wedge \ldots \wedge v_p \rangle = \det\left[\{\langle u_i, v_j\rangle\}_{i,j}\right] .$$

Throughout the book the norm on $\Lambda^p \mathbb{R}^d$ will be the one associated with this scalar product. Note that $\{e_{i_1} \wedge \ldots \wedge e_{i_p}, 1 \leq i_1 < i_2 < \ldots < i_p \leq d\}$ is an orthonormal basis of $\Lambda^p \mathbb{R}^d$, and that $\|u_1 \wedge u_2 \wedge \ldots \wedge u_p\|$ is the p-dimensional volume of the parallelogram generated by $u_1, \ldots, u_p$. We have, for any non zero x, y in $\mathbb{R}^d$,

$$\boxed{\delta(\bar{x}, \bar{y}) = \frac{\|x \wedge y\|}{\|x\| \; \|y\|}} \qquad (9)$$

If M is in $Gl(d,\mathbb{R})$ we define an automorphism $\Lambda^p M$ of $\Lambda^p \mathbb{R}^d$ by

$$(\Lambda^p M)(u_1 \wedge \ldots \wedge u_p) = Mu_1 \wedge Mu_2 \wedge \ldots \wedge Mu_p .$$

$\Lambda^p M$ will denote either the linear mapping or its matrix in the basis $(e_{i_1} \wedge e_{i_2} \wedge \dots \wedge e_{i_p})$. We set

$$\|\Lambda^p M\| = \operatorname{Sup} \{ \|(\Lambda^p M)w\| , \ w \in \Lambda^p \mathbb{R}^d, \ \|w\| = 1\} \quad .$$

Note that $\Lambda^p (MN) = (\Lambda^p M)(\Lambda^p N)$ so that $\|\Lambda^p MN\| \leq \|\Lambda^p M\| \, \|\Lambda^p N\|$.
If K is an orthogonal matrix

$$<Ku_1 \wedge \dots \wedge Ku_p, Kv_1 \wedge \dots \wedge Kv_p> = \det\{<Ku_i,Kv_j>\} = \det\{<u_i,v_j>\},$$

hence $\Lambda^p K$ is orthogonal.

The importance of exterior powers in the study of random matrices comes from (9) and the following.

LEMMA 5.3. *Let* M *be a matrix in* $Gl(d,\mathbb{R})$, *and* $a_1(M) \geq a_2(M) \geq \dots \geq a_d(M) > 0$ *the square roots of the eigenvalues of* M^*M. *Then for any* p, $1 \leq p \leq d$,

$$\|\Lambda^p M\| = a_1(M)\, a_2(M) \dots a_p(M) \ .$$

Proof : Write a polar decomposition $M = KAU$ with $K, U \in O(d)$ and $A = \operatorname{diag}(a_1(M), \dots, a_d(M))$. Since $\Lambda^p K$ and $\Lambda^p U$ are isometries of $\Lambda^p \mathbb{R}^d$, $\|\Lambda^p M\| = \|(\Lambda^p K)(\Lambda^p A)(\Lambda^p U)\| = \|\Lambda^p A\|$. But

$$(\Lambda^p A)(e_{i_1} \wedge \dots \wedge e_{i_p}) = a_{i_1}(M) \dots a_{i_p}(M) e_{i_1} \wedge \dots \wedge e_{i_p}$$

whence

$$\|\Lambda^p A\| = \operatorname{Sup}\{a_{i_1}(M) \dots a_{i_p}(M) \ ; \ 1 \leq i_1 < i_2 < \dots < i_p \leq d\} = a_1(M) \dots a_p(M).$$

LEMMA 5.4. *For* M *in* $Gl(d,\mathbb{R})$ *set* $\ell(M) = \operatorname{Max}(\operatorname{Log}^+ \|M\|, \operatorname{Log}^+ \|M^{-1}\|)$. *Then for any* p, $1 \leq p \leq d$, *and any unit vector* w *of* $\Lambda^p \mathbb{R}^d$,

$$|\operatorname{Log} \|\Lambda^p M\| \,| \leq p\, \ell(M) \quad \text{and} \quad |\operatorname{Log} \|\Lambda^p M w\|| \leq p\, \ell(M) \ .$$

Proof : With the notations of the foregoing lemma we have

$$\|\Lambda^p M\| = a_1(M) \dots a_p(M) \leq a_1(M)^p \leq \|M\|^p$$

so that $\operatorname{Log} \|\Lambda^p M\| \leq p \operatorname{Log} \|M\| \leq p \operatorname{Log}^+ \|M\| \leq p\, \ell(M)$.
Since $\|\Lambda^p M^{-1} \Lambda^p M\| = \|\Lambda^p \operatorname{Id}\| = 1$ we also have

$$\|\Lambda^p M\|^{-1} \leq \|\Lambda^p M^{-1}\| \leq \|M^{-1}\|^p$$

so that $- \operatorname{Log} \|\Lambda^p M\| \leq p \operatorname{Log} \|M^{-1}\| \leq p \operatorname{Log}^+ \|M^{-1}\| \leq p\, \ell(M)$.

Therefore $\qquad |\operatorname{Log} \|\Lambda^p M\| \,| \leq p\, \ell(M)$.

Since , for $\|w\| = 1$, we have $\|\Lambda^p M w\| \leq \|\Lambda^p M\|$ and $\|\Lambda^p M w\|^{-1} \leq \|\Lambda^p M^{-1}\|$ we also have

$$|\mathrm{Log}\ \|\Lambda^p M\, w\|\ | \leq p\ \ell(M)\ .$$

Remark that if $|\det M| = 1$, $\ell(M) \leq d\ \mathrm{Log}\ \|M\|$, (see Remark 3.5).

We can now define the d Lyapunov exponents associated to a sequence of random matrices.

<u>DEFINITION 5.5.</u> *Let* $Y_1, Y_2, \ldots$ *be i.i.d. random matrices in* $Gl(d,\mathbb{R})$ *with* $\mathbb{E}(\mathrm{Log}^+\|Y_1\|) < \infty$. *The Lyapunov exponents* $\gamma_1, \gamma_2, \ldots, \gamma_d$ *associated to* (Y_n) *are defined inductively by* $\gamma = \gamma_1$ *and for* $p \geq 2$,

$$\sum_{i=1}^{p} \gamma_i = \lim_{n\to\infty} \frac{1}{n}\ \mathbb{E}(\mathrm{Log}\ \ \|\Lambda^p\ Y_n \ldots Y_1\|)\ .$$

In other words $\sum_{i=1}^{p} \gamma_i$ is the upper Lyapunov exponent associated with the matrices $(\Lambda^p Y_n)$. As such it exists and is equal to the almost sure limit of $\frac{1}{n}\ \mathrm{Log}\ \|\Lambda^p Y_n \ldots Y_1\|$ (see Chapter I). The definition is ambiguous if for some p $\sum_{i=1}^{p-1} \gamma_i = -\infty$, in this case we put $\gamma_p = \gamma_{p+1} = \ldots = \gamma_d = -\infty$.

Note that if $\mathbb{E}(\mathrm{Log}^+\ \|Y_1^{-1}\|)$ is also finite, Lemma 5.4 implies that for any p

$$\left|\frac{1}{p} \sum_{i=1}^{p} \gamma_i\right| \leq \mathrm{Sup}\ \mathbb{E}(\mathrm{Log}^+\|Y_1\|\ ,\ \mathrm{Log}^+\ \|Y_1^{-1}\|) < \infty$$

so that all the exponents are finite.

<u>PROPOSITION 5.6.</u> *If* $a_1(n) \geq a_2(n) \geq \ldots \geq a_d(n) > 0$ *are the square roots of the eigenvalues of* $(Y_n \ldots Y_1)^*(Y_n \ldots Y_1)$, *then with probability one,*

$$\gamma_p = \lim \frac{1}{n}\ \mathbb{E}\ (\mathrm{Log}\ a_p(n)) = \lim \frac{1}{n}\ \mathrm{Log}\ a_p(n)\ .$$

<u>Proof</u> : If $\gamma_1 + \ldots + \gamma_{p-1} = -\infty$, $a_p(n)^{p-1} \leq a_1(n) \ldots a_{p-1}(n)$, implies

$$(p-1)\ \lim \frac{1}{n}\ \mathrm{Log}\ a_p(n) \leq \lim \frac{1}{n}\ \mathrm{Log}\ a_1(n) \ldots a_{p-1}(n)$$

which yields, using Lemma 5.3

$$(p-1) \lim \frac{1}{n} \mathrm{Log}\ a_p(n) \leq \lim \frac{1}{n} \log\ \|\Lambda^p\ Y_n \ldots Y_1\| = \gamma_1 + \ldots + \gamma_{p-1} = -\infty$$

so $\gamma_p = -\infty = \lim \frac{1}{n} \text{Log } a_p(n)$.

If $\gamma_1 + \ldots + \gamma_{p-1} \neq -\infty$,

$$\lim \frac{1}{n} \text{Log } a_p(n) = \lim \frac{1}{n} \text{Log } \frac{a_1(n)\ldots a_p(n)}{a_1(n)\ldots a_{p-1}(n)}$$

$$= \lim \frac{1}{n} \{\text{Log } \|\Lambda^p Y_n \ldots Y_1\| - \text{Log } \|\Lambda^{p-1} Y_n \ldots Y_1\|\}$$

$$= \gamma_p .$$

It is thus clear that $\gamma_1 \geq \gamma_2 \geq \ldots \geq \gamma_d$, which justifies the fact that we have called $\gamma = \gamma_1$ the upper Lyapunov exponent.

We have $|\det(Y_n \ldots Y_1)| = a_1(n)a_2(n)\ldots a_d(n)$, therefore

$$\gamma_1 + \gamma_2 + \ldots + \gamma_d = \mathbb{E}(\text{Log } |\det Y_1|)$$

and all the exponents are finite if and only if $\text{Log}|\det Y_1|$ is integrable.

III.6. Comparison of the top Lyapunov exponents and Furstenberg's theorem

We now prove the main results of this chapter. Under the assumptions $\mathbb{E}(\text{Log}^+ \|Y_1\|) < \infty$ and T_μ strongly irreducible we determine following Guivarc'h and Raugi the least $p > 1$ such that $\gamma_1 > \gamma_p$. In particular we recover Furstenberg's theorem. As before $\gamma_1 \geq \gamma_2 \geq \ldots \geq \gamma_d$ are the Lyapunov exponents associated with μ. The following is maybe the principal result of this book. It appeared in [34].

THEOREM 6.1. *Let* μ *be a probability measure on* $Gl(d,\mathbb{R})$ *such that* $\int \text{Log}^+ \|M\| \, d\mu(M)$ *is finite. We suppose* T_μ *irreducible. Then* $\gamma_1 > \gamma_2$ *if and only if* T_μ *is strongly irreducible and contracting.*

Proof : Suppose that T_μ is strongly irreducible and contracting (i.e. the index of T_μ is one). From Proposition 3.2. we know that for any x in $\mathbb{R}^d$, if $S_n = Y_n \ldots Y_1$, where $\{Y_n, n \geq 1\}$ are i.i.d. matrices with law μ, then

$$\operatorname*{Inf}_{n} \frac{\|S_n x\|}{\|S_n\|} \neq 0 \quad \text{and} \quad \lim_{n\to\infty} \frac{a_2(n)}{\|S_n\|} = \lim_{n\to\infty} \frac{\|\Lambda^2 S_n\|}{\|S_n\|^2} = 0 .$$

This implies that for each non zero matrix M, a.s.

$$\lim_{n\to\infty} \frac{\|S_n M\|^2}{\|\Lambda^2 S_n\|} = +\infty \tag{10}$$

We shall use Lemma II.2.2 to deduce from (10) that $\gamma_1 > \gamma_2$. Consider each matrix as an endomorphism. From the proof of the theorem of Furstenberg and Kesten (Theorem I.4.1) we know that there exists a μ-invariant distribution ν_1 on B_1, the set of norm-one endomorphisms of $\mathbb{R}^d$, such that

$$\lim_{n\to\infty} \frac{1}{n} \operatorname{Log} \|S_n M\| = \gamma_1$$

for $\mathbb{P} \otimes \nu_1$-almost all (ω, M). In the same way, working with $(\Lambda^2 Y_i)$ there exists a μ-invariant distribution ν_2 on B_2, the set of norm-one endomorphisms of $\Lambda^2 \mathbb{R}^d$, such that

$$\lim_{n\to\infty} \frac{1}{n} \operatorname{Log} \|\Lambda^2 S_n N\| = \gamma_1 + \gamma_2$$

for $\mathbb{P} \otimes \nu_2$-almost all (ω, N).

Consider $B = B_1 \times B_2$. It is a $Gl(d,\mathbb{R})$-space if we put

$$Y\cdot(M,N) = \left(\frac{YM}{\|YM\|}, \frac{(\Lambda^2 Y)N}{\|(\Lambda^2 Y)N\|}\right), \qquad Y \in Gl(d,\mathbb{R}),\ M \in B_1,\ N \in B_2.$$

By Lemma I.3.5 it is clear that each limit point ν of $\{\frac{1}{n} \sum_{i=1}^{n} \mu^i * (\nu_1 \otimes \nu_2)\ ;\ n \geq 1\}$ is a μ-invariant probability measure on $B_1 \times B_2$.

Let $\sigma : Gl(d,\mathbb{R}) \times B \longrightarrow \mathbb{R}$ be the cocycle defined by

$$\sigma(Y, (M,N)) = \operatorname{Log} \frac{\|YM\|^2}{\|\Lambda^2 Y N\|} .$$

Since the projections of ν on B_1 and B_2 are ν_1 and ν_2 we have, $\mathbb{P} \otimes \nu$ a.s.

$$\begin{aligned} \lim_{n\to\infty} \frac{1}{n} \sigma(S_n, (M,N)) &= \lim_{n\to\infty} \frac{2}{n} \operatorname{Log} \|S_n M\| - \lim_{n\to\infty} \frac{1}{n} \operatorname{Log} \|\Lambda^2 S_n N\| \\ &= 2\gamma_1 - (\gamma_1 + \gamma_2) \\ &= \gamma_1 - \gamma_2 . \end{aligned}$$

On the other hand, by (10),

$$\lim_{n\to\infty} \sigma(S_n, (M,N)) \geq \lim_{n\to\infty} \operatorname{Log} \frac{\|S_n M\|^2}{\|\Lambda^2 S_n\|} = \infty ,$$

and $\gamma_1 > \gamma_2$ by Lemma II.2.2.

Since a direct proof of the converse would be tedious, we make use of the fact that when G_μ is irreducible, for any non zero x in $\mathbb{R}^d$

$$\lim_{n\to\infty} \frac{1}{n} \operatorname{Log} \|Y_n \ldots Y_1 x\| = \gamma_1 \qquad \text{a.s.} .$$

We will show this in Proposition 7.2 . Taking it for granted we see that if $\gamma_1 > \gamma_2$, for each unit vectors x, y

$$\lim_{n\to\infty} \delta(S_n(\omega)\cdot\bar{x}, S_n(\omega)\cdot\bar{y}) \leq \lim_{n\to\infty} \frac{\|\Lambda^2 S_n(\omega)\|}{\|S_n(\omega)x\| \, \|S_n(\omega)y\|} \leq 0 \qquad \text{a.s.}$$

This in turn implies that T_μ is strongly irreducible and contracting (see Proposition 4.4).

More generally we have

PROPOSITION 6.2. *Let* μ *be a probability measure on* $Gl(d,\mathbb{R})$ *such that* $\int \operatorname{Log}^+ \|M\| \, d\mu(M)$ *is finite. If* T_μ *is strongly irreducible and if its index is* p *then*

$$\gamma_1 = \gamma_2 = \ldots = \gamma_p > \gamma_{p+1} .$$

Proof : It is obvious from Lemma 1.4 that $\gamma_1 = \gamma_2 = \ldots \gamma_p$. By Proposition 3.2 for any non zero matrix M , a.s.

$$\lim_{n\to\infty} \frac{\|S_n M\|^{p+1}}{\|\Lambda^{p+1} S_n\|} = \lim_{n\to\infty} \frac{\|S_n\|}{a_{p+1}(n)} = \infty .$$

From this relation one deduces that $\gamma_p > \gamma_{p+1}$ using Lemma II.2.2 by a proof similar to the proof of Theorem 6.1 .

In particular we obtain the following result of [21] .

THEOREM 6.3. (Furstenberg's theorem) *If* μ *is a probability measure on* $\{M \in Gl(d,\mathbb{R}) , |\det M| = 1\}$ *such that*

(a) $\int \operatorname{Log} \|M\| \, d\mu(M) < \infty$.

(b) G_μ *is strongly irreducible and non compact* .

Then the upper Lyapunov exponent is positive $(\gamma > 0)$.

Proof : By Proposition 1.7, the index p of T_μ satisfies $p < d$. Therefore by Proposition 6.2 (recall that if G_μ is strongly irreducible then T_μ is also strongly irreducible) $\gamma_p > \gamma_{p+1}$. But the definition of the Lyapunov exponents gives

$$\begin{aligned}\gamma_1 + \gamma_2 + \dots + \gamma_d &= \lim_{n\to\infty} \frac{1}{n} \operatorname{Log} \|\Lambda^d S_n\| \\ &= \lim_{n\to\infty} \frac{1}{n} \operatorname{Log} |\det S_n| \\ &= 0 .\end{aligned}$$

If γ were not positive then, since $\gamma = \gamma_1$, for $1 \leqq i \leqq p$, $\gamma_i \leqq \gamma_1 \leqq 0$ and for $p+1 \leqq i \leqq d$, $\gamma_i < \gamma_p \leqq 0$. This would imply that $\sum_{i=1}^{d} \gamma_i < 0$ which is absurd.

Our analysis of the central limit theorem rests upon the following result of Le Page [49]. Recall that

$$\ell(M) = \operatorname{Sup}(\operatorname{Log}^+ \|M\| , \operatorname{Log}^+ \|M^{-1}\|) .$$

PROPOSITION 6.4. *Suppose that* $\{Y_n, n \geqq 1\}$ *are i.i.d. random matrices in* $Gl(d,\mathbb{R})$ *with common distribution* μ. *If* $\mathbb{E}\{\ell(Y_1)\}$ *is finite and if* T_μ *is strongly irreducible and contracting then, for* $S_n = Y_n \dots Y_1$,

(i) For any $\bar{x}, \bar{y}$ *in* $P(\mathbb{R}^d)$,

$$\lim_{n\to\infty} \frac{1}{n} \operatorname{Log} \frac{\delta(S_n\cdot\bar{x}, S_n\cdot\bar{y})}{\delta(\bar{x},\bar{y})} < 0 , \qquad \text{a.s.},$$

(ii)

$$\overline{\lim_{n\to\infty}} \sup_{\bar{x},\bar{y}} \frac{1}{n} \mathbb{E}\left(\operatorname{Log} \frac{\delta(S_n\cdot\bar{x}, S_n\cdot\bar{y})}{\delta(\bar{x},\bar{y})}\right) < 0 .$$

Proof : We have already noticed that under these hypotheses γ_1 and γ_2 are finite. By (9), for any x, y in $\mathbb{R}^d - \{0\}$,

$$\frac{\delta(S_n\cdot\bar{x}, S_n\cdot\bar{y})}{\delta(\bar{x},\bar{y})} \leqq \frac{\|S_n x\| \, \|S_n y\| \, \|x\| \, \|y\|}{\|S_n x\| \, \|S_n y\| \, \|x \wedge y\|} \leqq \|\Lambda^2 S_n\| \frac{\|x\| \, \|y\|}{\|S_n x\| \, \|S_n y\|} .$$

Since $\frac{1}{n} \operatorname{Log} \|\Lambda^2 S_n\|$ converges a.s. to $\gamma_1 + \gamma_2$ and $\frac{1}{n} \operatorname{Log} \frac{\|S_n x\|}{\|x\|}$ converges to γ_1 (see Corollary 3.4), we have

$$\lim_{n\to\infty} \frac{1}{n} \operatorname{Log} \frac{\delta(S_n\cdot\bar{x}, S_n\cdot\bar{y})}{\delta(\bar{x},\bar{y})} \leqq \gamma_1 + \gamma_2 - 2\gamma_1 = \gamma_2 - \gamma_1 .$$

By Theorem 6.1, $\gamma_2 - \gamma_1 < 0$, proving (i). We also have by Corollary 3.4,

$$\overline{\lim_{n\to\infty}} \ \operatorname{Sup}_{\bar{x},\bar{y}} \frac{1}{n} \mathbb{E}\{\operatorname{Log} \frac{\delta(S_n\cdot\bar{x}, S_n\cdot\bar{y})}{\delta(\bar{x},\bar{y})}\}$$

$$\leq \lim_{n\to\infty} \frac{1}{n} \mathbb{E}(\operatorname{Log} \|\Lambda^2 S_n\|) + 2 \lim_{n\to\infty} \{\operatorname{Sup}_x \frac{1}{n} \operatorname{Log} \frac{\|x\|}{\|S_n x\|}\}$$

$$\leq \gamma_2 - \gamma_1 < 0 \quad .$$

This completes the proof.

Exercise 6.5. Let $Y_1, Y_2, \ldots$ be a sequence of i.i.d. positive matrices of order d. With the notations of Remark 4.7 prove that

(i) For any $\bar{x}, \bar{y}$ in C, $\lim_{n\to\infty} \theta(Y_n \ldots Y_1\cdot\bar{x}, Y_n \ldots Y_1\cdot\bar{y}) = 0$.

(ii) For any continuous function f with compact support on C

$$\lim_{n\to\infty} |\mathbb{E}\{f(Y_n \ldots Y_1\cdot\bar{x}\} - \mathbb{E}\{f(Y_n \ldots Y_1\cdot\bar{y})\}| = 0 \ .$$

(iii) Deduce from (ii) that there exists a unique μ-invariant distribution on C.

(iv) Show that

$$\overline{\lim_{n\to\infty}} \ \operatorname{Sup}_{\bar{x},\bar{y}\in C} \frac{1}{n} \mathbb{E}(\operatorname{Log} \frac{\theta(Y_n \ldots Y_1\cdot\bar{x}, Y_n \ldots Y_1\cdot\bar{y})}{\theta(\bar{x},\bar{y})}) < 0 \ .$$

III.7. Complements. The irreducible case

So far we have most often assumed that G_μ is strongly irreducible. In this section we consider the case where G_μ is merely irreducible. We first prove a technical lemma which is useful to deal with this situation in various contexts. We then apply it to give a necessary and sufficient condition ensuring that there are at least two distinct Lyapunov exponents. (Recall that when G_μ is not strongly irreducible then $\gamma_1 = \gamma_2$ by Theorem 6.1). This is related to the amenability of G_μ. Therefore we obtain as a by-product a complete description of all closed amenable subgroups of $Gl(d,\mathbb{R})$.

LEMMA 7.1. *Let* $S_n = Y_n \ldots Y_1$ *where* $\{Y_n, n \geq 1\}$ *are i.i.d. random elements of* $Gl(d,\mathbb{R})$ *with common distribution* μ. *Suppose that* G_μ *is irreducible. Then there exists a direct sum decomposition* $\mathbb{R}^d = V_1 \oplus V_2 \oplus \ldots \oplus V_k$ *and a closed normal subgroup* H *of* G_μ *with the following properties* :

(i) All the V_i*'s have the same dimension, say* r .

(ii) The group G_μ/H *is finite.*

(iii) For any $1 \leqq i \leqq k$ *and* M *in* H , $M(V_i) = V_i$.

(iv) The distribution of $\tau = \inf\{n \geqq 1\ ,\ S_n \in H\}$ *has an exponentially decaying tail. For any* $1 \leqq i \leqq k$ *let* $S_\tau^{(i)}$ *denote an* $r \times r$ *matrix which represents the endomorphism of* V_i *obtained by restriction of* S_τ *to* V_i *, in a fixed basis of* V_i *. Then, if* $\lambda(i)$ *is the distribution of* $S_\tau^{(i)}$, $G_{\lambda(i)}$ *is a strongly irreducible subgroup of* $Gl(r,\mathbb{R})$.

(v) If moreover $\mathbb{E}(\mathrm{Log}^+ \|Y_1\|)$ *is finite then, for* $1 \leqq i \leqq k$, *the upper Lyapunov exponent* $\gamma^{(i)}$ *associated to* $S_\tau^{(i)}$ *is equal to* $\mathbb{E}(\tau)\gamma$.

<u>Proof</u> : Consider (see Lemma 4.5) subspaces $V_1,\dots,V_\ell$ of $\mathbb{R}^d$ with the same dimension r such that for $i \neq j$, $V_i \cap V_j = \{0\}$ and for $1 \leqq i \leqq \ell$,

$$\{M\,V_i\ ;\ M \in G_\mu\} = \{V_1,\dots,V_\ell\}\ . \tag{11}$$

We shall choose r minimal among the integers for which such subspaces exist. If $r = d$, G_μ is strongly irreducible and the lemma is obvious (take $H = G_\mu$). Therefore we suppose $r < d$. In the same way that we have proved the assertion $V_i \cap V_j = \{0\}$ in Lemma 4.5 one sees that for any $i_1,\dots,i_p$, j, either $V_j \subset \bigoplus_{s=1}^{p} V_{i_s}$ or $V_j \cap \bigoplus_{s=1}^{p} V_{i_s} = \{0\}$.

Since G_μ is irreducible the linear span of $\bigcup_{i=1}^{\ell} V_i$ is $\mathbb{R}^d$ and we can renumber the V_i's in such a way that for some k

$$V_1 \oplus V_2 \oplus \dots \oplus V_k = \mathbb{R}^d\ . \tag{12}$$

Consider now the map $\psi : G_\mu \longrightarrow P_\ell$, where P_ℓ is the permutation group over ℓ letters, defined by

$$\{\Psi(M)\}(i) = j \qquad \text{if} \qquad M(V_i) = V_j\ .$$

It is clear that Ψ is a group homomorphism. Its kernel

$$H = \{M \in G_\mu\ ;\ M(V_i) = V_i\ ,\quad i = 1,\dots,\ell\}$$

is thus a closed normal subgroup of G_μ and G_μ/H is finite.

If we write each matrix Y_i in a different fixed basis all the properties we are looking for are preserved. We may therefore suppose for convenience that if $\{e_1,\dots,e_d\}$ is the canonical basis of $\mathbb{R}^d$,

$\{e_1,\dots,e_r\}$ spans V_1, $\{e_{r+1},\dots,e_{2r}\}$ spans V_2 and so on. We can write any matrix M of H as

$$M = \begin{bmatrix} M^{(1)} & & 0 \\ & M^{(2)} & \\ & & \\ 0 & & M^{(k)} \end{bmatrix} \tag{13}$$

where each $M^{(i)}$ is a $r \times r$ matrix .

We define inductively $\tau(1) = \tau$, $\tau(n+1) = \inf\{m > \tau(n); S_m S_{\tau(n)}^{-1} \in H\}$. It is easy to see that $\tau(1)$, $\tau(2) - \tau(1),\dots,\tau(n+1) - \tau(n)$ are i.i.d. random variables. Notice that since τ is the first time when the finite states Markov chain $\Psi(S_n)$ returns to the identity, its distribution has an exponentially decaying tail. It is also clear that $S_{\tau(1)}$, $S_{\tau(2)}S_{\tau(1)}^{-1},\dots,S_{\tau(n)}S_{\tau(n-1)}^{-1}$ are independent random matrices in H with a common distribution λ . Making use of the stopping time property of τ , one sees that if $\mathbb{E}(\mathrm{Log}^+ \|Y_1\|)$ is finite then $\mathbb{E}(\mathrm{Log}^+\|S_\tau\|)$ is also finite.

We shall now prove that for each $1 \leq m \leq k$, $G_{\lambda(m)}$ is a strongly irreducible subgroup of $Gl(r,\mathbb{R})$. For this we first remark that $T_\lambda = T_\mu \cap H$ (since G_μ/H is finite, H is open in G_μ and M is in $T_\mu \cap H$ if and only if for any neighborhood O of M in H , $\mathbb{P}(\exists\, n > 0,\ S_n \in O) \neq 0$, i.e. $\mathbb{P}(\exists\, n > 0\ ,\ S_{\tau(n)} \in O) \neq 0$. This latter condition is equivalent to $M \in T_\lambda$). On the other hand there exist $t_1,t_2,\dots,t_p$ in T_μ such that

$$G_\mu = t_1^{-1}H \cup t_2^{-1}H \cup \dots \cup t_p^{-1}H$$

for, since $\Psi(T_\mu^{-1})$ is a finite semigroup in the permutation group P_ℓ , $\Psi(T_\mu^{-1})$ is a subgroup and $\Psi^{-1}(\Psi(T_\mu^{-1}))$ must be a closed subgroup which contains T_μ . Therefore $\Psi(T_\mu^{-1}) = \Psi(G_\mu)$ and it suffices to choose $t_1,\dots,t_p$ in T_μ with $\Psi(\{t_1^{-1},\dots,t_p^{-1}\}) = \Psi(G_\mu)$.

Suppose that $G_{\lambda(m)}$ is not strongly irreducible for some $m \in \{1,\dots,k\}$. Then we can find proper subspaces $W_1,\dots,W_s$ of V_m such that, for any M in G_λ ,

$$M\Big(\bigcup_{i=1}^{s} W_i\Big) = \bigcup_{i=1}^{s} W_i \quad .$$

For any matrix A in T_μ there exists some index j such that $t_jA \in H$. This implies that $t_jA \in T_\lambda$ (since $T_\lambda = T_\mu \cap H$) and that

$(t_j A)(W_1) \subset \bigcup_{i=1}^{s} W_i$. Therefore $AW_1 \subset \bigcup_{i,j} t_j^{-1} W_i$ and we can find $M_1, \ldots, M_q$ such that for each A in T_μ

$$A\{M_1 W_1, \ldots, M_q W_1\} = \{M_1 W_1, \ldots, M_q W_1\}$$

in contradiction with the minimal assumption on r .

Finally we prove (v). Using (iv) and Corollary 3.4 we see that for any $x_i \in V_i$, $x_i \neq 0$,

$$\lim \frac{1}{n} \operatorname{Log} \|S^{(i)}_{\tau(n)}\| = \lim \frac{1}{n} \operatorname{Log} \|S^{(i)}_{\tau(n)} x_i\| = \gamma^{(i)} . \qquad (14)$$

This leads to

$$\lim \frac{1}{n} \operatorname{Log} \|S_{\tau(n)}\| = \operatorname{Sup}_i \lim \frac{1}{n} \operatorname{Log} \|S^{(i)}_{\tau(n)}\| = \operatorname{Sup}_i \gamma^{(i)}$$

and to, if x_i is the orthogonal projection of x on V_i ,

$$\lim \frac{1}{n} \operatorname{Log} \|S_{\tau(n)} x\| = \operatorname{Sup}_i \lim \frac{1}{n} \operatorname{Log} \|S^{(i)}_{\tau(n)} x_i\|$$
$$= \operatorname{Sup}_{\{i;x_i \neq 0\}} \gamma^{(i)} .$$

If $\mathbb{E}(\tau) = \alpha$, making use of the usual law of large numbers we have $\lim \frac{\tau(n)}{n} = \alpha$ so that by (14) , for any $x \neq 0$ in $\mathbb{R}^d$,

$$\lim_{n\to\infty} \frac{1}{\tau(n)} \operatorname{Log} \|S_{\tau(n)} x\| = \operatorname{Sup}_{\{i,x_i \neq 0\}} \gamma^{(i)} .$$

For any positive integer n define θ_n by the formula

$$\tau(\theta_n) \leqq n < \tau(\theta_n + 1) .$$

It is readily seen that

$$\lim_{n\to\infty} \frac{\tau(\theta_n)}{n} = 1 \qquad \text{a.s.}$$

from which we deduce that for any x in $\mathbb{R}^d$, $x \neq 0$,

$$\lim_{n\to\infty} \frac{1}{n} \operatorname{Log} \|S_n x\| = \lim_{n\to\infty} \frac{1}{\tau(\theta_n)} \operatorname{Log} \|S_{\tau(\theta_n)} x\| = \operatorname{Sup}_{\{i;x_i \neq 0\}} \frac{\gamma^{(i)}}{\alpha} \qquad (15)$$

(use the relations

$$\operatorname{Log} \|S_{\tau(\theta_n+1)} x\| - \sum_{i=n+1}^{\tau(\theta_n+1)} \operatorname{Log}^+ \|Y_i\| \leqq \operatorname{Log} \|S_n x\|$$

and
$$\operatorname{Log} \|S_n x\| \leqq \operatorname{Log} \|S_{\tau(\theta_n)} x\| + \sum_{i=\tau(\theta_n)+1}^{n} \operatorname{Log}^+ \|Y_i\|) .$$

Now suppose that for some $1 \leqq i \leqq k$, $\gamma^{(i)} \neq \alpha\gamma$. Then

$$W = \{x \in \mathbb{R}^d ; \lim \frac{1}{n} \operatorname{Log} \|S_n x\| < \gamma \qquad \text{a.s.}\}$$

is a linear subspace of $\mathbb{R}^d$ different from $\{0\}$. Since $Y_1(\omega)W = W$ a.s. , $MW = W$ for each M of G_μ . The irreducibility assumption on G_μ implies that $W = \mathbb{R}^d$. We thus can find a basis $v_1,\ldots,v_d$ in W and this leads to a contradiction for if $\|M\| = \operatorname{Sup}\{ \|Mv_j\| ; 1 \leq j \leq d\}$ we find that $\lim \frac{1}{n} \operatorname{Log} \|S_n\| < \gamma$ (recall that this limit does not depend on the chosen norm). The proof is complete.

The end of the foregoing proof shows that for each $x \neq 0$, $\lim \frac{1}{n} \operatorname{Log} \|S_n x\| = \gamma$ (see (15)). One readily deduces from this the following result of Furstenberg (see the proof of Corollary 3.4, (ii)). Actually this proposition can be proved directly in a short way using some arguments of Markov chain theory (see in particular Furstenberg and Kifer [26]).

PROPOSITION 7.2. *Let* $S_n = Y_n \ldots Y_1$ *where* $\{Y_n, n \geq 1\}$ *are i.i.d. random elements of* $Gl(d,\mathbb{R})$ *with common distribution* μ . *Assume that* G_μ *is irreducible and that* $\mathbb{E}(\operatorname{Log}^+\|Y_1\|)$ *is finite. Then*

(i) For any $x \neq 0$

$$\lim_{n\to\infty} \frac{1}{n} \operatorname{Log} \|S_n x\| = \gamma \qquad a.s. .$$

(ii) For any μ*-invariant distribution* ν *on* $P(\mathbb{R}^d)$,

$$\gamma = \iint \operatorname{Log} \frac{\|Mx\|}{\|x\|} \, d\mu(M) \, d\nu(\bar{x}) .$$

Let us state without proof an important consequence of this proposition (see Furstenberg and Kifer [26] or Hennion [35]).

COROLLARY 7.3. *Let* (μ_n) *be a sequence of probability measures on* $Gl(d,\mathbb{R})$ *with upper Lyapunov exponents* $\gamma(\mu_n)$ *such that*

$$\int_{\|g\|\geq N} \operatorname{Log}^+ \|g\| \, d\mu_n(g) + \int_{\|g^{-1}\|\geq N} \operatorname{Log}^+ \|g^{-1}\| \, d\mu_n(g)$$

converges to zero as $N \longrightarrow \infty$ *uniformly in* n . *Then if* μ_n *converges weakly to the probability measure* μ *and if* G_μ *is irreducible then* $\lim_{n\to\infty} \gamma(\mu_n)$ *exists and is equal to the upper Lyapunov exponent associated with* μ .

Let us notice that when each G_{μ_n} is irreducible then the proof is easy : if ν_n is a μ_n-invariant probability measure on $P(\mathbb{R}^d)$ then each limit point ν of (ν_n) is μ-invariant and

$$\lim_{n\to\infty} \gamma(\mu_n) = \lim_{n\to\infty} \int \text{Log} \frac{\|Mx\|}{\|x\|} d\mu_n(M)\, d\nu_n(x)$$
$$= \int \text{Log} \frac{\|Mx\|}{\|x\|} d\mu(M)\, d\nu(x) = \gamma(\mu) .$$

But the conclusion of corollary can be false when G_μ is not irreducible (consider the case where $\mu_n\{\begin{pmatrix}2 & 0\\ 0 & 2^{-1}\end{pmatrix}\} = 1 - \frac{1}{n}$ and $\mu_n\{\begin{pmatrix}0 & 1\\ 1 & 0\end{pmatrix}\} = \frac{1}{n}$) .

We now return to the main line of this section and state its main result, namely

THEOREM 7.4. *Let* μ *be a probability measure on* $Gl(d,\mathbb{R})$ *such that* G_μ *is irreducible and such that* $\int \text{Log}^+ \|M\|\, d\mu(M)$ *is finite. Then* $\gamma_1 = \gamma_d$ *if and only if there exists a closed normal abelian subgroup* A *of* G_μ *such that* G_μ/A *is compact.*

We shall see in the proof below that $\gamma_1 = \gamma_d$ if and only if, with the notation (13) ,

$$\{(|\det M^{(1)}|^{-1/r} M^{(1)},\ldots,|\det M^{(k)}|^{-1/r} M^{(k)}) \ ; \ M \in H\}$$

is contained in a compact subgroup of $\{Gl(r,\mathbb{R})\}^k$. This condition is easily tractable on concrete examples. We thus have a complete description of the probability measures for which $\gamma_1 = \gamma_d$.

Proof : Consider the subgroup H and the stopping time τ which are given in Lemma 7.1 . We first prove that $G_\lambda = H$, where λ is the distribution of S_τ .

We have seen in the proof of this lemma that $G_\mu = T_\mu^{-1} H$ and that $T_\mu \cap H = T_\lambda$ so that $T_\mu \cap H$ is contained in G_λ . This first implies that T_μ is contained in $T_\mu^{-1} G_\lambda$ (because if M is in T_μ , we can find N in T_μ and A in H such that $M = N^{-1} A$, using $G = T_\mu^{-1} H$. Thus $A = NM$ is in $T_\mu \cap H$ so that A is in G_λ). Secondly, since the identity matrix is in the group G_λ , T_μ^{-1} is contained in $T_\mu^{-1} G_\lambda$. Therefore $T_\mu^{-1} G_\lambda$ is a semigroup which contains T_μ and T_μ^{-1} , which yields that G_μ is contained in $T_\mu^{-1} G_\lambda$. Pick now an A in H and write $A = N^{-1}M$ with N in T_μ and M in G_λ. Since $N = A^{-1}M$ is in $T_\mu \cap H$, N is in G_λ and A itself is in G_λ. This proves that $H = G_\lambda$.

We now prove the theorem itself. Without loss of generality we may assume that μ is carried by $\{M \in Gl(d,\mathbb{R})\ ;\ |\det M| = 1\}$ and, in that case, $\gamma_1 = \gamma_d$ is equivalent to $\gamma = 0$.

Suppose that for some i, $1 \leq i \leq k$, the set

$$\{|\det M^{(i)}|^{-1/r} M^{(i)}\ ;\ M \in H\}$$

is not confined in a compact subgroup of $Gl(d,\mathbb{R})$ (we use the notation (13)). Then since $H = G_\lambda$, we have by (iv) of Lemma 7.1 and by Furstenberg's theorem 6.3,

$$\gamma^{(i)} > \frac{1}{r}\,\mathbb{E}(\mathrm{Log}\,|\det S_\tau^{(i)}|)\ .$$

Therefore, since all the exponents $\gamma^{(j)}$, $1 \leq j \leq k$, are equal to $\mathbb{E}(\tau)\gamma$,

$$rk\,\mathbb{E}(\tau)\gamma > \sum_{i=1}^{k} \mathbb{E}(\mathrm{Log}\,|\det S_\tau^{i}|) = \mathbb{E}(\mathrm{Log}\,|\det S_\tau|) = 0\ ,$$

and $\gamma \neq 0$.

Otherwise consider the homomorphism $\phi : H \longrightarrow \{Gl(r,\mathbb{R})\}^k$ defined by

$$\phi(M) = (|\det M^{(1)}|^{-1/r} M^{(1)},\ldots,|\det M^{(k)}|^{-1/r} M^{(k)})\ .$$

If its image is contained in a compact subgroup, it is obvious that for any i, $1 \leq i \leq k$,

$$\gamma^{(i)} = \frac{1}{r}\,\mathbb{E}(\mathrm{Log}\,|\det S_\tau^{(i)}|)$$

so that

$$rk\,\mathbb{E}(\tau)\gamma = \mathbb{E}(\mathrm{Log}\,|\det S_\tau|) = 0$$

and $\gamma = 0$. Let A denote the kernel of ϕ. It is clear that A is an abelian closed subgroup of G_μ such that G_μ/A is compact and it is easy to check that A is normal in G_μ. Finally we sketch the proof of the fact that if A is an arbitrary closed abelian normal subgroup of G_μ such that G_μ/A is compact then $\gamma_1 = \gamma_d$. Let $\pi : G_\mu \longrightarrow G_\mu/A$ be the canonical map. $H/H\cap A$ is isomorphic to $\pi(H)$ hence it is compact. Let

$$A_1 = \{M^{(1)} \in Gl(d,\mathbb{R}) \quad ; \quad M \in H \cap A\}$$

$$H_1 = \{M^{(1)} \in Gl(r,\mathbb{R}) \quad ; \quad M \in H\}$$

then H_1/A_1 is compact and A_1 is abelian. Using the Markov Kakutani theorem and the Haar measure on H_1/A_1, we can construct a probability measure m on $P(\mathbb{R}^r)$ such that, for each M in H

$$M^{(1)} m = m .$$

But this implies (Use Exercises 2.4 and 3.6) that

$$\{|\det M^{(1)}|^{-1/r} M^{(1)} \; ; \; M \in H\}$$

is compact. One then deduces as above that $\gamma_1 = \gamma_d$.

As a consequence we obtain the following result of Furstenberg .

<u>COROLLARY 7.5</u>. *With the hypotheses of Theorem 7.4 ,* $\gamma_1 = \gamma_d$ *if and only if there exists a probability measure* m *on* $P(\mathbb{R}^d)$ *such that* $Mm = m$ *for each* M *in* G_μ .

<u>Sketch of proof</u> : If $\gamma_1 = \gamma_d$, let A be the kernel of ϕ constructed in the proof above. For any x in V_1

$$M\bar{x} = \bar{x} \qquad \text{for any } M \text{ in } A .$$

If ρ is the Haar measure on G_μ/A we can define m by

$$m = \int g\, \delta_{\bar{x}}\, d\rho(\pi(g))$$

(where $\pi : G_\mu \longrightarrow G_\mu/A$ is the canonical projection).

Conversely if there is some measure m such that $Mm = m$ for each M in G_μ then $Mm = m$ for each M in H . If $m(\overline{V_2 \oplus \ldots \oplus V_k}) = 0$ we can consider the image m_1 of m on $\bar{V}_1$ under the orthogonal projection. We have $M^{(1)} m_1 = m_1$ for $M \in H$ and this implies (see the end of the proof above) that $\gamma_1 = \gamma_d$. If $m(\overline{V_2 \oplus \ldots \oplus V_k}) \neq 0$ the restriction m' of m on $\overline{V_2 \oplus \ldots \oplus V_k}$ satisfies also $Mm' = m'$ for each M in H and we argue in the same way with m' considering if possible its projection on $\bar{V}_2$, and so forth.

From the foregoing it is natural to introduce the following class of groups (the reader is refered to Greenleaf [29] and Glasner [27] for details. It is shown in the latter that the definition we give is equivalent with the more common one).

DEFINITION 7.6. *A topological group* G *is said to be amenable if there exists on each compact* G-*space* B *a probability measure* m *such that* gm = m *for each* g *in* G .

If there exists a closed normal abelian subgroup of G such that G/A is compact then G is amenable. Therefore it is clear that we can express the foregoing results in the following way.

COROLLARY 7.7. *Let* G *be an irreducible closed subgroup of* $Gl(d,\mathbb{R})$. *The following are equivalent :*

(i) G *is amenable .*

(ii) There exists a closed normal abelian subgroup A *of* G *such that* G/A *is compact.*

(iii) For each probability measure μ *on* G *with* $G_\mu = G$, *all the Lyapunov exponents associated with* μ *are equal.*

(iv) There exists a probability measure μ *on* G *with* $G_\mu = G$ *such that all the Lyapunov exponents associated with* μ *are equal.*

One can find in Guivarc'h [31] a direct proof of the equivalence (i) $\Longleftrightarrow$ (iv).

It is easy to deduce from the equivalence (i) $\Longleftrightarrow$ (ii) that an arbitrary closed subgroup G of $Gl(d,\mathbb{R})$ (or $Gl(d,\mathbb{C})$) is amenable if an only if there exists a nested sequence

$$\{e\} = H_o \subset H_1 \subset \ldots \subset H_n = G$$

of closed subgroups H_i of G such that H_i is normal in H_{i+1} and such that H_{i+1}/H_i has a cocompact abelian normal closed subgroup for any $0 \leqq i < n$.

To conclude, note that when G is irreducible and amenable there exists an open normal subgroup H of G such that G/H is finite and such that H can be imbedded in some $\mathbb{R}^\ell \times K$ where K is compact. This implies easily that there exists a recurrent random walk on G with distribution μ satisfying $G_\mu = G$ if and only if G is amenable and the rank of the subgroup A given in the corollary is smaller than 3 (we say that $S_n = Y_n \ldots Y_1$ is recurrent if for almost all ω, $(S_n(\omega))$ returns infinitely often in each neighborhood of the identity matrix).

CHAPTER IV

COMPARISON OF LYAPUNOV EXPONENTS AND BOUNDARIES

In the preceding chapter we have given a criterion ensuring that the two upper Lyapunov exponents are distinct. It will give us all we need for the study of limit theorems. But a sharp study of the behaviour at infinity of the random products S_n requires a precise knowledge of the relations between all the exponents. For instance they provide all the limit values of $\frac{1}{n}$ Log $\|S_n(\omega)x\|$, when ω is kept fixed and x runs through $\mathbb{R}^d$ (Osseledec's theorem) and determine the possible boundaries towards which S_n converges.

Therefore we digress from our study of limit theorems and prove the theorem of Guivarc'h and Raugi which gives us a condition ensuring that two given Lyapunov exponents are distinct. We then study in all detail some important examples. In particular we consider the case of symplectic matrices which is often encountered in concrete applications (see e.g. Chapter IV of Part B). Finally we proceed with a short introduction to Furstenberg's theory of boundaries and establish its connection with the previous results.

This chapter will not be needed in the sequel of this part of the book (but has important applications in part B).

IV.1. A criterion ensuring that Lyapunov exponents are distinct

We first generalize the definitions given in Chapter III.

DEFINITION 1.1. *Let* T *be a subset of* $Gl(d,\mathbb{R})$ *and* p *be an integer in* $\{1,\dots,d-1\}$. *We say that*

(i) T *is* p*-strongly irreducible if there does not exist a finite union* L *of proper linear subspaces of* $\Lambda^p \mathbb{R}^d$ *such that, for all* M *in* T , $(\Lambda^p M)(L)$ *is contained in* L .

(ii) T *is* p*-contracting if there exists a sequence* $\{M_n, n \geq 1\}$ *in* T *such that* $\|\Lambda^p M_n\|^{-1} \Lambda^p M_n$ *converges to a rank-one matrix.*

In other words T is p-strongly irreducible (resp. p-contracting) if $\{\Lambda^p M ; M \in T\}$ is a strongly irreducible (resp. contracting) subset of the set of linear automorphisms of $\Lambda^p \mathbb{R}^d$. We shall give in the next section some criteria with which one can check these properties on examples.

The following important result is taken from Guivarc'h and Raugi [34] .

THEOREM 1.2. *Let* $S_n = Y_n \dots Y_1$ *where* $Y_1, Y_2, \dots$ *are i.i.d. random matrices of* $Gl(d,\mathbb{R})$ *with common distribution* μ . *Suppose that* $\mathbb{E}(\text{Log}^+ \|Y_1\|)$ *is finite and that, for some* p *in* $\{1,\dots,p-1\}$, T_μ *is* p*-strongly irreducible and* p*-contracting. Then, if* $\gamma_1 \geqq \gamma_2 \geqq \dots \geqq \gamma_d$ *are the Lyapunov exponents associated with* μ , *the following hold*

(i) When $\gamma_p \neq -\infty$, $\gamma_p > \gamma_{p+1}$.

(ii) There exists a unique μ*-invariant distribution* ν_p *on* $P(\Lambda^p \mathbb{R}^d)$ *and*

$$\sum_{i=1}^{p} \gamma_i = \iint \text{Log} \frac{\|\Lambda^p M x\|}{\|x\|} \, d\mu(M) \, d\nu_p(\bar{x}) .$$

(iii) For any non zero x *in* $\Lambda^p \mathbb{R}^d$,

$$\lim_{n\to\infty} \frac{1}{n} \text{Log} \|\Lambda^p S_n x\| = \sum_{i=1}^{p} \gamma_i \quad , \quad a.s. \quad .$$

Proof : Let q be the dimension of $\Lambda^p \mathbb{R}^d$. We may view each $\Lambda^p Y_i$ as a matrix in $Gl(q,\mathbb{R})$. If $\mu(q)$ is its distribution, the hypotheses entail that $T_{\mu(q)}$ is a strongly irreducible and contracting subset of $Gl(q,\mathbb{R})$. Therefore, if $\lambda_1 \geqq \lambda_2$ denote the two upper Lyapunov exponents associated with $\{\Lambda^p Y_n, n \geqq 1\}$ we obtain by applying

Corollary III.3.4, Theorem III.4.3 and III.6.1 to this sequence that

(a) $\lambda_1 > \lambda_2$.

(b) There exists a unique $\mu(q)$-invariant distribution ν_p on $P(\Lambda^p \mathbb{R}^d)$ and

$$\lambda_1 = \iint \mathrm{Log} \frac{\|Mx\|}{\|x\|} \, d\mu(q)(M) \, d\nu_p(\bar{x}) .$$

(c) For any non zero x in $\Lambda^p \mathbb{R}^d$,

$$\lim \frac{1}{n} \mathrm{Log} \|\Lambda^p S_n x\| = \lambda_1 , \qquad \text{a.s.} .$$

It is obvious that the theorem will follow from these assertions once we have proved that

$$\lambda_1 = \sum_{i=1}^{p} \gamma_i \quad \text{and} \quad \lambda_2 = \sum_{i=1}^{p-1} \gamma_i + \gamma_{p+1} .$$

A short way to verify this is to write a polar decomposition of S_n. If $S_n = K_n A_n U_n$ with K_n, U_n in $O(d)$ and $A_n = \mathrm{diag}(a_1(n),\ldots,a_d(n))$, $a_1(n) \geq a_2(n) \geq \ldots \geq a_d(n) > 0$, we have $\Lambda^p S_n = (\Lambda^p K_n)(\Lambda^p A_n)(\Lambda^p U_n)$. Let $\{e_i, 1 \leq i \leq d\}$ denote as usual the canonical basis of $\mathbb{R}^d$. In the basis $\{e_{i_1} \wedge \ldots \wedge e_{i_p} ; 1 \leq i_1 < \ldots < i_p \leq d\}$ of $\Lambda^p \mathbb{R}^d$, the matrices $\Lambda^p K_n$ and $\Lambda^p U_n$ are orthogonal and

$$\Lambda^p A_n = \mathrm{diag}\{a_{i_1}(n) a_{i_2}(n) \ldots a_{i_p}(n) ; 1 \leq i_1 < i_2 < \ldots < i_p \leq d\} .$$

Therefore the two top eigenvalues of $(\Lambda^p S_n)^* \Lambda^p S_n$ are $\{a_1(n)a_2(n)\ldots a_{p-1}(n)a_p(n)\}^2$ and $\{a_1(n)a_2(n)\ldots a_{p-1}(n)a_{p+1}(n)\}^2$. This yields that (see Proposition III.5.6)

$$\lambda_1 = \lim_{n\to\infty} \frac{1}{n} \mathrm{Log}\, a_1(n)a_2(n)\ldots a_{p-1}(n)a_p(n) = \sum_{i=1}^{p} \gamma_i$$

and

$$\lambda_2 = \lim_{n\to\infty} \frac{1}{n} \mathrm{Log}\, a_1(n)a_2(n)\ldots a_{p-1}(n)a_{p+1}(n) = \sum_{i=1}^{p-1} \gamma_i + \gamma_{p+1} .$$

The proof is complete.

Exercise 1.6 provides an example where T_μ is strongly irreducible and 2-contracting and where $\gamma_2 = \gamma_3$.

As a consequence of this theorem let us prove

<u>COROLLARY 1.3</u>. *Suppose that* T_μ *is* p-*strongly irreducible and*

p-contracting for any p , $1 \leq p \leq d-1$. *Then one can find a polar decomposition of* $S_n = Y_n \ldots Y_1$, *namely*

$$S_n = K_n A_n U_n$$

with K_n, $U_n \in O(d)$ *and* $A_n = \operatorname{diag}(a_1(n),\ldots,a_d(n))$, *such that* U_n *converges almost surely to some orthogonal matrix* U_∞ . *Moreover if* $\mathbb{E}(\operatorname{Log}^+\|Y_1\|) < \infty$, *then* $\{S_n^* S_n\}^{1/2n}$ *converges a.s. to* $U_\infty^* \operatorname{diag}(\exp(\gamma_1),\ldots,\exp(\gamma_d))U_\infty$ and $\gamma_1 > \gamma_2 > \ldots > \gamma_d$.

Proof : Consider any polar decomposition $S_n = \tilde{K}_n A_n \tilde{U}_n$ of S_n and write for any p $\Lambda^p S_n = (\Lambda^p \tilde{K}_n)(\Lambda^p A_n)(\Lambda^p \tilde{U}_n)$. If we apply (a) of Proposition III.2.2 to the sequence $\{\Lambda^p Y_n,\ n \geq 1\}$ we see that the direction of $\tilde{U}_n^* e_1 \wedge \ldots \wedge \tilde{U}_n^* e_p$ converges a.s. . Choose a unit vector x in $\mathbb{R}^d$ such that the direction of $\tilde{U}_n^* e_1$ converges to $\bar{x}$. It is clear that one can find $\varepsilon_n^1 = \varepsilon_n^1(\omega)$ with values in $\{1,-1\}$ such that $\tilde{U}_n^* \varepsilon_n^1 e_1$ actually converges to x . In the same way, since $\Lambda^2 \tilde{U}_n^* \overline{(e_1 \wedge e_2)}$ converges a.s. in $P(\Lambda^2 \mathbb{R}^d)$ one can find $\varepsilon_n^2 \in \{1,-1\}$ such that $\Lambda^2 \tilde{U}_n^* \varepsilon_n^1 \varepsilon_n^2 (e_1 \wedge e_2)$ converges. With this procedure we find $(\varepsilon_n^1, \varepsilon_n^2, \ldots, \varepsilon_n^d) \in \{1,-1\}^d$ such that $\Lambda^p \tilde{U}_n^* \varepsilon_n^1 \ldots \varepsilon_n^p (e_1 \wedge \ldots \wedge e_p)$ converges for any p . Let $U_n = \operatorname{diag}(\varepsilon_n^1,\ldots,\varepsilon_n^d)\tilde{U}_n$ and $K_n = \tilde{K}_n \operatorname{diag}(\varepsilon_n^1,\ldots,\varepsilon_n^d)$. It is clear that $S_n = K_n A_n U_n$ is a polar decomposition and U_n converges to some U_∞ . By Theorem 1.2, when $\mathbb{E}(\operatorname{Log}^+\|Y_1\|) < \infty$ then $\gamma_1 > \gamma_2 > \ldots > \gamma_d$ and (see Proposition III.5.6) $A_n^{1/n}$ converges to $A_\infty = \operatorname{diag}(\exp(\gamma_1),\ldots,\exp(\gamma_d))$. By definition

$$(S_n^* S_n)^{1/2n} = (U_n^* A_n^* K_n^* K_n A_n U_n)^{1/2n} = U_n^*(A_n^{1/n})U_n$$

thus this sequence converges a.s. to $U_\infty^* A_\infty U_\infty$.

Exercise 1.4. [56]. Let $Y_1, Y_2, \ldots$ be i.i.d. matrices in $Gl(d,\mathbb{R})$ with $\mathbb{E}(\operatorname{Log}^+\|Y_1\|) < \infty$.

(i) Suppose that for any orthogonal matrix K , $K Y_1 K^{-1}$ and Y_1 have the same distribution. Show that

$$\gamma_1 + \ldots + \gamma_p = \mathbb{E}(\operatorname{Log}\|Y_1 e_1 \wedge Y_1 e_2 \wedge \ldots \wedge Y_1 e_p\|)$$

(see I.2).

(ii) We suppose that the d^2 entries of Y_1 are independent Gaussian variables with mean zero and variance one. Show that the hypothesis of (i) is satisfied. Fix an integer p , $1 \leq p < d$. Let P

be the (random) projection on the orthogonal of the subspace spanned by $Y_1 e_1, \ldots, Y_1 e_{p-1}$ and Q be the orthogonal projection onto the span of $e_1, \ldots, e_{d-p+1}$. Show that

$$\gamma_p = \mathbb{E}(\operatorname{Log} \|P\, Y_1\, e_p\|) = \mathbb{E}(\operatorname{Log}\|Q\, Y_1\, e_p\|) .$$

Deduce that $\gamma_p = \frac{1}{2}\,\mathbb{E}\,(\operatorname{Log}\,|\sum_{i=1}^{d-p+1} (Y_1)^2_{i,p}|) = \frac{1}{2}\,\{\operatorname{Log} 2 + \frac{\Gamma'(d-p+1/2)}{\Gamma(d-p+1/2)}\}$.

Exercise 1.5. Suppose $\gamma_1 > \gamma_2 > \ldots > \gamma_d$. Show that T_μ is p-contracting for any p.

Exercise 1.6. With the basis $\{e_1 = \begin{pmatrix}1 & 0\\ 0 & 0\end{pmatrix}, e_2 = \begin{pmatrix}0 & 0\\ 1 & 0\end{pmatrix}, e_3 = \begin{pmatrix}0 & 1\\ 0 & 0\end{pmatrix}, e_4 = \begin{pmatrix}0 & 0\\ 0 & 1\end{pmatrix}\}$ one can identify the vector space $M(2,\mathbb{R})$ of 2×2 matrices with $\mathbb{R}^4$. Define $\Psi : Sl(2,\mathbb{R}) \times Sl(2,\mathbb{R}) \longrightarrow Gl(4,\mathbb{R})$ by requiring that, for A, B in $Sl(2,\mathbb{R})$, $\Psi(A,B)$ is the matrix which represents the linear endomorphism $\Psi(A,B)(M) = A M B^{-1}$ of $M(2,\mathbb{R})$.

(i) Show that Ψ is a group homomorphism.

(ii) Let G be the group $\Psi(Sl(2,\mathbb{R}) \times Sl(2,\mathbb{R}))$. Show that G is closed in $Gl(4,\mathbb{R})$ and p-contracting for $p = 1, 2, 3$. (One can use Corollary 2.2 and the image of a pair of diagonal matrices). Prove that G is strongly irreducible.

(iii) Let μ_1, μ_2 be two probability measures on $Sl(2,\mathbb{R})$. If λ_1, λ_2 are the top Lyapunov exponents associated with μ_1 and μ_2 prove that the Lyapunov exponents associated with the image μ of $\mu_1 \otimes \mu_2$ under Ψ are $\lambda_1\lambda_2$, $\lambda_1^{-1}\lambda_2$, $\lambda_1\lambda_2^{-1}$, $\lambda_1^{-1}\lambda_2^{-1}$.

(iv) Show that one can have $T_\mu = G$ and $\lambda_1^{-1}\lambda_2 = \lambda_1\lambda_2^{-1}$.

IV.2. Some examples

We first give some criteria which can be used to check that a set is p-contracting. We then study two important examples where all the Lyapunov exponents are distinct.

PROPOSITION 2.1. *Given a subset* T *of* $Gl(d,\mathbb{R})$ *and an integer* p, $1 \leq p < d$, *the following statements are equivalent :*

(i) T *is* p-*contracting* .

(ii) There exists a sequence (M_n) *in* T *such that, if* $a_1(n) \geq a_2(n) \geq \ldots \geq a_d(n) > 0$ *are the square roots of the eigenvalues of* $M_n^* M_n$ *, then* $\lim_{n\to\infty} \{a_p(n)\}^{-1} a_{p+1}(n) = 0$.

(iii) There is a compact subset C *of* $Gl(d,\mathbb{R})$ *such that* $\{PMQ\ ;\ P \in C\ ,\ M \in T\ ,\ Q \in C\}$ *is* p-*contracting.*

Proof : If (i) holds one can find (M_n) in T such that $\|\Lambda^p M_n\|^{-1} \Lambda^p M_n$ converges to a rank-one matrix. Write a polar decomposition $M_n = K_n A_n U_n$. We have $\Lambda^p A_n = \mathrm{diag}(a_1(n)a_2(n)\ldots a_p(n),\ a_1(n)a_2(n)\ldots a_{p-1}(n)a_{p+1}(n),\ldots)$ so that $\|\Lambda^p M_n\|^{-1} \Lambda^p A_n = \mathrm{diag}(1\ ,\ \dfrac{a_{p+1}(n)}{a_p(n)}\ ,\ \ldots)$, where the dots indicate positive numbers smaller than $a_p(n)^{-1} a_{p+1}(n)$. Since each limit point of $\Lambda^p K_n$ and $\Lambda^p U_n$ is an orthogonal matrix, $\|\Lambda^p M_n\|^{-1} \Lambda^p A_n$ converges to a rank-one matrix, i.e. $\lim a_p(n)^{-1} a_{p+1}(n) = 0$. This proves that (i) implies (ii), and (i) follows from (ii) by the same reasoning. Remark that

$$\frac{a_p(n)}{a_{p+1}(n)} = \frac{\|\Lambda^p M_n\|^2}{\|\Lambda^{p-1} M_n\|\ \|\Lambda^{p+1} M_n\|}\ .$$

Thus, if P_n and Q_n are in a compact set, there exists $\alpha > 0$ such that

$$\frac{1}{\alpha}\frac{a_p(n)}{a_{p+1}(n)} \leq \frac{\|\Lambda^p P_n M_n Q_n\|^2}{\|\Lambda^{p-1} P_n M_n Q_n\|\ \|\Lambda^{p+1} P_n M_n Q_n\|} \leq \alpha \frac{a_p(n)}{a_{p+1}(n)}$$

for each n . This implies that (ii) and (iii) are equivalent.

The following corollary is particularly useful (Guivarc'h has announced that the converse holds under an irreducibility assumption on T). Recall that an eigenvalue λ of a matrix M is called simple if $\mathrm{Ker}(M-\lambda\,\mathrm{Id})$ is one-dimensional and equal to $\mathrm{Ker}(M-\lambda\,\mathrm{Id})^2$. We shall say that λ is dominating if $|\lambda| > |\lambda'|$ for any other eigenvalue λ'.

COROLLARY 2.2. *A semigroup* T *in* $Gl(d,\mathbb{R})$ *which contains a matrix with a simple dominating eigenvalue is contracting. If there exists in* T *a matrix with* d *eigenvalues of distinct moduli, then* T *is* p-*contracting for each* $p\ ,\ 1 \leq p \leq d-1$.

Proof : Let M be a matrix in T with a simple dominating eigenvalue λ . It follows from the Jordan decomposition that $\|M^{2n}\|^{-1} M^{2n}$ converges to a projection matrix with range $\mathrm{Ker}(M-\lambda\,\mathrm{Id})$. Since this space is one dimensional, $\{M^n , n \geq 1\}$ is contracting (see Proposition 2.1).

If the d eigenvalues of M have distinct moduli, then $\Lambda^p M$ has a dominating simple eigenvalue for any p, $1 \leq p \leq d-1$. Therefore T is p-contracting.

Tutubalin was the first to show that when μ has a density on $Sl(d,\mathbb{R})$ with respect to the Haar measure, all the Lyapunov exponents are distinct (see Sazonov and Tutubalin [63], Tutubalin [69]). He assumed moreover that $\mathbb{E}(\mathrm{Log}^2 \|Y_1\|)$ is finite, and Raugi (see [59]) dropped this condition. But, when μ has a density, T_μ is open in $Sl(d,\mathbb{R})$. The following thus shows that we recover their results.

PROPOSITION 2.3. *If* T_μ *contains an open set of* $Sl(d,\mathbb{R})$, *then* T_μ *is* p*-strongly irreducible and* p*-contracting for any* $p \in \{1,\dots,d-1\}$.

Proof : Let V be an open set in $Sl(d,\mathbb{R})$ contained in T_μ. We will show that, for n large enough, $V^n = \{M_1 M_2 \dots M_n \; ; \; M_i \in V , 1 \leq i \leq n\}$ contains a matrix whose eigenvalues have distincts moduli. Since V^n is contained in T_μ this will prove, by Corollary 2.2, that T_μ is p-contracting for any p. Let $M \in V$, with the Jordan decomposition we can find $Q \in Gl(d,\mathbb{R})$ such that

$$Q\,M\,Q^{-1} = \begin{bmatrix} A_1 & & * \\ & A_2 & \\ 0 & & A_r \end{bmatrix}$$

where, for each $i \in \{1,\dots,r\}$, A_i is either a 1×1 matrix λ_i, $\lambda_i \in \mathbb{R}$, or a 2×2 matrix which can be written

$$A_i = \lambda_i \, R(\theta_i) \quad \text{with} \quad R(\theta) = \begin{pmatrix} \cos\theta & -\sin\theta \\ \sin\theta & \cos\theta \end{pmatrix} .$$

If $\varepsilon_1, \varepsilon_2, \dots, \varepsilon_r$ are real numbers close enough to 1 with $\varepsilon_1 \dots \varepsilon_r = 1$, then

$$M' = \begin{bmatrix} \varepsilon_1 A_1 & & * \\ & \varepsilon_2 A_2 & \\ 0 & & \varepsilon_r A_r \end{bmatrix}$$

is still in the open set $Q\,V\,Q^{-1}$ and we can choose $\varepsilon_1, \dots, \varepsilon_r$ in such a way that all the eigenvalues of M' are distinct. Therefore one can find $P \in Gl(d,\mathbb{R})$ such that $P\,V\,P^{-1}$ contains a matrix

$$\begin{bmatrix} B_1 & & 0 \\ & B_2 & \\ 0 & & B_r \end{bmatrix} \qquad (1)$$

with $B_i = \beta_i$ or $\beta_i R(\alpha_i)$. We first suppress the rotation terms. If for instance $B_1 = \beta_1 R(\alpha_1)$, since $P V P^{-1}$ is open we can find $\varepsilon > 0$ such that if $|\theta - \alpha_1| < \varepsilon$, then

$$\begin{bmatrix} \beta_1 R(\theta) & & 0 \\ & B_2 & \\ 0 & & B_r \end{bmatrix}$$

is still in $P V P^{-1}$. If n is a positive integer such that $2n\varepsilon > 2\pi$ we can find $\theta_1,\ldots,\theta_n$ such that $|\theta_i - \alpha_i| < \varepsilon$ and $R(\theta_1)R(\theta_2) \ldots R(\theta_n) = \mathrm{Id}$. This implies that

$$\begin{bmatrix} \beta_1^n & 0 & & 0 \\ 0 & \beta_1^n & & \\ & & B_2^n & \\ & 0 & & B_r^n \end{bmatrix} \quad \text{is in } P V^n P^{-1} .$$

If we carry on this procedure we find an integer m such that $P V^m P^{-1}$ contains a diagonal matrix. We then perturb slightly this matrix to obtain an element of $P V^m P^{-1}$ which has d eigenvalues of distinct modulus.

Finally we prove that T_μ , or equivalently G_μ , is p-strongly irreducible. Since $Sl(d,\mathbb{R})$ is connected, the only subgroup of $Sl(d,\mathbb{R})$ which contains V is $Sl(d,\mathbb{R})$ itself. Therefore $G_\mu = Sl(d,\mathbb{R})$ and it is easy to check that $Sl(d,\mathbb{R})$ is p-strongly irreducible (use e.g. Exercise 2.9).

Remark 2.4. It is of course sufficient to require that $\{|\det M|^{-1/d} M ; M \in T_\mu\}$ contains an open set of $\{M \in Gl(d,\mathbb{R}) ; |\det M| = 1\}$ to obtain the conclusions of the proposition.

We have met in II.5.1 and II.7 probability measures μ on $Sl(d,\mathbb{R})$ which satisfy $\varepsilon_K * \mu * \varepsilon_{K^{-1}} = \mu$ for any K in $SO(d)$. It is then clear that $KT_\mu K^{-1} = T_\mu$ for any K . Therefore the following shows that unless μ is supported by $SO(d)$ the conclusions of Corollary 1.3 hold for μ (see an interpretation of this in Section 4). Actually, using some analysis one can deduce this result from the preceding proposition. But we prefer to prove it directly since the method can be used in other contexts.

PROPOSITION 2.5. *Let* T *be a semigroup in* $Gl(d,\mathbb{R})$. *If there is a matrix* M *such that*

(i) $\|M\|^{-1} M$ *is not orthogonal* .

(ii) *For any* $K \in SO(d)$, $K M K^{-1}$ *is in* T ,

then T *is* p-*contracting and* p-*strongly irreducible for each* p , $1 \leq p < d$.

Proof : If T' is the smallest semigroup which contains $\{K M K^{-1}$, $K \in SO(d)\}$, $KT'K^{-1} = T'$ for any $K \in SO(d)$ and T' is contained in T. Without loss of generality we will suppose that $T' = T$. Using the polar decomposition it is clear that one can write $M = U_1 B U_2$ with U_1, U_2 in $SO(d)$ and $B = \mathrm{diag}(b_1,\ldots,b_d)$ where b_i are real numbers. Assumption (i) implies that there exists j such that $|b_j| > |b_i|$ for any $i \neq j$. We claim that for any $K_1,\ldots,K_n$ in $SO(d)$ there exists $K \in SO(d)$ such that

$$K(K_1 B K_1^{-1})(K_2 B K_2^{-1}) \ldots (K_n B K_n^{-1}) \in T . \qquad (2)$$

For proving this by induction it suffices to show that if $P \in T$ and $K \in SO(d)$ there is some $K' \in SO(d)$ such that $K'PKBK^{-1} \in T$. But $U_1 K^{-1} P K U_1^{-1} \in T$ and $U_1 B U_2 \in T$ whence

$$K U_2 (U_1 K^{-1} P K U_1^{-1})(U_1 B U_2) U_2^{-1} K^{-1} = K U_2 U_1 K^{-1} P K B K^{-1}$$

is in T .

Since $|b_j| > |b_i|$ for $i \neq j$ it is clear that for m large enough and for a convenient choice of $K_1,\ldots,K_m$ among the permutation matrices, one can find $A = (K_1 B K_1^{-1}) \ldots (K_m B K_m^{-1})$ such that

$$A = \mathrm{diag}(a_1,\ldots,a_d) \quad \text{with} \quad a_1 > a_2 > \ldots > a_d > 0 .$$

By (2) KA is in T for some K in $SO(d)$. Let us show that, for any positive integer n , $K^n A^n$ is in T . This is true for $n = 1$ and if $K^{n-1}A^{n-1} \in T$, since $AK = K^{-1}(KA)K \in T$, then $K^{n-1}A^{n-1}AK$ is in T and $K^n A^n = K(K^{n-1} A^n K)K^{-1}$ is in T . By Proposition 2.1 it is now clear that $\{K^n A^n , n \geq 1\}$, hence T , is p-contracting for any p .

We now prove that T is p-strongly irreducible. This amounts to say that the group G generated by T is p-strongly irreducible. Since K^2A^2 and KA are in T, $K^2AK^{-1} \varepsilon G$ and $K^3A = K(K^2AK^{-1}) \varepsilon G$ (it is obvious that for each U in $SO(d)$, $UGU^{-1} = G$). This implies that $A^{-3}K^{-3}K^3A \varepsilon G$ so A^2 is in G.

Suppose there exist proper subspaces $V_1,\dots,V_r$ of $\Lambda^p\mathbb{R}^d$ such that for any $i \varepsilon \{1,\dots,r\}$ and Y in G, $\Lambda^pY(V_i) = V_j$ for some j. Since A^2 induces a bijection of $\{1,\dots,r\}$ (defined by $A^2(i) = j$ iff $(\Lambda^pA^2)(V_i) = V_j$) we can find an integer m such that A^m keeps fixed each of these points, that is, writing D for A^m,

$$\Lambda^pD(V_i) = V_i \qquad , \qquad 1 \leq i \leq r .$$

Since $SO(d)$ is connected this implies in particular that

$$\Lambda^p(U\,D^n\,U^{-1})(V_1) = V_1$$

for any rotation U and positive integer n. Notice that $\lim_{n\to\infty} \|\Lambda^pD^n\|^{-1}\Lambda^pD^n$ is the orthogonal projection P on $e_1\Lambda e_2\Lambda\dots\Lambda e_p$ and therefore,

$$(\Lambda^pU)P(\Lambda^pU^{-1})V_1 = V_1 .$$

This yields $V_1 = \Lambda^p\mathbb{R}^d$; for notational convenience we just prove this for $p = 2$. If $w = \sum_{i<j} \lambda_{i,j}\, e_i \Lambda e_j$ is in V_1 and $\lambda_{i,j} \neq 0$, consider a rotation matrix U which satisfies $U(e_1) = e_i$, $U(e_2) = \pm e_j$. Then $(\Lambda^pU)P(\Lambda^pU^{-1})w = \lambda_{i,j}\, e_i \Lambda e_j$ is in V_1. Therefore V_1 must be spanned by a subfamily of $\{e_i \Lambda e_j\ ,\ 1 \leq i < j \leq d\}$. The same must hold for each $(\Lambda^pU)V_1$ where $U \varepsilon SO(d)$, which implies that $V_1 = \Lambda^2\mathbb{R}^d$ (if for instance $e_1 \Lambda e_2 \varepsilon V_1$ and $e_1 \Lambda e_3 \notin V_1$, then choosing U such that $U(e_i) = e_i$, $i \neq 2,3$, $U(e_2) = \dfrac{e_2+e_3}{\sqrt{2}}$, $U(e_3) = \dfrac{e_2-e_3}{\sqrt{2}}$, we see that $e_1 \Lambda (e_2+e_3)$ is in $(\Lambda^2U)(V_1)$, so that $e_1 \Lambda e_2$ and $e_1 \Lambda e_3$ are in $(\Lambda^pU)(V_1)$ but $e_1 \Lambda (e_2-e_3) = \sqrt{2}(\Lambda^2U)(e_1 \Lambda e_3)$ is not in $(\Lambda^pU)(V_1)$, a contradiction). The proof is complete.

Exercise 2.6. Define the map $\phi : \Lambda^p\mathbb{R}^d \longrightarrow \Lambda^{d-p}\mathbb{R}^d$ by the relation

$$u \Lambda v = \langle\phi(u),v\rangle$$

for any $u \varepsilon \Lambda^p\mathbb{R}^d$, $v \varepsilon \Lambda^{d-p}\mathbb{R}^d$. Show that ϕ is a linear homomorphism. Prove that, for any M in $Gl(d,\mathbb{R})$ and any u in $\Lambda^p\mathbb{R}^d$,

$$(\det M)\ \phi(\Lambda^pM\,u) = (\Lambda^{d-p}M^*)^{-1}\phi(u) .$$

Show that if T is a p-strongly irreducible subgroup of $Gl(d,\mathbb{R})$, then T is $(d-p)$-strongly irreducible.

Exercise 2.7. If T is a p-contracting subset of $Gl(d,\mathbb{R})$, prove that $T^{-1} = \{M ; M^{-1} \in T\}$ is $(d-p)$-contracting.

Exercise 2.8. Let μ be a probability measure on $Sl(d,\mathbb{R})$ such that $\varepsilon_k * \mu * \varepsilon_{k^{-1}} = \mu$ for any k in $SO(d)$. Let m_d^p denote the distribution of $K\cdot\overline{e_1 \wedge e_2 \wedge \ldots \wedge e_p}$ on $P(\Lambda^p \mathbb{R}^d)$ where K is a random matrix whose law is the Haar measure on $SO(d)$. Prove that m_d^p is μ-invariant. Show that when μ is not confined on $SO(d)$, m_d^p is the only μ-invariant distribution on $P(\Lambda^p \mathbb{R}^d)$.

Exercise 2.9. Let T be a connected subset of $Gl(d,\mathbb{R})$. Show that, if p is an integer $(1 \leq p < d)$, T is p-strongly irreducible if and only if T is p-irreducible (i.e. there is no proper subspace V of $\Lambda^p \mathbb{R}^d$ such that $\Lambda^p M(V) = V$ for any M in T). Hint : for any p, $\{\Lambda^p M , M \in T\}$ is connected thus it suffices to consider the case $p = 1$. When $p = 1$ use the fact that if T is not strongly irreducible there exists some q such that for an $\bar{w}$ in $P(\Lambda^q \mathbb{R}^d)$ $T\cdot\bar{w}$ is finite.

IV.3. The case of symplectic matrices

Let $Sp(d,\mathbb{R})$ denote the group of order $2d$ symplectic matrices, i.e. matrices M in $Gl(2d,\mathbb{R})$ which satisfy

$$\sum_{i=1}^{d} Me_i \wedge Me_{i+d} = \sum_{i=1}^{d} e_i \wedge e_{i+d} .$$

It is clear that the action of $Sp(d,\mathbb{R})$ on $\Lambda^2 \mathbb{R}^{2d}$ is not irreducible since the space spanned by $\sum_{i=1}^{d} e_i \wedge e_{i+d}$ is invariant. Thus Proposition 1.2 is not suited for the study of Lyapunov exponents associated with random symplectic matrices. Since these matrices occur in many applications (see e.g. Chapter IV of Part B, and the references in Tutubalin [68]) it is worth proving a criterion adapted to this situation. Before stating it let us recall some basic facts.

It is immediate to see that if

$$J = \begin{pmatrix} 0 & I \\ -I & 0 \end{pmatrix}$$

where I is the identity matrix of $Gl(d,\mathbb{R})$, then $Sp(d,\mathbb{R}) = \{M \in Gl(2d,\mathbb{R})\ ;\ M^* J M = J\}$, (recall that M^* denotes the transpose of M).

LEMMA 3.1. *Given a matrix* M *in* $Sp(d,\mathbb{R})$, *then*

(i) M^* *is in* $Sp(d,\mathbb{R})$.

(ii) If λ *is an eigenvalue of* M *so is* $\frac{1}{\lambda}$.

(iii) There exist two orthogonal matrices K *and* U *in* $Sp(d,\mathbb{R})$ *and a diagonal matrix* $A = \mathrm{diag}(a_1,\dots,a_d, a_1^{-1},\dots,a_d^{-1})$ *with* $a_1 \geq a_2 \geq \dots \geq a_d \geq 1$ *such that* $M = K A U$.

Proof : We have $M^*JM = J$ so that $M^{-1}J^{-1}(M^*)^{-1} = J^{-1}$. But $J^{-1} = -J$ thus $M^{-1}J(M^*)^{-1} = J$ and $MJM^* = J$, proving (i). If λ is an eigenvalue of M with eigenvector v we have

$$M^*(Jv) = \frac{1}{\lambda} M^* J Mv = \frac{1}{\lambda}(Jv)$$

and $\frac{1}{\lambda}$ is an eigenvalue of M^*. To verify (iii) note that the above relation implies that we can find an orthonormal basis of $\mathbb{R}^{2d}$ of the form $\{f_1,\dots,f_d, Jf_1,\dots,Jf_d\}$ such that, if $a_1^2 \geq a_2^2 \geq \dots \geq a_d^2 \geq 1$ are the eigenvalues of M^*M greater than one, then $M^*M f_i = a_i^2 f_i$ and $M^*MJ f_i = a_i^{-2} J f_i$. Let U denote the orthogonal matrix defined by the relations

$$U f_i = e_i\ , \quad U J f_i = -(e_{i+d})\ , \qquad i = 1,2,\dots,d\ .$$

We have

$$(U^* J U) f_i = U^* J e_i = -U^* e_{i+d} = J f_i$$

and
$$(U^* J U)(J f_i) = -U^* J e_{i+d} = -U^* e_i = -f_i = J(J f_i)\ .$$

Therefore $U^* J U = U$ and U is symplectic. If $A = \mathrm{diag}(a_1,\dots,a_d, a_1^{-1},\dots,a_d^{-1})$ and $K = M U^{-1} A^{-1}$, A and thus K are symplectic and $M = K A U$. Since K is easily seen to be orthogonal, the proof is complete.

From this lemma one readily deduces

PROPOSITION 3.2. *If* $\gamma_1 \geq \gamma_2 \geq \dots \geq \gamma_{2d}$ *are the Lyapunov exponents associated with symplectic random matrices, then*

$$\gamma_{d+i-1} = -\gamma_i \qquad \text{for} \quad i = 1,\dots,d .$$

For any $p \in \{1,\dots,d\}$ let L_p be the subspace of $\Lambda^p \mathbb{R}^{2d}$ spanned by $\{Me_1 \wedge \dots \wedge Me_p \ , \ M \in Sp(d,\mathbb{R})\}$, which is sometimes called the p-Lagrangian submanifold (it is known that $u_1 \wedge u_2 \wedge \dots \wedge u_p$ is in L_p iff $u_i^* J u_j = 0$ for any $1 \leq i, j \leq p$, but we won't use it). Notice that $L_1 = \mathbb{R}^{2d}$.

DEFINITION 3.3. *We say that a subset* T *of* $Sp(d,\mathbb{R})$ *is* L_p*-strongly irreducible if there does not exist a finite union* W *of proper linear subspaces of* L_p *such that* $(\Lambda^p M)(W) = W$ *for any* M *in* T.

The main result about random products of symplectic matrices is

PROPOSITION 3.4. *Let* $\{Y_n, n \geq 1\}$ *be a sequence of i.i.d. random symplectic matrices of order* $2d$ *and* p *be an integer in* $\{1,\dots,d\}$. *Suppose that* T_μ *is p-contracting and* L_p*-strongly irreducible and that* $\mathbb{E}(\operatorname{Log} \|Y_1\|)$ *is finite. Then the following hold*

(i) $\gamma_p > \gamma_{p+1}$.

(ii) For any non zero x *in* L_p,

$$\lim_{n\to\infty} \frac{1}{n} \operatorname{Log} \|\Lambda^p \, Y_n \dots Y_1 \, x\| = \sum_{i=1}^{p} \gamma_i .$$

(iii) There exists a unique μ*-invariant probability distribution* ν_p *on* $P(L_p) = \{\bar{x} \in P(\Lambda^p \mathbb{R}^{2d}) \ ; \ x \in L_p\}$ *and*

$$\iint \operatorname{Log} \frac{\|\Lambda^p \, M \, x\|}{\|x\|} \, d\mu(M) \, d\nu_p(\bar{x}) = \sum_{i=1}^{p} \gamma_i .$$

Proof : Let k be the dimension of L_p and $\{f_1,\dots,f_k\}$ a basis of L_p to be chosen later on. For any M in $Sp(d,\mathbb{R})$ let $\overline{M}$ denote the matrix in $Gl(k,\mathbb{R})$ whose entries are

$$\overline{M}_{i,j} = \langle \Lambda^p M \, f_i \, , \, f_j \rangle \qquad 1 \leq i, j \leq k .$$

The proof of the proposition is broken in three steps.

First step : We shall show that for any M in $Sp(d,\mathbb{R})$, if $\phi(M) = \Lambda^p M$, then $\|\phi(M)\| = \|\overline{M}\|$ and $\|\Lambda^2 \phi(M)\| = \|\Lambda^2 \, \overline{M}\|$.

Let us first prove this when $p < d$. Let $f_1 = e_1 \wedge e_2 \wedge \dots \wedge e_p$ and $f_2 = e_1 \wedge e_2 \wedge \dots \wedge e_{p-1} \wedge e_{p+1}$. Since all the matrices

$$Y = \begin{bmatrix} B & 0 \\ C & B^{*-1} \end{bmatrix} \tag{3}$$

with B, C in $Gl(d,\mathbb{R})$ and C symmetric are in $Sp(d,\mathbb{R})$, we can find such a Y for which $f_2 = \Lambda^p Y f_1$. This shows that f_1 and f_2 are in L_p. If we write $M = KAU$ with K, U in $Sp(d,\mathbb{R}) \cap O(d)$ and $A = \mathrm{diag}(a_1,\dots,a_d, a_1^{-1},\dots,a_d^{-1})$, $a_1 \geq a_2 \geq \dots \geq a_d \geq 1$, we have, since $p < d$,

$$\|\phi(M)\| = a_1\dots a_p = \|(\Lambda^p A)(f_1)\| = \|\overline{A} f_1\| \leq \|\overline{A}\| \leq \|\Lambda^p A\| = \|\phi(M)\|$$

so that $\|\phi(M)\| = \|\overline{A}\|$, but $\|\overline{A}\| = \|\overline{K}\,\overline{A}\,\overline{U}\| = \|\overline{M}\|$ and $\|\phi(M)\| = \|\overline{M}\|$.
On the other hand

$$\begin{aligned} \|\Lambda^2\ \phi(M)\| &= (a_1\dots a_p)(a_1\dots a_{p-1}a_{p+1}) = \|\phi(A)f_1 \wedge \phi(A)f_2\| \\ &= \|\overline{A}(f_1 \wedge f_2)\| \leq \|\Lambda^2\ \overline{A}\| \leq \|\Lambda^2\ \phi(M)\| \end{aligned} \tag{4}$$

so that $\|\Lambda^2\ \phi(M)\| = \|\Lambda^2\ \overline{A}\| = \|\Lambda^2\ \overline{K}\,\overline{A}\,\overline{U}\| = \|\Lambda^2\ \overline{M}\|$, proving the claim when $p < d$. If $p = d$ we choose $f_1 = e_1 \wedge \dots \wedge e_d$ but $f_2 = e_1 \wedge \dots \wedge e_{d-1} \wedge e_{2d}$. If Y is the matrix of the form (3) with $B = Id$ and C such that $C_{i,j} = 0$ except $C_{d,d} = 1$ we see that $\Lambda^p Y f_1 = f_1 + f_2$. Therefore f_2 is in L_p. The same proof as above works with this new choice because (4) is replaced by

$$\|\Lambda^2\ \phi(M)\| = (a_1\dots a_d)(a_1\dots a_{d-1}\frac{1}{a_d}) = \|\phi(A)f_1 \wedge \phi(A)f_2\|.$$

Second step : Let $\overline{\mu}$ denote the distribution of $\overline{Y}_1$. Let us show that $T_{\overline{\mu}}$ is a strongly irreducible and contracting subset of $Gl(k,\mathbb{R})$. Since T_μ is L_p-strongly irreducible, it is clear that $T_{\overline{\mu}}$ is strongly irreducible. On the other hand we know (see Proposition 2.1) that, since T_μ is p-contracting, there exists a sequence (M_n) in T_μ such that $\lim_{n\to\infty} \|\phi(M_n)\|^2\ \|\Lambda^2\ \phi(M_n)\|^{-1} = \infty$. This implies that $\lim_{n\to\infty} \|\overline{M}_n\|^2\ \|\Lambda^2\ \overline{M}_n\|^{-1} = \infty$, so that $T_{\overline{\mu}}$ is contracting.

Third step : We now prove the proposition. It follows readily from the statement proved in the first step and from the definition of the Lyapunov exponents (see especially Proposition III.5.6) that the two upper Lyapunov exponents associated with $(\Lambda^p\ Y_n)$ and with $(\overline{Y}_n)$ are the same. Therefore (see e.g. the proof of Theorem 1.2) they are equal to $\sum_{i=1}^{p} \gamma_i$ and $\sum_{i=1}^{p-1} \gamma_i + \gamma_{p+1}$. The proposition is thus an immediate consequence of Theorem 1.2 applied to the sequence $(\overline{Y}_n)$ (with $p = 1$ in this theorem).

The fact that this proposition is suitable for studying symplectic random matrices is clear from the proof we give of the following result of Guivarc'h and Raugi [34].

PROPOSITION 3.5. *If* μ *is a probability measure on* $Sp(d,\mathbb{R})$ *such that* T_μ *contains an open subset of* $Sp(d,\mathbb{R})$ *then all the Lyapunov exponents are distinct.*

Proof : Taking into account Proposition 3.2 it suffices to show that $\gamma_i > \gamma_{i+1}$ for $i = 1,2,\ldots,p$. For this we verify that the assumptions of Proposition 3.4 are valid for any $p \in \{1,\ldots,d\}$.

a. We first prove that T_μ is L_p-strongly irreducible. This is equivalent to the fact that G_μ is L_p-strongly irreducible. But $Sp(d,\mathbb{R})$ being connected, $G_\mu = Sp(d,\mathbb{R})$ (for G_μ contains an open set by hypothesis). Applying Exercise 2.9 we see that we have to show that there is no proper subspace V of L_p such that $(\Lambda^p M)(V) \subset V$ for any M in $Sp(d,\mathbb{R})$. Suppose there exists such a V . If $A = \operatorname{diag}(a_1,\ldots,a_d, a_1^{-1},\ldots,a_d^{-1})$, $a_1 > a_2 > \ldots > a_d > 1$, the relation

$$(\Lambda^p A^n)(V) \subset V \qquad , \qquad \text{for any } n > 0 ,$$

implies either that $e_1 \wedge \ldots \wedge e_p$ is in V , but then

$$\{M e_1 \wedge \ldots \wedge M e_p \ , M \in Sp(d,\mathbb{R})\} \subset V$$

and V is not proper, or that $e_1 \wedge \ldots \wedge e_p$ is in the orthogonal W of V . But then by (i) of Lemma 3.1, if $w \in W$ and $v \in V$

$$<\Lambda^p M w \ , v> = <w, \Lambda^p M^* v> = 0$$

for any M in $Sp(d,\mathbb{R})$. Therefore $M e_1 \wedge \ldots \wedge M e_p$ is in W for any M and $L_p = W$. This contradicts also the fact that V is proper.

b. We now show that T_μ is p-contracting. For this it suffices to verify that T_μ contains a matrix M with $2d$ eigenvalues whose moduli are distinct (see Corollary 2.2). We first show that if U is an open subset contained in T_μ , then there is in U a matrix with distinct, eventually complex, eigenvalues. We may of course suppose that -1 is not an eigenvalue of the matrices in U . In that case the mapping $\psi : U \longrightarrow M(2d,\mathbb{R})$ defined by $\psi(M) = (Id-M)(Id+M)^{-1}$ sends U onto an open subset of $S = \{M \in M(2d,\mathbb{R}) \ , \ M^*J + JM = 0\}$.

Since the eigenvectors of M and $\psi(M)$ are the same it suffices to show that each open subset of S contains a matrix with distinct eigenvalues. For each M in S let $P(M)$ be the resultant of the caracteristic polynomial associated with M and its derivative (see e.g. Lang [45]). M has a double eigenvalue iff $P(M) = 0$. But $P(\cdot)$ is a polynomial on the vector space S. It can be zero on an open set if and only if it is the zero polynomial. This would imply that each matrix of S has a double eigenvalue, which is false since $\mathrm{diag}(a_1,\dots,a_d, -a_1,\dots,-a_d)$ is in S for any $a_1,\dots,a_d$.

So we have found a matrix M in U with distinct eigenvalues $\lambda_1,\dots,\lambda_d, \frac{1}{\lambda_1},\dots,\frac{1}{\lambda_d}$. Suppose that $\lambda_{2k+1},\dots,\lambda_d$ are the real elements of $\{\lambda_1,\dots,\lambda_d\}$. Let $\{f_1,f_2,\dots,f_d, h_1,\dots,h_d\}$ be a basis of $\mathbb{R}^{2d}$ such that

$$M(f_{2j+1} + i\, f_{2j+2}) = \lambda_{2j+1}(f_{2j+1} + i\, f_{2j+2}) \ , \qquad j = 0,\dots,k-1 \ ;$$

$$M(h_{2j+1} + i\, h_{2j+2}) = \lambda_{2j+1}^{-1}(h_{2j+1} + i\, h_{2j+2}) \ , \qquad j = 0,\dots,k-1 \ ;$$

$$M\, f_j = \lambda_j\, f_j \ , \qquad M\, h_j = \lambda_j^{-1}\, h_j \ , \qquad j = 2k+1,\dots,d \ .$$

It is immediate to deduce from these relations that for any $r,s,\ r \neq s$,

$$f_r^* J f_s = 0 \ , \qquad f_r^* J h_s = 0$$

and

$$f_r^* J f_r = 0 \ , \qquad f_r^* J h_r = \alpha_r \neq 0 \ .$$

If we replace h_r by $\alpha_r^{-1} h_r$ this means that the matrix P defined by $P(e_i) = f_i$ and $P(e_{i+d}) = h_{i+d}$, $1 \leq i \leq d$, is symplectic. Therefore if M' is the matrix which represents M in the basis $\{f_1,\dots,f_d, h_1,\dots,h_d\}$, $M = P^{-1} M' P$, M' is in the open subset $P U P^{-1}$ of $Sp(d,\mathbb{R})$. Moreover M' can be written as

$$M' = \begin{bmatrix} B_1 & & & & & 0 \\ & \ddots & & & & \\ & & B_d & & & \\ & & & B_1^{-1} & & \\ & & & & \ddots & \\ 0 & & & & & B_d^{-1} \end{bmatrix}$$

where B_i is either a real number β_i, or a 2×2 matrix $\beta_i\, R(\theta_i)$. Working as in the proof of Proposition 2.3 (with obvious modifications) we see that for some n, $P U^n P^{-1}$, and therefore U^n, contains a matrix with 2d eigenvalues whose moduli are distinct.

Exercise 3.6. : Show that if T_μ contains an open set of $Sp(d,\mathbb{R})$ then T_μ is strongly irreducible.

IV.4. μ-boundaries

We now return to the study of general random matrices. We shall be interpreting the results of Section 1 as informations on the behaviour of products $Y_1 \ldots Y_n$ towards boundaries of $Gl(d,\mathbb{R})$.

Consider integers $1 \leq p_1 < p_2 < \ldots < p_k < d$. Let $F(p_1,\ldots,p_k)$ denote the subset

$$\{(\overline{x_1 \wedge x_2 \wedge \ldots \wedge x_{p_1}}, \overline{x_1 \wedge \ldots \wedge x_{p_1} \wedge \ldots \wedge x_{p_2}}, \ldots, \overline{x_1 \wedge \ldots \wedge x_{p_1} \wedge \ldots \wedge x_{p_k}}) \; ; \; x_i \in \mathbb{R}^d\}$$

of $P(\Lambda^{p_1} \mathbb{R}^d) \times P(\Lambda^{p_2} \mathbb{R}^d) \times \ldots \times P(\Lambda^{p_k} \mathbb{R}^d)$. It follows from Lemma III.5.1 that $F(p_1,\ldots,p_k)$ may be identified with the set of k-uples $(E_1,E_2,\ldots,E_k)$ where E_i is a p_i-dimensional subspace of $\mathbb{R}^d$ and where

$$E_1 \subset E_2 \subset \ldots \subset E_k .$$

For instance $F(1) = P(\mathbb{R}^d)$ and $F(r)$ is the set of r-dimensional subspaces of $\mathbb{R}^d$. $F = F(1,2,3,\ldots,d-1)$ is called the flag manifold.

It is clear that each $F(p_1,\ldots,p_k)$ is a $Gl(d,\mathbb{R})$-space for the obvious action. Note that $F(p_1,\ldots,p_k)$ is the orbit under $O(d)$ of

$$e(p_1,\ldots,p_k) = (\overline{e_1 \wedge \ldots \wedge e_{p_1}}, \overline{e_1 \wedge \ldots \wedge e_{p_1} \wedge \ldots \wedge e_{p_2}}, \ldots, \overline{e_1 \wedge \ldots \wedge e_{p_k}}).$$

As such it is compact. At last note that

$$S = \{M \in Gl(d,\mathbb{R}) \; ; \; M \cdot e(1,2,\ldots,d-1) = e(1,2,\ldots,d-1)\}$$

is the group of upper triangular matrices and F is the homogeneous space $Gl(d,\mathbb{R})/_S$.

PROPOSITION 4.1. *Let* $Y_1,Y_2,\ldots$ *be independent matrices in* $Gl(d,\mathbb{R})$ *with the same distribution* μ. *We suppose that, for any* $p \in \{1,\ldots,d-1\}$, T_μ *is* p-*strongly irreducible. Let* $1 \leq p_1 < p_2 < \ldots < p_k < d$ *be all the integers* p *such that* T_μ *is* p-*contracting. Then*

(i) There exists a unique μ*-invariant distribution* ν *on* $F(p_1,\dots,p_k)$.

(ii) With probability one the measures $Y_1\dots Y_n\nu$ *on* $F(p_1,\dots,p_k)$ *converge weakly to a Dirac measure.*

(iii) If $1 \leq q_1 < q_2 < \dots < q_\ell < d$ *and* $\{q_1,\dots,q_\ell\}$ *is not contained in* $\{p_1,\dots,p_k\}$ *, then for any* μ*-invariant distribution* ν *on* $F(q_1,q_2,\dots,q_\ell)$ *, the probability that* $Y_1\dots Y_n\nu$ *converges to a Dirac measure is zero.*

Proof : If we apply Theorem III.4.3 to the sequence $\{\Lambda^{p_i} Y_n^*, n \geq 1\}$ we know that there exists $\overline{Z}_i$ in $P(\Lambda^{p_i} \mathbb{R}^d)$ such that $\Lambda^{p_i} Y_1\dots Y_n\cdot\overline{x}$ converges in probability to $\overline{Z}_i$, uniformly in $\overline{x} \in P(\Lambda^{p_i} \mathbb{R}^d)$. This implies that there is a subsequence (n_j) such that for any $\overline{x}$,

$$\lim_{n\to\infty} \Lambda^{p_i} Y_1\dots Y_{n_j}\cdot\overline{x} = \overline{Z}_i \qquad \text{a.s.} \quad .$$

Therefore $\overline{Z}_i$ is a decomposable p_i-vector. Repeating this for each p_i we find a point Z in $F(p_1,\dots,p_k)$ such that

$$\lim_{n\to\infty} Y_1\dots Y_{n_j}\cdot\xi = Z \tag{5}$$

for any ξ in $F(p_1,\dots,p_k)$, with probability one. If ν is an μ-invariant probability measure on $F(p_1,\dots,p_k)$, (5) holds for ν-almost all ξ with probability one, by Fubini's theorem. Therefore

$$\lim_{j\to\infty} Y_1\dots Y_{n_j}\nu = \delta_Z , \qquad \text{a.s.} \quad .$$

A compacity argument yields

$$\lim_{n\to\infty} Y_1\dots Y_n\nu = \delta_Z$$

this proves (ii) but also implies (i) (ν must be the distribution of Z).

To prove (iii) suppose that $q_i \notin \{p_1,\dots,p_k\}$ and that, with probability $\neq 0$, $\lim Y_1\dots Y_n\nu = \delta_Z$. If $\overline{\nu}$ (resp. $\overline{Z}$) is the projection of ν (resp. Z) on $P(\Lambda^{q_i} \mathbb{R}^d)$, then with non zero probability

$$\lim \Lambda^{q_i} Y_1\dots Y_n\overline{\nu} = \delta_{\overline{Z}} \quad .$$

But $\overline{\nu}$ is proper (see Proposition III.2.3) and this would imply (see Exercise III.1.12) that T_μ is q_i-contracting. This is a contradiction.

Consider now symplectic matrices of order $2d$. Put, for integers $1 \leq p_1 < p_2 < \ldots < p_k \leq d$,

$$L(p_1,\ldots,p_k) = \{(M(e_1\wedge\ldots\wedge e_{p_1}),M(e_1\wedge\ldots\wedge e_{p_1}\wedge\ldots\wedge e_{p_2}),\ldots,M(e_1\wedge\ldots\wedge e_{p_k}));$$
$$M \in Sp(d,\mathbb{R})\}.$$

Notice that $L(p)$ is the p-Lagrangian submanifold. Using the proof of Proposition 3.4 one shows as above the following

PROPOSITION 4.2. *Let* $\{Y_n, n \geq 1\}$ *be i.i.d. random matrices in* $Sp(d,\mathbb{R})$ *with common distribution* μ . *We suppose that for any* $p \in \{1,\ldots,d\}$, T_μ *is* L_p*-strongly irreducible and denote by* $1 \leq p_1 < \ldots < p_k \leq d$ *all the integers* $p \in \{1,\ldots,d\}$ *such that* T_μ *is* p*-contracting. Then*

(i) There exists a unique μ*-invariant distribution* ν *on* $L(p_1,\ldots,p_k)$.

(ii) $Y_1\ldots Y_n\nu$ *converges to a Dirac measure, a.s.* .

(iii) If $1 \leq q_1 < q_2 < \ldots < q_\ell \leq d$ *and* $\{q_1,\ldots,q_\ell\}$ *is not contained in* $\{p_1,\ldots,p_k\}$ *then for any* μ*-invariant distribution* ν *on* $L(q_1,\ldots,q_\ell)$*, a.s.* , $Y_1\ldots Y_n\nu$ *does not converge weakly to a Dirac measure.*

Exercise 4.3. : Show that, under the hypotheses of Proposition 4.1,

(i) There exists a unique μ-invariant distribution on

$$P(\Lambda^{p_1}\mathbb{R}^d) \times P(\Lambda^{p_2}\mathbb{R}^d) \times \ldots \times P(\Lambda^{p_k}\mathbb{R}^d).$$

(ii) If $Y_1\ldots Y_n = K_n A_n U_n$ is a polar decomposition then $U_n\cdot e(p_1,\ldots,p_k)$ converges a.s. in $F(p_1,\ldots,p_k)$.

We shall now define, following Furstenberg, the μ-boundaries. We begin by a somewhat informal discussion.

Let G denote the group $Gl(d,\mathbb{R})$. If $Y_1,Y_2,\ldots$ are independent random elements of G with a common distribution μ we put $Z_n = Y_1\ldots Y_n$ and, for g in G , $Z_o^g = g$, $Z_n^g = g\,Y_1\ldots Y_n$. It is readily seen that $\{Z_n^g, n \geq 0\}$ is a Markov chain on G. Proposition 4.1 tells us something about the behaviour at infinity of this Markov chain. Namely we have found G-spaces B such that if ν is a μ-invariant distribution on B , $Y_1 \dots Y_n \nu \xrightarrow{w} \delta_Z$ almost surely. This yields that $Z_n^g \nu \longrightarrow \delta_{g \cdot Z}$. Let $M_1(B)$ denote the set of probability measures on B and define a map $\Psi : G \cup B \longrightarrow M_1(B)$ by $\Psi(g) = g\nu$ if $g \in G$, $\Psi(x) = \delta_x$ if $x \in B$. Following Furstenberg [71] consider the coarsest topology $\mathcal{T}$ on $G \cup B$ which induces on G and B their own topology and for which Ψ is continuous. It is not difficult to see that with this topology

(a) G is open and dense in $G \cup B$.

(b) With probability one, Z_n^g converges to $g.Z$ in $G \cup B$.

Thus the pair $(B, \mathcal{T})$ appears as a kind of boundary of G associated with μ .

Proposition 4.1 provides us with such boundaries towards which Z_n^g converges a.s. . If it was, in some way, natural to introduce the G-spaces $F(p_1, \dots, p_k)$ one can wonder if we have exhausted all possible "boundaries" associated with μ . In other words, are there other $(G \cup B, \mathcal{T})$ for which (a) and (b) hold which give us more information ? We want to show that the answer to this latter question is essentially no. This will not be needed later so we shall be rather sketchy. (The interested reader should consult the works of Furstenberg [20], [22], [23], [24] to have a full account of these problems). We first state this question in a more tractable form.

Suppose that B is a compact G-space such that for some topology $\mathcal{T}$ on $G \cup B$, (a) and (b) hold. It is natural to impose that $G \cup B$ is itself a G-space and in this case (b) is equivalent to (b') : Z_n converges to Z a.s. . For any p , $\{Y_{p+1} \dots Y_n , n > p\}$ is just a copy of $\{Z_n, n \geq 1\}$ so that $Y_{p+1} \dots Y_n$ converges a.s. to some $Z(p)$. We have $Y_1 Z(1) = Z$ so the distribution, say ν , of Z is μ-invariant. We will suppose that in G , Z_n goes to infinity (that's the usual case) so that $Z \in B$. For any continuous function f on B , if $\mathcal{F}_p$ is the σ-algebra generated by $Y_1, \dots, Y_p$,

$$\mathbb{E}^{\mathcal{F}_p}\{f(Z)\} = \mathbb{E}^{\mathcal{F}_p}\{f(Z_p \cdot Z(p+1))\} = \int_B f(Z_p \cdot x) \, d\nu(x) .$$

The martingale convergence theorem yields

$$\lim_{p \to \infty} \int_B f(Z_p \cdot \bar{x}) \, d\nu(x) = f(Z) \qquad \text{a.s.}$$

and (if B is separable), $Y_1 \dots Y_n \nu \xrightarrow{w} \delta_Z$ with probability one.

The conclusion of this discussion is that it is equivalent to consider a topology $\mathscr{C}$ on $G \cup B$ for which (a) and (b) hold or to specify a μ-invariant probability measure ν on B such that

(a') $\{\delta_x, x \in B\}$ is contained in the closure of $\{g\nu, g \in G\}$.

(b') $Y_1 \ldots Y_n \nu$ converges weakly to some δ_Z, with probability one.

Condition (a') takes the place of the density assumption of G in $G \cup B$. A pair (B,ν) for which (a') and (b') hold is called a μ-boundary.

Given a probability measure μ on a group G it can be difficult to describe all μ-boundaries. But it is worth describing compact G-spaces B such that (B,ν) is a μ-boundary for each μ-invariant distribution ν on B . In fact this is easy and these spaces are important. We have the following (amenable groups have been defined in III.7.5).

PROPOSITION 4.4. *Let* G *be a second countable locally compact group,* μ *a probability measure on* G *and* H *a closed (not necessarily normal) amenable subgroup of* G *such that* $G_{/H}$ *is compact. If* B *is a compact metric* G*-space such that* (B,ν) *is a* μ*-boundary for any* μ*-invariant distribution* ν *, then there exists a closed subgroup* H' *of* G *which contains* H *such that* $B = G_{/H'}$.

Proof : Since H is amenable there exists a probability measure m on B with, for any h in H

$$hm = m .$$

Let (n_i) be a subsequence such that $\lim \frac{1}{n_i} \sum_{j=1}^{n_i} \mu^j \dot{*} m = \nu$ exists. By Lemma I.3.4, ν is μ-invariant. Let us show that there is a probability measure λ on G such that $\nu = \lambda \dot{*} m$. Let $\pi : G \longrightarrow G_{/H}$ denote the canonical projection. Since $G_{/H}$ is compact we can suppose that $\pi(\frac{1}{n_i} \sum_{j=1}^{n_i} \mu^j)$ converges to some probability measure ρ on $G_{/H}$. If λ is a measure on G with $\pi(\lambda) = \rho$, $\nu = \lambda \dot{*} m$.

Since (B,ν) is a μ-boundary there exists a sequence g_n in G such that $g_n \nu$ converges to some δ_z . Let V_p be a decreasing sequence of neighborhoods of z in B with $\bigcap_p V_p = \{ z \}$. For n large

$$g_n \nu (V_p) = \int gm(V_p) \, d(\varepsilon_{g_n} * \lambda)(g) \geq 1 - \frac{1}{p} .$$

So we can find g'_p in G such that $g'_p m(V_p) \geq 1 - \frac{1}{p}$. This yields $\lim g'_p m = \delta_z$. Write $G = KH$ for some compact set K in G . If $g'_p = k_p h_p$, $k_p \in K$, $h_p \in H$, we may suppose that k_p converges to some k . Taking into account the relation

$$g'_p m = k_p h_p m = k_p m$$

we obtain $km = \delta_z$ and m is supported by $x_1 = k^{-1} \cdot z$. If $H' = \{g \in G \ , \ g \cdot x_1 = x_1\}$, H' is a closed subgroup of G which contains H and $G_{/H'}$ can be identified with $G \cdot x_1$. But $G \cdot x_1 = K \cdot x_1$ is compact so it carries a μ-invariant probability measure ν . Condition (a') implies that $G \cdot x_1$ is equal to B . The proof is complete.

If, as in this proposition, (B,ν) is a μ-boundary for every μ-invariant distribution ν, ν must be unique. For if ν_1 and ν_2 are μ-invariant, $\nu_3 = \frac{\nu_1 + \nu_2}{2}$ is also μ-invariant and if $Y_1 \ldots Y_n \nu_i \longrightarrow \delta_{Z(i)}$ for, $i = 1,2,3$, $Z(1) = Z(2) = Z(3)$ and $\nu_1 = \nu_2$. This class of boundaries is thus exactly the one of boundaries which carry a unique μ-invariant distribution. Consider the case where $G = Gl(d,\mathbb{R})$. We have seen that if S is the subgroup of uppertriangular matrices $G_{/S}$ is the flag manifold. S, being solvable, is amenable and $G_{/S}$ is compact. So $G_{/S}$ is, by Proposition 4.1, a μ-boundary. Moreover one can show that $\{F(p_1,\ldots,p_k)$; $1 \leq p_1 < p_2 < \ldots < p_k < d$, $k = 1,\ldots,d-1\}$ exhausts all the $G_{/H'}$ where H' is a closed subgroup of G which contains S . Therefore Proposition 4.1 describes, when T_μ is p-strongly irreducible for each p , all the boundaries (B,ν) which carry a unique μ-invariant probability measure. With the notations of this proposition $F(p_1,\ldots,p_k)$ is the maximal μ-boundary with this property, the others being its equivariant quotients.

Now look at the case of matrices in $Sp(d,\mathbb{R})$. Let $L = L(1,2,\ldots,p)$. It is easy to see that if

$$S = \{M \in Sp(d,\mathbb{R}) \ ; \ M \cdot \overline{e_1 \Lambda \ldots \Lambda e_k} = \overline{e_1 \Lambda \ldots \Lambda e_k} \ , \ k = 1,\ldots,d\}$$

then S is solvable (hence amenable) and that $Sp(d,\mathbb{R})_{/S}$ is equal to L . Since one can show that all the equivariant quotients of L are the $L(p_1,\ldots,p_k)$, $1 \leq p_1 < p_2 < \ldots < p_k \leq d$, Proposition 4.2 describes all the μ-boundaries which carry a unique μ-invariant distribution, when T_μ is L_p-strongly irreducible for each p .

The reader who is aware of Lie group theory may guess that if we deal with a general semi-simple Lie group G with compact center, the possible μ-boundaries carrying a unique μ-invariant distribution will to be found among the $G_{/P}$ where P runs through the set of parabolic subgroups of G. This approach is developed in Guivarc'h and Raugi [34].

Remark 4.4. In many cases one can show that under the hypotheses of Proposition 4.1, $F(p_1,\dots,p_k)$ is actually maximal among all μ-boundaries. In this case it can be identified with the Poisson space associated with μ (i.e. each μ-harmonic bounded function on G has a unique integral representation on this space). This is proved, under the assumption $T_\mu = G_\mu$, by Furstenberg (see [20]) when μ is absolutely continuous on $Sl(d,\mathbb{R})$ and by Ledrappier (see [47]) when G_μ is discrete.

Remark 4.5. Suppose that μ is a probability measure on $Sl(d,\mathbb{R})$ such that $\varepsilon_k * \mu * \varepsilon_{k^{-1}} = \mu$ for any k in $SO(d)$. Let D denote the quotient space $Sl(d,\mathbb{R})_{/SO(d)}$ and $\pi : Sl(d,\mathbb{R}) \longrightarrow D$ the canonical map. Exactly as in II.7 one sees that for any g in $Sl(d,\mathbb{R})$, $\pi(gY_1\dots Y_n)$ is a Markov chain on D. By Proposition 2.5 we know that if μ is not confined on $SO(d)$, the flag manifold F itself is a μ-boundary. For any ξ in F let $m_F = m_K * \delta_\xi$ (where m_K is the normalized Haar measure on $SO(d)$). It is clear that m_F does not depend upon ξ and is the unique μ-invariant measure. Therefore $gY_1\dots Y_n m_F$ converges to a Dirac measure on F. In this case one may interpret this as a convergence of the Markov chain in a natural compactification of D. Namely let $\overline{D}$ be the closure of $\{g m_F ,$ $g \in Sl(d,\mathbb{R})\}$ in the space of probability measures on F. Since $g m_F = g' m_F$ if and only if $\pi(g) = \pi(g')$ (easy exercise) D can be seen as a dense subset of $\overline{D}$. On the other hand F is clearly identified with a closed subset of $\overline{D}$. Therefore F is a part of the boundary of D in its compactification $\overline{D}$, and $\pi(gY_1\dots Y_n)$ converges to a point of F. When $d = 2$, D can be identified with the unit disk $\{z \in \mathbb{C} ; |z| < 1\}$ and $\overline{D}$ with $\{z \in \mathbb{C} ; |z| \leq 1\}$. We have thus generalized the results of Section II.7.

CHAPTER V

CENTRAL LIMIT THEOREM AND RELATED RESULTS

In 1970 Kaijser showed in some particular but typical cases that the contractive action of the random products $S_n = Y_n \dots Y_1$ on $P(\mathbb{R}^d)$ implies that $\operatorname{Log} \|S_n x\|$ suitably normalized converges in distribution to a gaussian law (see [38], [39], [40]). This idea was later fully developed by Le Page in [49] where he proved that, loosely speaking, $\operatorname{Log} \|S_n x\|$ behaves like a sum of i.i.d. real random variables and satisfies analogues of the main classical limit theorems. We shall give here a detailed introduction to the important work of Le Page and present, with some improvements, his main results (namely, the central limit theorem with speed of convergence and an estimate of the large deviations of $\operatorname{Log} \|S_n x\|$). We also study the tightness of $\{S_n x, n \geq 1\}$ without moment assumption. This chapter ends with an application to linear stochastic differential equations.

V.1. Introduction

Let us explain the main ideas leading to a proof of the central limit theorem for $\operatorname{Log} \|S_n x\|$, where $S_n = Y_n \dots Y_1$ is the product of i.i.d. random matrices in $Gl(d,\mathbb{R})$. They derive from the fact that if $E = Gl(d,\mathbb{R}) \times P(\mathbb{R}^d)$, for any direction $\bar{x}$, the sequence $X_0 = (Y_1, \bar{x})$, $X_1 = (Y_2, Y_1 \cdot \bar{x})$, ..., $X_n = (Y_{n+1}, S_n \cdot \bar{x})$ is a Markov chain on E. Notice that if $\sigma : E \longrightarrow \mathbb{R}$ is defined by

$$\sigma(M,\bar{x}) = \text{Log}\,\frac{\|Mx\|}{\|x\|}$$

then, for $\|x\| = 1$,

$$\text{Log}\,\|S_n x\| = \sigma(X_{n-1}) + \sigma(X_{n-2}) + \ldots + \sigma(X_o) .$$

Therefore $\text{Log}\,\|S_n x\|$ is just an additive functional of the Markov chain (X_n) and we may try to apply the general devices for studying such objects.

In fact (at least) two methods are useful. The first one, due to Gordin and Lifschitz, relies upon martingale theory. We shall see (cf. complements 2.8) that it gives a short proof of the central limit theorems. The second method, due to Nagaev ([55]) and first used for random products by Tutubalin ([66]), rests on perturbation theory for operators. It is technically more involved but gives the best estimates. Since it may also be applied to prove other results (e.g. large deviations) we shall develop it at length.

For a general Markov chain (X_n) on E the idea is to introduce operators $\{P(z),\ z \in \mathbb{C}\}$ acting on some space of functions ϕ on E by the formula

$$[P(z)\phi](y) = \mathbb{E}_y\{e^{z\,\sigma(X_1)}\,\phi(X_1)\} . \tag{1}$$

The Markov property implies that

$$[P^n(z)\phi](y) = \mathbb{E}_y\{e^{z\{\sigma(X_1)+\ldots+\sigma(X_n)\}}\,\phi(X_n)\} .$$

So that, if $\mathbb{1}$ is the function identically equal to one,

$$\mathbb{E}_y\{e^{z\{\sigma(X_1)+\ldots+\sigma(X_n)\}}\} = [P^n(z)\ \mathbb{1}]\,(y) .$$

Thus these operators $P(z)$ play the role of the usual Fourier-Laplace transform.

In our specific model, Tutubalin noticed that since for $y = (M,\bar{x})$, $(P(z)\phi)(y)$ depends only on $M\cdot\bar{x}$, it suffices to consider the functions ϕ on E which can be written

$$\phi(M,\bar{x}) = f(M\cdot\bar{x}) .$$

If we put $(T(z)f)(M\cdot\bar{x}) = (P(z)\phi)(M,\bar{x})$ we obtain, if $\|x\| = 1$,

$$(T(z)f)(\bar{x}) = \mathbb{E}\{e^{z\,\text{Log}\,\|Y_1 x\|}\,f(Y_1\cdot\bar{x})\} \tag{2}$$

and (see lemma 5.2)

$$(T^n(z)f)(\bar{x}) = \mathbb{E}\{e^{z\,\text{Log}\,\|S_n x\|}\,f(S_n\cdot\bar{x})\} .$$

The point now is to find a space $\mathbb{L}$ of functions on $P(\mathbb{R}^d)$ which contains $\mathbb{1}$, such that $T(z)$ acts on $\mathbb{L}$ in such a way that a thorough analysis of the iterates of $T(z)$ is available (at least for $|z|$ small). Nagaev and Tutubalin supposed that (X_n) satisfies the Döblin condition and used for $\mathbb{L}$ the space of Borel bounded functions. But this hypothesis is too stringent for random products. The idea of Le Page, suggested by particular cases studied by Kaijser, is to choose $\mathbb{L}$ to be a space of Hölder continuous functions.

We shall first see that if $T(0)$ acts on such a space, then $T(0) = N + Q$ where N is a rank-one projection, Q an operator with spectral radius smaller than one and $NQ = QN = 0$. This is really the fundamental result. From this we shall deduce, using some easy facts of perturbation theory, that for z small enough, $T(z)$ is a bounded operator on $\mathbb{L}$ which can be written

$$T(z) = \lambda(z)N(z) + Q(z)$$

where $\lambda(z)$ is in $\mathbb{C}$, $N(z)$ is a rank-one projection, $N(z)Q(z) = Q(z)N(z) = 0$, and the spectral radius of $Q(z)$ is smaller than $|\lambda(z)|$. This implies for instance that $\mathbb{E}\{e^{i\frac{t}{\sqrt{n}}\operatorname{Log}\|S_n x\|}\}$ is equivalent to $\lambda(\frac{it}{\sqrt{n}})^n$ as $n \longrightarrow \infty$.

It is then easy to prove the analogues of the classical limit theorems for sums of i.i.d. real random variables which depend only upon the behaviour of the Fourier-Laplace transform near zero, by mimicking the usual proofs.

V.2. Exponential convergence to the invariant measure

Let $Y_1, Y_2, \ldots$ be independent random matrices in $Gl(d,\mathbb{R})$ with common distribution μ. We have seen in Theorem III.4.3 that if T_μ is strongly irreducible and contracting then there exists a unique μ-invariant distribution ν on $P(\mathbb{R}^d)$. Moreover, for all continuous functions f on $P(\mathbb{R}^d)$ and $\bar{x}$ in $P(\mathbb{R}^d)$

$$\mathbb{E}\{f(S_n \cdot \bar{x})\} \longrightarrow \int f \, d\nu$$

where $S_n = Y_n \ldots Y_1$.

We shall improve this result under a moment condition on μ.

DEFINITION 2.1. *Let* $\ell(M) = \mathrm{Sup}\{\mathrm{Log}^+ \|M\| , \mathrm{Log}^+ \|M^{-1}\|\}$ *for* M *in* $Gl(d,\mathbb{R})$. *We say that a probability measure* μ *on* $Gl(d,\mathbb{R})$ *has*

(i) a moment of order p *if* $\int \ell(M)^p \, d\mu(M) < \infty$;

(ii) an exponential moment if for some $\tau > 0$, $\int \exp \tau\ell(M) \, d\mu(M)$ *is finite.*

For the reader's convenience we restate Lemma III.5.4, adding the obvious inequality (i).

LEMMA 2.2. *(i) For any* M_1, M_2 *in* $Gl(d,\mathbb{R})$

$$\ell(M_1 M_2) \leq \ell(M_1) + \ell(M_2) .$$

(ii) For any integer p, $1 \leq p < d$, M *in* $Gl(d,\mathbb{R})$ *and* w *in* $\Lambda^p \mathbb{R}^d$,

$$|\mathrm{Log}\,\|\Lambda^p M\|\,| \leq p\,\ell(M)$$

and

$$|\mathrm{Log}\,\|\Lambda^p Mw\|\,| \leq p\,\ell(M)\,\|w\| .$$

Note that by (i) the convolution powers of μ have at least as many moments as μ has.

We first prove a general lemma on cocycles. We use the notations of I.3 ; in particular μ^n denotes the nth convolution power of μ.

LEMMA 2.3. *Let* T *be a topological semigroup acting on a set* B *and* s *be an additive cocycle on* $T \times B$. *Consider a probability measure* μ *on* T *such that*

(i) for $r(g) = \sup_{x\in B} |s(g,x)|$, *there is some* $\tau > 0$ *such that* $\int_T \exp \tau\, r(g) d\mu(g)$ *is finite ;*

(ii) for some positive integer p, $\sup_{x\in B} \int s(g,x) d\mu^p(g) < 0$. *Then* $\lim_{n\to\infty} \frac{1}{n} \mathrm{Log}\, \{\sup_x \int e^{\alpha s(g,x)} d\mu^n(g)\} < 0$ *for* $\alpha > 0$ *small enough.*

Proof : For each α in $(0,\tau)$ set $a_n = \mathrm{Log}\{\sup_x \int e^{\alpha s(g,x)} d\mu^n(g)\}$. Since s is a cocycle we have for all positive integers n, m,

$$\begin{aligned}\int e^{\alpha s(g,x)} d\mu^{n+m}(g) &= \int e^{\alpha s(g_1 g_2,x)} d\mu^n(g_1) d\mu^m(g_2) \\ &= \int e^{\alpha s(g_1, g_2\cdot x)}\, e^{\alpha s(g_2,x)} d\mu^n(g_1)\, d\mu^m(g_2)\end{aligned}$$

$$\leq \int e^{\alpha s(g_2,x)} \{\operatorname{Sup}_y \int e^{\alpha s(g_1,y)} d\mu^n(g_1)\} d\mu^m(g_2)$$

$$\leq \{\operatorname{Sup}_y \int e^{\alpha s(g_1,y)} d\mu^n(g_1)\}\{\operatorname{Sup}_x \int e^{\alpha s(g_2,x)} d\mu^m(g_2)\}$$

so that the sequence $a_n = \operatorname{Log}\{\operatorname{Sup}_x \int e^{\alpha s(g,x)} d\mu^n(g)\}$ is subadditive. This implies that

$$\lim_{n\to\infty} \frac{1}{n} \operatorname{Log}\{\operatorname{Sup}_x \int e^{\alpha s(g,x)} d\mu^n(g)\} = \inf_n \frac{1}{n} a_n .$$

We claim (and this will prove the lemma) that for α small enough,

$$\operatorname{Sup}_x \int e^{\alpha s(g,x)} d\mu^p(g) < 1 .$$

From the relation

$$e^x \leq 1 + x + \frac{x^2}{2} e^{|x|} ,$$

we deduce that

$$e^{\alpha s(g,x)} \leq 1 + \alpha s(g,x) + \frac{\alpha^2}{2} r(g)^2 e^{\alpha r(g)}$$

and

$$\operatorname{Sup}_x \int e^{\alpha s(g,x)} d\mu^p(g) \leq 1 + \alpha \operatorname{Sup}_x \int s(g,x) d\mu^p(g) + \frac{\alpha^2}{2} \int r(g)^2 e^{\alpha r(g)} d\mu^p(g)$$

Therefore the claim follows readily from assumption (ii).

PROPOSITION 2.3. *Let* μ *be a probability measure on* $Gl(d,\mathbb{R})$ *such that* T_μ *is strongly irreducible and contracting. If for some* $\tau > 0$, $\int e^{\tau \ell(M)} d\mu(M)$ *is finite then there exists* $\alpha_0 > 0$ *such that for each* α *in* $(0, \alpha_0]$,

$$\lim_{n\to\infty} \{\operatorname*{Sup}_{\substack{\bar{x},\bar{y}\in P(\mathbb{R}^d)\\ \bar{x}\neq\bar{y}}} \int \{\frac{\delta(M\cdot\bar{x},M\cdot\bar{y})}{\delta(\bar{x},\bar{y})}\}^\alpha d\mu^n(M)\}^{1/n} < 1 .$$

Proof : Let $B = \{(\bar{x},\bar{y}) \; ; \; \bar{x} \in P(\mathbb{R}^d), \bar{y} \in P(\mathbb{R}^d), \bar{x} \neq \bar{y}\}$. It is clear that B is a $Gl(d,\mathbb{R})$-space under the action

$$M\cdot(\bar{x},\bar{y}) = (M\bar{x} , M\bar{y}) \quad , \qquad M \in Gl(d,\mathbb{R}) , \; (\bar{x},\bar{y}) \in B .$$

The relation

$$s(M,(\bar{x},\bar{y})) = \operatorname{Log} \frac{\delta(M\cdot\bar{x},M\cdot\bar{y})}{\delta(\bar{x},\bar{y})}$$

defines a cocycle on $Gl(d,\mathbb{R}) \times B$. By Lemma 2.2

$$\operatorname*{Sup}_{(\bar{x},\bar{y})\in B} s(M,(\bar{x},\bar{y})) = \operatorname*{Sup}_{(\bar{x},\bar{y})\in B} \operatorname{Log} \frac{\|Mx\wedge My\|}{\|x\wedge y\|}\frac{\|x\|\ \|y\|}{\|Mx\|\ \|My\|}$$

$$\leq 4\,\ell(M)\ .$$

Since we have seen (Proposition III.6.4) that for some $p > 0$

$$\operatorname*{Sup}_{x,y} \int s(M,(\bar{x},\bar{y}))d\mu^p(M) = \operatorname*{Sup}_{x,y} \mathbb{E}\,\{\operatorname{Log} \frac{\delta(S_p\cdot\bar{x},S_p\cdot\bar{y})}{\delta(\bar{x},\bar{y})}\} < 0$$

the proposition follows from the preceding lemma.

We now introduce the space of Hölder-continuous functions with which we shall work.

DEFINITION 2.4. *Given* $\alpha > 0$ *we set for any continuous function* f *on* $P(\mathbb{R}^d)$

$$|f|_\infty = \operatorname*{Sup}_{\bar{x}\in P(\mathbb{R}^d)} |f(\bar{x})| \quad \text{and} \quad m_\alpha(f) = \operatorname*{Sup}_{\substack{\bar{x},\bar{y}\in P(\mathbb{R}^d)\\ \bar{x}\neq\bar{y}}} \frac{|f(\bar{x}) - f(\bar{y})|}{\delta(\bar{x},\bar{y})^\alpha}\ .$$

We define $\mathbb{L}(\alpha)$ *as the set of continuous functions* f *on* $P(\mathbb{R}^d)$ *for which* $\|f\|_\alpha = |f|_\infty + m_\alpha(f)$ *is finite.*

It is readily seen that $\mathbb{L}(\alpha)$, with the norm $\|\cdot\|_\alpha$, is a Banach space.

The following theorem, due to Le Page [49] , is basic. Note that, with its notations, we have for any continuous function f on $P(\mathbb{R}^d)$,

$$\mathbb{E}\,\{f(S_n\cdot\bar{x})\} = \int f(M\cdot\bar{x})d\mu^n(M) = P^n f(\bar{x})\ .$$

Therefore it shows that, in some sense, the distribution of $S_n\cdot\bar{x}$ converges to ν exponentially fast.

THEOREM 2.5. *Let* μ *be a probability measure on* $Gl(d,\mathbb{R})$ *such that* T_μ *is strongly irreducible and contracting, and let* ν *be the* μ*-invariant probability measure on* $P(\mathbb{R}^d)$ *. Suppose that for some* $\tau > 0$*.* $\int \exp \tau\,\ell(M)d\mu(M)$ *if finite. Then there exists* $\alpha_0 > 0$ *such that for* $0 < \alpha \leq \alpha_0$ *, the operators* P *et* N *defined on* $\mathbb{L}(\alpha)$ *by*

$$Pf(\bar{x}) = \int f(M\cdot\bar{x})d\mu(M)$$

$$Nf(\bar{x}) = \int f(\bar{y})\ d\nu(\bar{y})$$

are bounded and satisfy

$$\lim_{n\to\infty} \|P^n - N\|_\alpha^{1/n} < 1 .$$

Proof : We have, using Lemma 2.2 ,

$$|Pf(\bar{x})| \leq |f|_\infty$$

and

$$\frac{|Pf(\bar{x}) - Pf(\bar{y})|}{\delta(\bar{x},\bar{y})^\alpha} \leq \int \frac{\delta(M\cdot\bar{x} , M\cdot\bar{y})^\alpha}{\delta(\bar{x},\bar{y})^\alpha} m_\alpha(f) d\mu(M)$$

$$\leq m_\alpha(f) \int e^{4\alpha \ell(M)} d\mu(M)$$

so that P is bounded on $\mathbb{L}(\alpha)$ for $0 < \alpha < \frac{\tau}{4}$. It is clear that N is bounded on each $\mathbb{L}(\alpha)$. For α_0 small enough we know by Proposition 2.3 that for each α in $(0, \alpha_0]$ there exist $C > 0$ and ρ, $0<\rho<1$ such that

$$\int \{\frac{\delta(M\cdot\bar{x} , M\cdot\bar{y})}{\delta(\bar{x},\bar{y})}\}^\alpha d\mu^n(M) \leq C \rho^n .$$

This implies that for f in $\mathbb{L}(\alpha)$

$$\frac{|(P^n - N)f(\bar{x}) - (P^n - N)f(\bar{y})|}{\delta(\bar{x},\bar{y})^\alpha} = \frac{|P^n f(\bar{x}) - P^n f(\bar{y})|}{\delta(\bar{x},\bar{y})^\alpha}$$

$$\leq \int m_\alpha(f) \{\frac{\delta(M\cdot\bar{x} , M\cdot\bar{y})}{\delta(\bar{x},\bar{y})}\}^\alpha d\mu^n(M)$$

$$\leq C \rho^n m_\alpha(f) .$$

On the other hand, since ν is μ-invariant.

$$|(P^n-N)f(\bar{x})| = |\int f(M\cdot\bar{x})d\mu^n(M) - \int f(M\cdot\bar{y})d\mu^n(M)d\nu(\bar{y})|$$

$$\leq \int |f(M\cdot\bar{x}) - f(M\cdot\bar{y})| d\mu^n(M) d\nu(\bar{y})$$

$$\leq m_\lambda(f) \int \delta(M\cdot\bar{x} , M\cdot\bar{y})^\alpha d\mu^n(M) d\nu(\bar{y})$$

and, since $\delta(\bar{x},\bar{y}) \leq 1$,

$$|(P^n-N)f|_\infty = \sup_x |(P^n-N)f(\bar{x})|$$

$$\leq m_\alpha(f) \int [\sup_{\bar{x},\bar{y}} \int \{\frac{\delta(M\cdot\bar{x} , M\cdot\bar{y})}{\delta(\bar{x},\bar{y})}\}^\alpha d\mu^n(M)] d\nu(\bar{y})$$

$$\leq C \rho^n m_\alpha(f) .$$

This proves that

$$\|(P^n-N)f\|_\alpha = m_\alpha\{(P^n-N)(f)\} + |(P^n-N)f|_\infty$$

$$\leq 2C\rho^n m_\alpha(f)$$

$$\leq 2C\rho^n \|f\|_\alpha$$

so that

$$\|P^n - N\|_\alpha \leq 2C\rho^n .$$

Noting that $P^n-N = (P-N)^n$ since $PN = NP = N$ and $N^2 = N$, the limit of $\|P^n-N\|^{1/n}$ exists by subadditivity. This completes the proof.

Things are simpler when μ has a density. In that case we may work with continuous functions and we don't need any assumption on the moments of μ. The following result due to Tutubalin ([66]), holds.

PROPOSITION 2.6. *Suppose that* μ *is a probability measure on* $Gl(d,\mathbb{R})$ *(or* $Sl(d,\mathbb{R})$*) which has a density with respect to the Haar measure. If* $\mathcal{C}$ *denotes the space of continuous functions on* $P(\mathbb{R}^d)$, *the conclusions of the theorem hold for* P *and* N *acting on* $\mathcal{C}$.

Proof : We first prove that P is a compact operator on $\mathcal{C}$, i.e. that $\{Pf , |f|_\infty \leq 1\}$ is a conditionally compact subset of $\mathcal{C}$. For any f in $\mathcal{C}$ and k in $SO(d)$, put

$$Tf(k) = Pf(k\cdot\bar{e}_1) .$$

Since $\{k\cdot\bar{e}_1 , k \in SO(d)\} = P(\mathbb{R}^d)$ it is sufficient to check that $\{Tf , |f|_\infty \leq 1\}$ is a conditionally compact subset of $\mathcal{C}(SO(d))$, the space of continuous functions on $SO(d)$. By the theorem of Arzela-Ascoli (see e.g. Dunford and Schwartz IV.6.7 , [17]) we have only to show that this set is equicontinuous. Let ϕ denote the density of μ w.r.t. the Haar measure dM on $Gl(d,\mathbb{R})$ (or $Sl(d,\mathbb{R})$). For k_1, k_2 in $SO(d)$ we have

$$|Tf(k_1) - Tf(k_2)| = \left|\int \{f(Mk_1\cdot\bar{e}_1) - f(Mk_2\cdot\bar{e}_1)\}\, \phi(M)\, dM\right|$$

$$\leq \int |f(M\cdot\bar{e}_1)|\, |\phi(Mk_1^{-1}) - \phi(Mk_2^{-1})|\, dM$$

$$\leq \int |\phi(Mk_1^{-1}) - \phi(Mk_2^{-1})|\, dM .$$

When k_2 tends to k_1, the right hand side converges to zero (this is obvious when ϕ is bounded and continuous ; in general approximate ϕ in L^1 by such functions). Since it does not depend upon f, $\{Tf\ ,\ |f| \leq 1\}$ is conditionally compact. This yields that P and $Q = P - N$ are compact operators.

By Theorem III.4.3 , we have

$$\lim_{n\to\infty} (P-N)^n f = \lim_{n\to\infty} (P^n - N)f = 0 \ .$$

Therefore the modulus of each eigenvalue of Q is smaller than one. Its spectral radius, namely $\lim \|P^n - N\|^{1/n}$, is thus smaller than one.

Remark : It is not difficult to see that when μ has a density, the Markov chain $X_n = (Y_{n+1}, Y_n \dots Y_1 \cdot \bar{x})$ on $Gl(d,\mathbb{R}) \times P(\mathbb{R}^d)$ satisfies the Doeblin condition (see e.g. Revuz [61], chapter VI, for the definition).

Complements 2.8. (A first proof of the central limit theorem)

We shall show how Theorem 2.5 enables us to use the general result of Gordin and Lifschitz ([28]) for proving a central limit theorem. In Section 5 we shall give stronger results under the same assumptions without recourse to the theory of Markov chains. Therefore the reader may skip these complements and we just sketch the arguments.

Let (X_n) be the Markov chain on $E = Gl(d,\mathbb{R}) \times P(\mathbb{R}^d)$ defined by $X_o = (Y_1, \bar{x}_o), \dots, X_n = (Y_{n+1}, Y_n \dots Y_1 \cdot \bar{x}_o), \dots$ where $Y_1, Y_2, \dots$ are i.i.d. matrices whose distribution μ satisfies the hypotheses of Theorem 2.5 . We suppose also that $\bar{x}_o$ is a random variable on $P(\mathbb{R}^d)$, independent of $\{Y_n,\ n \geq 1\}$, with distribution ν . The transition probability R associated with this Markov chain is defined by, if A is a Borel set in E ,

$$R(M,\bar{x}),A) = \mathbb{E}\,((Y_1, M\cdot\bar{x}) \in A) \ .$$

It is easily seen that $\mu \otimes \nu$ is the unique invariant distribution for (X_n) so that (X_n) is a stationary ergodic sequence (see Breiman 7.3, [10]).

Let $\sigma(M,\bar{x}) = \mathrm{Log} \dfrac{\|Mx\|}{\|x\|}$. We want to show that

$$\mathrm{Log}\, \|S_n x_o\| - \mathrm{Log}\, \|x_o\| = \sum_{i=o}^{n-1} \sigma(X_i)$$

satisfies a central limit theorem. We may always suppose that the upper

Lyapunov exponent γ is equal to zero (by looking at $e^{-n\gamma} S_n$). In this case, if $\phi(\bar{x}) = \int \sigma(M,\bar{x}) d\mu(M)$, then $\int \phi(\bar{x}) d\nu(\bar{x}) = 0$. By the theorem of Gordin and Lifschitz it suffices to check that the equation

$$(I - R)\psi = \sigma$$

has a solution in $L^2(\mu \otimes \nu)$. But, for any $n \geq 1$,

$$\begin{aligned} R^n \sigma(M,\bar{x}) &= \mathbb{E}\{\sigma(Y_n, Y_{n-1} \ldots Y_1 M \cdot \bar{x})\} \\ &= \mathbb{E}\{\phi(Y_{n-1} \ldots Y_1 M \cdot \bar{x})\} \\ &= P^{n-1} \phi(M \cdot \bar{x}) . \end{aligned}$$

For $\alpha > 0$ small enough, ϕ is in $\mathbb{L}(\alpha)$ (see Lemma 4.2), and

$$\sup_{M,x} |R^n \sigma(M,\bar{x})| \leq \|P^{n-1} \phi\|_\alpha .$$

Using Theorem 2.5 and the fact that $N\phi = \int \phi \, d\nu = 0$ we see that this quantity converges to zero exponentially fast.

This proves that $\psi = \sum_{n=o}^{\infty} R^n \sigma(M,\bar{x})$ converges in $L^2(\mu \otimes \nu)$ and of course $(I-R)\psi = \sigma$. By [28], $\frac{1}{\sqrt{n}} \operatorname{Log} \|S_n x_o\|$ converges in distribution to the gaussian law $N(0,a^2)$, with $a^2 = \|\psi\|^2 - \|R\psi\|^2$.

Suppose now that $a^2 = 0$. Then $\|R\psi\| = \|\psi\|$, which implies that

$$\mathbb{E}(\{R\psi(X_o) - \psi(X_1)\}^2) = 0$$

and

$$R\psi(X_o) = \psi(X_1) \qquad \mathbb{P} \text{ a.s.} .$$

But $\psi(M,\bar{x}) = \sigma(M,\bar{x}) + \sum_{n=o}^{\infty} P^n \phi(M \cdot \bar{x})$. Hence, for $f(\bar{x}) = \sum_{n=o}^{\infty} P^n \phi(\bar{x})$, $\psi(M,\bar{x}) = \sigma(M,\bar{x}) + f(M \cdot \bar{x})$ and $R\psi(M,\bar{x}) = f(M \cdot \bar{x})$. We thus obtain

$$f(Y_1 \cdot \bar{x}_o) = \sigma(Y_2, Y_1 \cdot \bar{x}_o) + f(Y_2 Y_1 \cdot \bar{x}_o) .$$

Since f and σ are continuous and since the distribution of $(Y_2, Y_1 \cdot \bar{x})$ is $\mu \otimes \nu$ this yields

$$\sigma(M,\bar{x}) = f(\bar{x}) - f(M \cdot \bar{x}) \qquad \mu \otimes \nu \quad \text{a.s.}$$

and more generally

$$\sigma(M,\bar{x}) = f(\bar{x}) - f(M \cdot \bar{x}) \qquad \mu^n \otimes \nu \quad \text{a.s.} .$$

Since the support of ν is not confined a proper projective subspace and since f is bounded this implies that for some $c > 0$, $1 \leq \|M\| \leq c$, for all M in T_μ. We shall see in Lemma 8.2 that this cannot be true under our assumptions, proving $a^2 \neq 0$.

Exercise 2.9. Show that if μ is a probability measure on the symplectic group $Sp(d,\mathbb{R})$ which has a density w.r.t. the Haar measure, then $\lim \|P^n - N\|^{1/n} < 1$, where P acts on $\mathcal{C}(P(\mathbb{R}^{2d}))$.

Exercise 2.10. Show that the conclusion of Proposition 2.6 holds if for some integer p, μ^p is not singular with respect to the Haar measure. (Hint : Use Lemma 2, Section VII.8.1 (p. 709) of Dunford and Schwartz [17]).

Exercise 2.11. Let μ be a probability measure on the set of positive matrices M in $M(d,\mathbb{R})$ which satisfy, for some fixed $c > 0$,

$$M_{i,j} \frac{1}{c} \leqq M_{k,\ell} \leqq c\, M_{i,j} \quad , \quad 1 \leqq i,j,k,\ell \leqq d .$$

Let $B_c = \{\bar{x} \in P(\mathbb{R}^d) \,;\, x = (x_1,\dots,x_d) \,,\, \sum_{i=1}^{d} x_i = 1 \,,\, \frac{1}{c} \leqq \frac{x_i}{x_j} \leqq c\}$.

Show that B_c is a bounded set for the metric θ introduced in III.4.7. Define $\mathbb{L}_\alpha(B_c)$ as the Banach space of order α Hölder functions on (B_c, θ).

Show that the formula

$$Pf(\bar{x}) = \int f(M\cdot\bar{x})\,d\mu(M)$$

defines a bounded operator on $\mathbb{L}_\alpha(B_c)$ and that if

$$Nf(\bar{x}) = \int f(\bar{y})\,d\nu(\bar{y})$$

where ν is the unique μ-invariant distribution on B_c then

$$\lim \|P^n - N\|_\alpha^{1/n} < 1$$

(Use Exercise III.6.5).

V.3. A lemma of perturbation theory

We recall some elementary facts about the spectral theory of bounded operators. Although the lemma we will be needing is well known we present its proof for the sake of completeness.

We refer to Dunford and Schwartz ([17], VII.3) for the proof of the following.

Let $T : E \longrightarrow E$ be a bounded operator on a complex Banach space E. The spectrum of T is

$$\sigma(T) = \{z \in \mathbb{C} \ ; \ zI - T \ \text{has no bounded inverse}\} .$$

Outside $\sigma(T)$ the resolvent $R(z,T) = (zI-T)^{-1}$ is analytic.

Let $F(T)$ be the set of complex valued functions which are analytic in some neighborhood of $\sigma(T)$. Let $f \in F(T)$ and U be a bounded open set whose boundary ∂U is a finite union of C^1 curves. If $\sigma(T)$ is contained in U and if f is analytic in a neighborhood of $\overline{U}$, we set

$$f(T) = \frac{1}{2\pi i} \int_{\partial U} f(z)R(z,T)dz .$$

It follows from the Cauchy integral theorem that $f(T)$ depends only on f and T.

We shall use the facts that for f, g in $F(T)$,

$$(\alpha f + \beta g)(T) = \alpha f(T) + \beta g(T) ,$$

$$(fg)(T) = f(T)\, g(T) ,$$

$$f(\sigma(T)) = \sigma(f(T)) .$$

<u>LEMMA 3.1</u>. *Let* U_1 *and* U_2 *be two open sets in* $\mathbb{C}$ *with* C^1 *boundary such that* $\overline{U}_1 \cap \overline{U}_2 = \emptyset$ *and such that* $\sigma(T)$ *is contained in the union of* U_1 *and* U_2. *Define*

$$N_1 = \frac{1}{2\pi i} \int_{\partial U_1} R(z,T)dz , \qquad N_2 = \frac{1}{2\pi i} \int_{\partial U_2} R(z,T)dz .$$

Then

(i) N_1 *and* N_2 *are two projections such that* $N_1 T = T N_1$, $N_1 + N_2 = I$ *and* $N_1 N_2 = N_2 N_1$.

(ii) For $i = 1,2$, $T(N_i(E))$ *is contained in* $N_i(E)$ *and the spectrum of the restriction of* T *to* $N_i(E)$ *is* $\sigma(T) \cap U_i$.

<u>Proof</u> : Let V_1 and V_2 be two non-intersecting neighborhoods of $\overline{U}_1$ and $\overline{U}_2$. For $i = 1,2$, let f_i be the indicator function of V_i. Then f_i is in $F(T)$ and

$$N_i = \frac{1}{2\pi i} \int_{\partial U_i} f_1(z)\, R(z,T)dz = f_i(T) .$$

Since $f_1 + f_2 = 1$ on a neighborhood of $\sigma(T)$, $f_1(T) + f_2(T) = I$. Moreover $f_1^2 = f_1$ and $f_1 f_2 = f_2 f_1$ so that (i) is clear.

For any complex a set $g_a(z) = z f_1(z) + a f_2(z)$. The relation $g_a(T) = TN_1 + aN_2$ yields that, if T_1 is the restriction of T to

$N_1(E)$,

$$\sigma(g_a(T)) = \sigma(T_1) \cup \{a\} .$$

But $\sigma(g_a(T)) = g_a(\sigma(T)) = (\sigma(T) \cap U_1) \cup \{a\}$ so that $\sigma(T) \cup \{a\}$ is contained in $\{\sigma(T) \cap U_1\} \cup \{a\}$ for any a . This proves (ii).

We consider now an analytic family of bounded operators $\{T(\xi) ; \xi \in V\}$, where V is an open set in $\mathbb{C}$. This means that for every $\xi_o \in V$ we can find $\varepsilon > 0$ and a sequence (A_n) of bounded operators such that

(i) $\sum_{n\geq o} \varepsilon^n \|A_n\| < \infty$,

(ii) for $|\xi-\xi_o| < \varepsilon$ and ξ in V , $T(\xi) = \sum_{n=o}^{\infty} (\xi-\xi_o)^n A_n$.

We will use the following (see Dunford and Schwartz VII.6 [17]).

LEMMA 3.2. *Let* V *be a neighborhood of* 0 *in* $\mathbb{C}$ *and* $\{T(\xi) ; \xi \in V\}$ *be an analytic family of bounded operators on the Banach space* E . *Suppose that for some rank-one operator* N

$$\rho = \overline{\lim_{n\to\infty}} \|T(0)^n - N\| < 1 .$$

Then we can find $\varepsilon > 0$ *such that for* $|\xi| < \varepsilon$,

$$T(\xi) = \lambda(\xi) N(\xi) + Q(\xi)$$

where

(i) $\lambda(\xi)$ *is the unique eigenvalue of maximal modulus of* $T(\xi)$.

(ii) $N(S)$ *is a rank-one projection such that* $N(\xi)Q(\xi) = Q(\xi)N(\xi) = 0$.

(iii) The mappings $\lambda(\cdot)$, $N(\cdot)$ *and* $Q(\cdot)$ *are analytic.*

(iv) $|\lambda(\xi)| > \frac{2+\rho}{3}$ *and for each* $p \in \mathbb{N}$ *there exists* $c > 0$ *such that for each* $n \in \mathbb{N}$,

$$\left\|\frac{d^p}{d\xi^p} Q^n(\xi)\right\| \leq c\left\{\frac{1+2\rho}{3}\right\}^n .$$

Proof : 1) First notice that since $\lim T(0)^n = N$, $NT(0) = T(0)N = N$. This implies that $T(0) \operatorname{Ker} N \subset \operatorname{Ker} N$ and $T(0)N(E) \subset N(E)$. The restriction of $T(0)$ to $N(E)$ is the identity and on $\operatorname{Ker} N$ $T(0) = T(0) - N$. Therefore the spectral radius of the restriction of $T(0)$ to $\operatorname{Ker} N$ is $\lim \|(T(0) - N)^n\|^{1/n} = \lim \|T(0)^n - N\|^{1/n} = \rho$.

This implies that

$$\sigma(T(0)) \subset B(0,\rho) \cup \{1\} .$$

Moreover if γ is a small circle around 1, then for x in Ker N , $R(z, T(0))x$ is analytic in the disk whose boundary is γ , so that by the Cauchy Theorem

$$\frac{1}{2\pi i} \int_\gamma R(z , T(0))x \, dz = 0 .$$

For x in $N(E)$, $R(z, T(0))x = (z-1)^{-1}x$ and

$$\frac{1}{2\pi i} \int_\gamma R(z, T(0))x \, dz = x .$$

Whence

$$N = \frac{1}{2\pi i} \int_\gamma R(z, T(0))dz .$$

2) When ξ is in V , write $R(z,\xi)$ for the resolvent $R(z,T(\xi))$. If $\|T(\xi) - T(0)\| < \|R(z,0)\|^{-1}$ then $\sum_n [\{T(\xi) - T(0)\} R(z,0)]^n$ is the inverse operator of

$$\{zI - T(\xi)\} R(z,0) = Id - \{T(\xi) - T(0)\} R(z,0) .$$

Thus

$$R(z,0)^{-1} \{zI - T(\xi)\}^{-1} = \sum_n [\{T(\xi) - T(0)\} R(z,0)]^n$$

and $$R(z,\xi) = \sum_{n=o}^{\infty} R(z,0) [\{T(\xi) - T(0)\} R(z,0)]^n$$

This shows in particular that $\xi \longrightarrow R(z,\xi)$ is analytic.

3) Let D_1 be the disk with center 1 and radius $\frac{1-\rho}{3}$, D_2 the disk with center 0 and radius $\frac{1+2\rho}{3}$, U an open set which contains $B(0,\rho)$ and $\{1\}$ such that $\overline{U}$ is contained in $D_1 \cup D_2$, and $M = \text{Sup}\{ \|R(z,0)\| , z \notin U\}$.

We shall suppose that $\|T(\xi) - T(0)\| < \frac{1}{M}$. In that case, by 2), $\sigma(T(\xi))$ is contained in U . Let γ_1 (resp. γ_2) be the boundary of D_1 (resp. D_2). We define

$$N(\xi) = \frac{1}{2\pi i} \int_{\gamma_1} R(z,\xi)dz .$$

Using 1), 2) and the preceding lemma we see that $\xi \longrightarrow N(\xi)$ is analytic, that $N(0) = N$ and that $N(\xi)$ is a projection. Moreover

$$Id - N(\xi) = \frac{1}{2\pi i} \int_{\gamma_2} R(z,\xi)dz .$$

Let us show that if $\|N(\xi) - N(0)\| < 1$ then the rank of $N(\xi)$ is one. Suppose that $N(\xi)x = x$ and $N(\xi)y = y$. Since the rank of $N(0)$ is one there exists some λ in $\mathbb{C}$ such that $N(0)(x - \lambda y) = 0$. We have $\|x - \lambda y\| = \|N(\xi)(x - \lambda y)\| = \|(N(\xi) - N(0))(x - \lambda y)\| \leq \|N(\xi) - N(0)\| \|x - \lambda y\|$ Thus $x = \lambda y$ when $\|N(\xi) - N(0)\| < 1$.

4) Choose $\varepsilon > 0$ such that for $|\xi| < \varepsilon$, $\|T(\xi) - T(0)\| < \frac{1}{M}$, $\|N(\xi) - N(0)\| < 1$ and $\xi \longrightarrow N(\xi)$ is analytic.

Let $x_o \in E$ and $x'_o \in E'$ be such that $\|x_o\| = 1$, $Nx_o = x_o$ and $\langle x_o, x'_o\rangle = 1$. If $|\xi| < \varepsilon$ then the rank of $N(\xi)$ is one and there exists $\lambda(\xi) \in \mathbb{C}$ such that $T(\xi)N(\xi)x_o = N(\xi)T(\xi)x_o = \lambda(\xi)N(\xi)x_o$. The formula

$$\lambda(\xi) = \frac{\langle T(\xi)N(\xi)x_o, x'_o\rangle}{\langle N(\xi)x_o, x'_o\rangle}$$

shows that $\lambda(\cdot)$ is analytic. We find the wanted decomposition choosing $Q(\xi) = T(\xi)(Id - N(\xi))$.

4) To prove (iv) first note that, for $|\xi| < \varepsilon$, $\lambda(\xi)$ is in U since it is an eigenvalue of $T(\xi)$. But $\lambda(0) = 1$ and λ is continuous, thus $|\lambda(\xi)| \geq \frac{2+\rho}{3}$.

Now for each n in $\mathbb{N}$

$$Q^n(\xi) = T(\xi)^n\ (Id - N(\xi)) = \frac{1}{2\pi i}\int_{\gamma_2} z^n R(z,\xi)dz .$$

Therefore

$$\frac{d^p}{d\xi^p} Q^n(\xi) = \frac{1}{2\pi i}\int_{\gamma_2} z^n \frac{d^p}{d\xi^p} R(z,\xi)dz$$

and

$$\left\|\frac{d^p}{d\xi^p} Q^n(\xi)\right\| \leq c\left\{\frac{1+2\rho}{3}\right\}^n$$

for $c = \frac{1}{2\pi} \operatorname{Sup}\left\{\left\|\frac{d^p}{d\xi^p} R(z,\xi)\right\| ;\ z \in \gamma_2 ,\ |\xi| < \varepsilon\right\}$.

The proof of the lemma is complete.

<u>Remark 3.3.</u> It is easy to modify the proof in order to show that if, for some $m \in \mathbb{N}$, the family $\{T(\xi),\ \xi \in V\}$ is merely of class C^m then an analogous result holds. (In the statement one has to replace analytic by C^m and the integer p in (iv) has to verify $0 \leq p \leq m$).

V.4. The Fourier-Laplace transform near 0

Given a probability measure μ on $Gl(d,\mathbb{R})$ we define formally

$$(T(z)f)(\bar{x}) = \int e^{z \operatorname{Log} \frac{\|Mx\|}{\|x\|}} f(M\cdot\bar{x})\, d\mu(M)$$

where $z \in \mathbb{C}$, $\bar{x} \in P(\mathbb{R}^d)$ and f is a complex-valued function on $P(\mathbb{R}^d)$. Note that $T(0)$ is the operator P studied in Section 2.

Our first task is to show that for α small enough, $T(\cdot)$ is an analytic family of bounded operators on $\mathbb{L}(\alpha)$ near zero. Namely,

PROPOSITION 4.1. *Let μ be a probability measure on $Gl(d,\mathbb{R})$ such that for some $\tau > 0$, $\int \exp \tau\, \ell(M)\, d\mu(M)$ is finite. For any α with $0 < \alpha < \frac{1}{2}\tau$ there exists $\eta > 0$ such that $\{T(z), |z| < \eta\}$ is an analytic family of bounded operators on $\mathbb{L}(\alpha)$.*

The proof is tedious but straightforward. It will require the following lemma in which x and y denote two unit vectors with directions $\bar{x}$ and $\bar{y}$.

LEMMA 4.2. *For $z \in \mathbb{C}$ and $0 < \alpha \leq 1$, there exists $c_1, c_2 > 0$ such that, for any M in $Gl(d,\mathbb{R})$*

$$(i) \quad \sup_{\bar{x},\bar{y}\in P(\mathbb{R}^d)} \frac{\left|e^{z \operatorname{Log}\|Mx\|} - e^{z \operatorname{Log}\|My\|}\right|}{\delta(\bar{x},\bar{y})^{\alpha}} \leq c_1 \exp\{[(1+\alpha)|\operatorname{Re} z| + 2\alpha]\, \ell(M)\};$$

$$(ii) \quad \sup_{\bar{x},\bar{y}\in P(\mathbb{R}^d)} \frac{\left|\operatorname{Log}\|Mx\| - \operatorname{Log}\|My\|\right|}{\delta(\bar{x},\bar{y})^{\alpha}} \leq c_2\, \ell(M)\, e^{2\alpha\, \ell(M)}.$$

Proof of the lemma : a) First notice that for $\|x\| = \|y\| = 1$

$$\delta(\bar{x},\bar{y})^2 = 1 - \langle x,y\rangle^2 = \|x-y\|^2 \left\{1 - \frac{\|x-y\|^2}{4}\right\}.$$

Replacing, if needed, y by $-y$ we may suppose that $\|x-y\|^2 \leq 2$. In this case

$$\frac{1}{\sqrt{2}}\|x-y\| \leq \delta(\bar{x},\bar{y}) \leq \|x-y\|.$$

b) We now prove (i). Let $F : \mathbb{R}^d \longrightarrow \mathbb{C}$ be defined by $F(x) = \exp(z \operatorname{Log}\|Mx\|)$. Its differential DF_x satisfies

$$DF_x(h) = z\, \|Mx\|^{z-2} \langle Mh, Mx\rangle \qquad \text{for} \quad h \in \mathbb{R}^d$$

and $\qquad \|DF_x\| \leq |z| \ \|M\| \ \|Mx\|^{\operatorname{Re} z-1}$.

Therefore, if $a = |\operatorname{Re} z|$, $\|x\| = 1$ and $\|y\| = 1$ then

$$|F(x) - F(y)| \leq \|x-y\| \sup_{0\leq\lambda\leq 1} \|DF_{\lambda x+(1-\lambda)y}\|$$

$$\leq |z| \ \|x-y\| \ e^{(a+2)\ell(M)}$$

and $$\frac{|F(x)-F(y)|}{\delta(\overline{x},\overline{y})} \leq c \ e^{(a+2)\ell(M)} \qquad \text{for} \quad c = 2\sqrt{2}\ |z| \ .$$

This implies that for $M = c \exp(a+2)\ell(M)$ and $N = 2 \sup_{\|x\|=1} |F(x)|$

$$\|F(x)-F(y)\| \leq M\,\delta(\overline{x},\overline{y}) \wedge N \ .$$

Whence, if $M\delta(\overline{x},\overline{y}) \leq 1$,

$$\|F(x) - F(y)\| \leq M\,\delta(\overline{x},\overline{y}) \leq M^{\alpha}\ \delta(\overline{x},\overline{y})^{\alpha}$$

and if $M\,\delta(\overline{x},\overline{y}) \geq 1$

$$\|F(x)-F(y)\| \leq N \leq N\ M^{\alpha}\ \delta(\overline{x},\overline{y})^{\alpha} \ .$$

This yields

$$\frac{|F(x)-F(y)|}{\delta(\overline{x},\overline{y})^{\alpha}} \leq 2\{c\ \exp(|\operatorname{Re} z|+2)\ \ell(M)\}^{\alpha}\ e^{|\operatorname{Re} z|\ \ell(M)}$$

proving (i).

c) (ii) is proved along the same line, using the fact that if $G(x) = \operatorname{Log}\|Mx\|$ then $\|DG_x\| \leq e^{2\ell(M)}\ \|x\|^{-1}$.

<u>Proof of the proposition</u> : a) We first show that for $0 \leq |\operatorname{Re} z| \leq \frac{\tau-2\alpha}{1+\alpha}$, $T(z)$ is a bounded operator on $\mathbb{L}(\alpha)$. If $f \in \mathbb{L}(\alpha)$ and x,y are unit vectors, then

$$|T(z)f(\overline{x})| = \left|\int e^{z \operatorname{Log}\|Mx\|}\ f(M\cdot\overline{x})d\mu(M)\right|$$

$$\leq |f|_{\infty} \int \exp\ |\operatorname{Re} z|\ \ell(M)\ d\mu(M) \leq c_3\ |f|_{\infty}$$

for some $c_3 > 0$ if $|\operatorname{Re} z| \leq \tau$.

On the other hand, by the previous lemma,

$$\left|\frac{T(z)f(\overline{x}) - T(z)f(\overline{y})}{\delta(\overline{x},\overline{y})^{\alpha}}\right| \leq \int |e^{z \operatorname{Log}\|Mx\|}|\ \frac{|f(M\cdot\overline{x}) - f(M\cdot\overline{y})|}{\delta(\overline{x},\overline{y})^{\alpha}}\ d\mu(M)$$

$$+ \int |f(M\cdot\overline{y})|\ \frac{|e^{z \operatorname{Log}\|Mx\|} - e^{z \operatorname{Log}\|My\|}|}{\delta(\overline{x},\overline{y})^{\alpha}}\ d\mu(M)$$

$$\leq m_\alpha(f) \int e^{|\mathrm{Re}\, z|\, \ell(M)}\, d\mu(M) + c_1\, |f|_\infty \int e^{((1+\alpha)\,|\mathrm{Re}\, z|+2)\ell(M)}\, d\mu(M)$$

$$\leq c_4\, (m_\alpha(f) + |f|_\infty)$$

for some constant $c_4 > 0$, as soon as $|\mathrm{Re}\, z|\,(1+\alpha) + 2\alpha \leq \tau$.
Hence, if $|\mathrm{Re}\, z| \leq \frac{\tau-2\alpha}{1+\alpha}$,

$$\|T(z)f\|_\alpha = |T(z)f|_\infty + m_\alpha(T(z)f) \leq (c_4+c_3)\, \|f\|_\alpha$$

and $T(z)$ is bounded.

b) For each $n \in \mathbb{N}$ let $T_n f(\bar{x}) = \int \{\mathrm{Log}\, \|Mx\|\}^n f(M\cdot\bar{x})\, d\mu(M)$, where x is a unit vector with direction $\bar{x}$.

Taking into account the relation

$$\frac{|(\mathrm{Log}\|Mx\|)^n - (\mathrm{Log}\|My\|)^n|}{\delta(\bar{x},\bar{y})^\alpha} \leq \frac{|\mathrm{Log}\|Mx\| - \mathrm{Log}\|My\||}{\delta(\bar{x},\bar{y})^\alpha} \sum_{i=o}^{n-1} |\mathrm{Log}\|Mx\||^i\, |\mathrm{Log}\|My\||^{n-1-i}$$

$$\leq c_2\, n\, \ell(M)^{n+1} \exp 2\alpha\, \ell(M) \ ,$$

we see that

$$\frac{|T_n f(\bar{x}) - T_n f(\bar{y})|}{\delta(\bar{x},\bar{y})^\alpha} \leq \int |\mathrm{Log}^n \|Mx\||\, \frac{|f(M\cdot\bar{x}) - f(M\cdot\bar{y})|}{\delta(\bar{x},\bar{y})^\alpha}\, d\mu(M)$$
$$+ \int f(M\cdot\bar{y}) \frac{|\mathrm{Log}^n\|Mx\| - \mathrm{Log}^n\|My\||}{\delta(\bar{x},\bar{y})^\alpha}\, d\mu(M)$$

$$\leq m_\alpha(f) \int |\mathrm{Log}^n\|Mx\||\, d\mu(M) + |f|_\infty c_2 n \int \ell(M)^{n+1} c^{2\alpha\ell(M)} d\mu(M)$$

On the other hand

$$|T_n f|_\infty \leq |f|_\infty \operatorname*{Sup}_x \int |\mathrm{Log}^n\|Mx\||\, d\mu(M) \leq |f|_\infty \int \ell(M)^n\, d\mu(M) \ .$$

This shows that

$$\|T_n\|_\alpha \leq c_2\, n \int \ell(M)^{n+1}\, e^{2\alpha\,\ell(M)}\, d\mu(M) + 2 \int \ell(M)^n\, d\mu(M) \ ,$$

and $\sum_n \frac{|z|^n}{n!} \|T_n\|_\alpha \leq c_2 \int e^{2\alpha\,\ell(M)} \sum_n \frac{n|z|^n\, \ell(M)^{n+1}}{n!}\, d\mu(M) + 2\int e^{|z|\,\ell(M)} d\mu(M)$

so that $\sum_n \frac{|z|^n}{n!} \|T_n\|_\alpha$ is finite if $2\alpha + |z| < \tau$.

Therefore $\{T(z)\ ,\ |z| < \eta\}$ is analytic when $\eta = \frac{\tau-2\alpha}{1+\alpha}$.

Theorem 2.5, together with Lemma 3.2 and Proposition 4.1, yields the following result of Le Page [49].

THEOREM 4.3. *Let* μ *be a probability measure on* $Gl(d,\mathbb{R})$ *such that* T_μ *is strongly irreducible and contracting. Suppose that for some* $\tau > 0$, $\int \exp \tau\, \ell(M)\, d\mu(M)$ *is finite. Then there exists* $\alpha_0 > 0$ *such that if* $0 < \alpha < \alpha_0$ *, then for some* $\varepsilon > 0$ *, the operators* $T(z)$, $|z| < \varepsilon$ *, are bounded on* $\mathbb{L}(\alpha)$.
Moreover, for $|z| < \varepsilon$,

$$T(z) = \lambda(z)\, N(z) + Q(z)$$

where (i) $\lambda(0) = 1$, $N(0)f(\bar{x}) = \int f\, d\nu$ *if* ν *is the unique* μ*-invariant distribution on* $P(\mathbb{R}^d)$;

(ii) $\rho = \lim \|T(0)^n - N(0)\|_\alpha^{1/n} < 1$;

(iii) $\lambda(z)$ *is the unique eigenvalue of maximum modulus of* $T(z)$, $N(z)$ *is a rank-one projection and* $N(z)Q(z) = Q(z)N(z) = 0$;

(iv) *The mappings* $\lambda(\cdot)$, $N(\cdot)$ *and* $Q(\cdot)$ *are analytic* ;

(v) $|\lambda(z)| \geq \frac{2+\rho}{3}$ *and for each* $p \in \mathbb{N}$ *there exists* $c > 0$ *such that for each* $n \in \mathbb{N}$,

$$\left\|\frac{d^p}{dz^p} Q^n(z)\right\| \leq c \left(\frac{1+2\rho}{3}\right)^n .$$

The following will be used for studying the speed of convergence in the central limit theorem. Under the hypotheses of the theorem, if $\mathbb{1}$ is the function identically equal to one,

COROLLARY 4.4. *Suppose that for some* $\alpha \in (0,\alpha_0]$, $\lambda'(0) = 0$ *and* $\lambda''(0) = a^2 > 0$. *Then there exist* $\beta > 0$, $c > 0$ *and* $0 < r < 1$ *such that for* $n \in \mathbb{N}$ *and* $|t| < \beta\sqrt{n}$

$$\left\|T^n\left(\frac{it}{a\sqrt{n}}\right)\mathbb{1} - e^{-t^2/2}\,\mathbb{1}\right\|_\alpha \leq \frac{c}{\sqrt{n}} \{(|t^3|+|t|)e^{-t^2/4} + |t|r^n\} .$$

Proof : We have, by Theorem 4.3, for $|t| < \varepsilon$

$$T^n(it) = \lambda^n(it)\, N(it) + Q^n(it) .$$

If we write $\| \ \|$ instead of $\| \ \|_\alpha$, this implies for $|t| < a\varepsilon\sqrt{n}$

$$\|T^n(\frac{it}{a\sqrt{n}})\mathbb{1} - e^{-t^2/2}\mathbb{1}\| \leq \|\lambda^n(\frac{it}{a\sqrt{n}})N(\frac{it}{a\sqrt{n}}) - e^{-t^2/2}N(0)\| + \|Q^n(\frac{it}{a\sqrt{n}}) - Q^n(0)\|$$

(we here use the relations $N(0)\mathbb{1} = \mathbb{1}$ and $Q(0)\mathbb{1} = Q(0)N(0)\mathbb{1} = 0$) .
For notational convenience we write the proof for $a = 1$.

On the one hand, by (v) of Theorem 4.2, there exist $c_1 > 0$ and $0 < r < 1$ such that

$$\begin{aligned}\|Q^n(\frac{it}{\sqrt{n}}) - Q^n(0)\| &\leq \frac{|t|}{\sqrt{n}} \sup_{|z|<\varepsilon} \|\frac{d}{dz}Q^n(z)\| \\ &\leq c_1 \frac{|t|}{\sqrt{n}} r^n .\end{aligned} \tag{3}$$

On the other hand

$$\begin{aligned}\|\lambda^n(\frac{it}{\sqrt{n}})N(\frac{it}{\sqrt{n}}) - e^{-t^2/2}N(0)\| &\leq |\lambda^n(\frac{it}{\sqrt{n}}) - e^{-t^2/2}| \, \|N(0)\| \\ &+ |\lambda^n(\frac{it}{\sqrt{n}})| \, \|N(\frac{it}{\sqrt{n}}) - N(0)\|\end{aligned}$$

but, $N(\cdot)$ being analytic, $\|N(\frac{it}{\sqrt{n}}) - N(0)\| \leq \frac{t}{\sqrt{n}} \sup_{|z|<\varepsilon} \|\frac{\partial}{\partial z}N(z)\| \leq \frac{tk}{\sqrt{n}}$ for some $k > 0$, and

$$\|\lambda^n(\frac{it}{\sqrt{n}})N(\frac{it}{\sqrt{n}}) - e^{-t^2/2}N(0)\| \leq \frac{tk}{\sqrt{n}} |\lambda^n(\frac{it}{\sqrt{n}})| + |\lambda^n(\frac{it}{\sqrt{n}}) - e^{-t^2/2}| . \tag{4}$$

Making use of the relations $\lambda(0) = 1$, $\lambda'(0) = 0$ and $\lambda''(0) = 1$ we obtain

$$\lambda(it) = 1 - \frac{t^2}{2} + t^3 o(t) = \exp\{-\frac{t^2}{2} + t^3 o(t)\}$$

and

$$\lambda^n(\frac{it}{\sqrt{n}}) = e^{-\frac{t^2}{2} + \frac{t^3}{\sqrt{n}} o(\frac{t}{\sqrt{n}})} .$$

This implies (since $|e^u - 1| \leq |u| \, e^{|u|}$) that

$$\begin{aligned}|\lambda^n(\frac{it}{\sqrt{n}}) - e^{-t^2/2}| &= e^{-t^2/2} \, |e^{\frac{t^3}{\sqrt{n}} o(\frac{t}{\sqrt{n}})} - 1| \\ &= e^{-t^2/2} \frac{|t|^3}{\sqrt{n}} |o(\frac{t}{\sqrt{n}})| \exp|\frac{t^3}{\sqrt{n}} o(\frac{t}{\sqrt{n}})| .\end{aligned}$$

Hence, for β small enough there exists $c_2 > 0$ such that, for $|t| < \beta\sqrt{n}$

$$|\lambda^n(\frac{it}{\sqrt{n}}) - e^{-t^2/2}| \leq c_2 e^{-t^2/4} \frac{|t|^3}{\sqrt{n}}$$

and $$|\lambda^n(\frac{it}{\sqrt{n}})| \leq c_2\, e^{-t^2/4} .$$

This yields, by (4),

$$\|\lambda^n(\frac{it}{\sqrt{n}})\, N(\frac{it}{\sqrt{n}}) - e^{-t^2/2}\, N(0)\| \leq \frac{c_2\, e^{-t^2/4}}{\sqrt{n}}(k|t|+|t|^3) \quad (5)$$

The lemma follows from (3) and (5).

Remark that the statements proved above are valid for matrices with complex entries.

V.5. Central limit theorem

Given a sequence $Y_1, Y_2, \ldots$ of independent random matrices in $Gl(d,\mathbb{R})$ with common distribution μ and $S_n = Y_n \ldots Y_1$ we intend to prove under various hypotheses that for $x \in \mathbb{R}^d$, $S_n x$ satisfies a central limit theorem. We shall always assume that T_μ is strongly irreducible and that μ has an exponential moment. We first show in Theorem 5.1 that if in addition T_μ is contracting, then both the norm and the direction of $S_n x$ can be normalized in such a way that they converge in distribution to a non degenerate limit. We also give the speed of convergence. We then prove, in Theorem 5.2, that the central limit theorem for $\text{Log}\,\|S_n x\|$ is also true when T_μ is not contracting. This is useful because it is often difficult to check whether T_μ is contracting or not.

The following important result is due to Le Page ([49]), the only improvement being $a^2 > 0$. Let $N(0,a^2)$ be the centered normal distribution with variance a^2:

THEOREM 5.1. (Central limit theorem, first form) : *Let* $S_n = Y_n \ldots Y_1$ *where* $\{Y_n, n \geq 1\}$ *are i.i.d. random matrices in* $Gl(d,\mathbb{R})$ *with law* μ. *Suppose that* T_μ *is strongly irreducible and contracting and that, for some* $\tau > 0$, $\int \exp \tau\, \ell(M)\, d\mu(M)$ *is finite. Let* γ *denote the upper Lyapunov exponent associated with* (S_n) *and* ν *be the unique* μ-*invariant distribution on* $P(\mathbb{R}^d)$.

Then there exists $a > 0$ *such that*

(i) for any $x \neq 0$, *the* $\mathbb{R} \times P(\mathbb{R}^d)$-*valued random variables*

$(\frac{1}{\sqrt{n}}\{\text{Log}\,\|S_n x\| - n\gamma\},\ S_n\cdot\bar{x})$ *converge in distribution to the product measure* $N(0,a^2)\otimes\nu$.

(ii) The random variables $(\frac{1}{\sqrt{n}}\{\text{Log}\,\|S_n\| - n\gamma\},\ S_n\cdot\bar{x})$ *converge in distribution to* $N(0,a^2)\otimes\nu$.

(iii) There exists $C>0$ *such that, for any* $t\in\mathbb{R}$,

$$\sup_{\|x\|=1}\left|\mathbb{P}\left(\frac{\text{Log}\,\|S_n x\| - n\gamma}{a\sqrt{n}} < t\right) - \frac{1}{\sqrt{2\pi}}\int_{-\infty}^{t}\exp - \frac{u^2}{2}\,du\right| \leq \frac{C}{\sqrt{n}} .$$

We first prove a lemma. Recall that $T(z)$ is defined by

$$T(z)\,f(\bar{x}) = \int e^{z\,\text{Log}\,\|Mx\|}\,f(M\cdot\bar{x})\,d\mu(M)$$

if f is a complex valued function on $P(\mathbb{R}^d)$ and $\|x\| = 1$. Using the notation of Theorem 4.3, and λ', λ'' for the first two derivatives of λ ,

LEMMA 5.2. *For each positive integer n and unit vector* x ,

$$T^n(z)\,f(\bar{x}) = \mathbb{E}\,\{e^{z\,\text{Log}\,\|S_n x\|}\,f(S_n\cdot\bar{x})\} . \tag{6}$$

If $\lambda(z)$ *is the eigenvalue of* $T(z)$ *acting on* $\mathbb{L}(\alpha)$ *with maximum modulus which appears in Theorem 4.3 (for* $0<\alpha<\alpha_0$ *and* $|z|<\varepsilon$*) then* $\lambda'(0)$ *is the upper Lyapunov exponent* γ *and for* $a^2=\lambda''(0)-\gamma^2$, $\mathbb{E}\{(\text{Log}\,\|S_n x\| - n\gamma)^2\} - n a^2$ *converges as* $n\longrightarrow\infty$ *to a finite limit. In particular* $a=0$ *if and only if the sequence* $\{\mathbb{E}(\text{Log}\,\|S_n x\| - n\gamma)^2,\ n\geq 1\}$ *is bounded.*

Proof : It is worth noting that (6) is only a consequence of the cocycle property of

$$\sigma(M,\bar{x}) = \text{Log}\,\frac{\|Mx\|}{\|x\|} .$$

We proceed by induction. If

$$T^{n-1}(z)f(\bar{x}) = \mathbb{E}\,\{e^{z\,\sigma(S_{n-1},\bar{x})}\,f(S_{n-1}\cdot\bar{x})\}$$

then

$$\begin{aligned} T^n(z)f(\bar{x}) &= T^{n-1}(z)\,\{T(z)f\}(\bar{x}) \\ &= \mathbb{E}\,\{e^{z\,\sigma(S_{n-1},\bar{x})}\,(T(z)f)(S_{n-1}\cdot\bar{x})\} \end{aligned}$$

$$= \mathbb{E}\{e^{z\sigma(S_{n-1},\bar{x})} e^{z\sigma(Y_n,S_{n-1}\cdot\bar{x})} f(Y_n\cdot(S_{n-1}\cdot\bar{x}))\}$$

$$= \mathbb{E}\{e^{z\sigma(S_n,\bar{x})} f(S_n\cdot\bar{x})\}$$

since $S_n = Y_n S_{n-1}$ and $\sigma(S_n,\bar{x}) = \sigma(Y_n, S_{n-1}\cdot\bar{x}) + \sigma(S_{n-1},\bar{x})$.

This proves (6). We now prove the rest of the lemma. If t is a small real number we can write by Theorem 4.2

$$T(t) = \lambda(t)N(t) + Q(t)$$

and

$$T^n(t) = \lambda^n(t)N(t) + Q^n(t) .$$

But for $\|x\| = 1$,

$$\mathbb{E}(e^{t \operatorname{Log} \|S_n x\|}) = T^n(t)\, \mathbb{1}(\bar{x}) .$$

We thus have

$$\frac{d}{dt}\mathbb{E}(e^{t \operatorname{Log} \|S_n x\|}) = (\frac{d}{dt}\lambda^n(t))N(t)\,\mathbb{1}(\bar{x}) + \lambda^n(t)\frac{d}{dt}(N(t)\,\mathbb{1}(\bar{x})) + \frac{d}{dt}\{Q^n(t)\,\mathbb{1}(\bar{x})\}$$

so that, using Theorem 4.2 ,

$$\gamma = \lim \frac{1}{n}\mathbb{E}(\operatorname{Log}\|S_n x\|) = \lim \frac{1}{n}\frac{d}{dt}\mathbb{E}(e^{t \operatorname{Log}\|S_n x\|})(0) = \lambda'(0).$$

Now

$$\frac{d^2}{dt^2}\mathbb{E}(e^{t \operatorname{Log}\|S_n x\|} e^{-nt\gamma})(0) = n(\lambda''(0)-\gamma^2)N(0)\,\mathbb{1}(\bar{x}) + \frac{d^2}{dt^2}\{N(t)\,\mathbb{1}(\bar{x})\}(0) + \frac{d^2}{dt^2}\{e^{-nt\gamma}Q^n(t)\,\mathbb{1}(\bar{x})\}(0)$$

Notice that, by (v) of Theorem 4.2 ,

$$\frac{d^2}{dt^2}\{e^{-nt\gamma}Q^n(t)\,\mathbb{1}(\bar{x})\}(0) = n^2\gamma^2 Q^n(0)\,\mathbb{1}(\bar{x}) - 2n\gamma\frac{d}{dt}\{Q^n(t)\,\mathbb{1}(\bar{x})\}(0) + \frac{d^2}{dt^2}\{Q^n(t)\,\mathbb{1}(\bar{x})\}(0)$$

converges to zero. Therefore, since $N(0)\,\mathbb{1} = \mathbb{1}$,

$$\mathbb{E}\{(\operatorname{Log}\|S_n x\| - n\gamma)^2\} - n\,(\lambda''(0) - \gamma^2) - \frac{d^2}{dt^2}\{N(t)\,\mathbb{1}(\bar{x})\}(0)$$

converges to zero. This proves the lemma.

We postpone the proof of the following lemma until section 8 (cf. Corollary 8.6).

LEMMA 5.3. *Under the hypotheses of Theorem 5.1 the sequence* $\mathbb{E}\{(\operatorname{Log}\|S_n x\| - n\gamma)^2\}$ *is not bounded.*

Proof of Theorem 5.1. : Although (i) follows readily from (iii) we shall prove it directly. By Theorem 4.3 and Lemma 5.2, for $\alpha > 0$ small enough, f in $\mathbb{L}(\alpha)$ and $\|x\| = 1$, we have

$$\mathbb{E}\{e^{\frac{it}{\sqrt{n}}(\text{Log}\,\|S_n x\| - n\gamma)} f(S_n\cdot\bar{x})\} = e^{-i\sqrt{n}\,t\,\gamma}\{\lambda(\tfrac{it}{\sqrt{n}})^n N(\tfrac{it}{\sqrt{n}}) f(\bar{x}) + Q^n(\tfrac{it}{\sqrt{n}}) f(\bar{x})\}$$

whenever t is real and n sufficiently large.

But $$\lim_{n\to\infty} e^{-i\sqrt{n}\,t\,\gamma}\,\lambda(\tfrac{it}{\sqrt{n}})^n = \exp - \frac{t^2 a^2}{2}$$

since $\lambda'(0) = \gamma$ and $\lambda''(0) = a^2 + \gamma^2$ by Lemma 5.2 . Notice that $a \neq 0$ since, by Lemma 5.3 , $\mathbb{E}(\text{Log}\,\|S_n x\| - n\gamma)^2$ is not bounded. On the other hand

$$\lim_{n\to\infty} N(\tfrac{it}{\sqrt{n}}) f(\bar{x}) = N(0) f(\bar{x}) = \int f\, d\nu$$

and $$\lim_{n\to\infty} Q^n(\tfrac{it}{\sqrt{n}})\, f(\bar{x}) = 0$$

(use (v) of Theorem 4.2 with $p = 0$). Therefore for any f in $\mathbb{L}(\alpha)$,

$$\lim_{n\to\infty} \mathbb{E}(e^{i\frac{t}{\sqrt{n}}(\text{Log}\,\|S_n x\| - n\gamma)} f(S_n\cdot\bar{x})\} = \exp(-\frac{t^2 a^2}{2}) \int f\, d\nu .$$

Taking into account that each continuous function on $P(\mathbb{R}^d)$ is a uniform limit of a sequence of elements of $\mathbb{L}(\alpha)$ we immediately see that

$$(\frac{1}{\sqrt{n}}\,(\text{Log}\,\|S_n x\| - n\gamma)\ ,\ S_n\cdot\bar{x})$$

converges in distribution to $N(0\,,a^2) \otimes \nu$.

To prove (ii) it suffices to use (i) and the fact that for any $x \neq 0$, $\text{Log}\,\|S_n x\| - \text{Log}\,\|S_n\|$ is a.s. bounded (see (a) of Proposition III.2.2).

We now prove (iii). We may suppose $\gamma = 0$. In this case $\lambda'(0) = 0$ and $\lambda''(0) = a^2 > 0$. Recall the Berry-Esseen inequality (see Lemma XVI.3.2 of Feller [18]) : there exists $M > 0$ such that for all $A > 0$ and real t

$$|\mathbb{P}(\frac{1}{\sqrt{n}}\,\text{Log}\,\|S_n x\| < a\,t) - \frac{1}{\sqrt{2\pi}}\int_{-\infty}^{t} e^{-\frac{u^2}{2}} du| \leq \frac{1}{\pi}\int_{-A}^{A} |\frac{\phi_n(u) - e^{-\frac{u^2}{2}}}{u}| du + \frac{M}{A}$$

where $\phi_n(u) = \mathbb{E}\{\exp \frac{iu}{a\sqrt{n}}\,\text{Log}\,\|S_n x\|\} = T^n(\frac{iu}{a\sqrt{n}})\,\mathbb{1}(\bar{x})$.

By Corollary 4.4 there are $\beta > 0$, $c > 0$ and $0 < r < 1$ such that when $|u| < \beta\sqrt{n}$

$$\sup_{\|x\|=1} |\phi_n(u) - e^{-u^2/2}| \leq \frac{c}{\sqrt{n}} \{(|u|^3+|u|)e^{-u^2/4} + |u|r^n\} .$$

Choosing $A = \beta\sqrt{n}$ we obtain

$$\sup_{\|x\|=1} |P(\mathrm{Log}\,\|S_n x\| < a\,t\,\sqrt{n}) - \frac{1}{\sqrt{2\pi}}\int_{-\infty}^{t} e^{-u^2/2}\,du|$$

$$\leq \frac{M}{\beta\sqrt{n}} + \frac{c}{\pi\sqrt{n}} \int_{-\beta\sqrt{n}}^{\beta\sqrt{n}} ((|u^2|+1)e^{-u^2/4} + r^n)du$$

$$\leq \frac{M}{\beta\sqrt{n}} + \frac{c}{\pi\sqrt{n}} \int_{-\infty}^{\infty} (|u|^2+1)\ e^{-u^2/4}du + \frac{2\beta}{\pi} r^n$$

from which (iii) follows.

In the next theorem (announced in [7]) we do not suppose that T_μ is contracting. In this case we cannot study the convergence in distribution of $S_n \cdot \bar{x}$, we do not even know if there is a unique μ-invariant distribution on $P(\mathbb{R}^d)$. Recall that G_μ is the smallest closed subgroup of $Gl(d,\mathbb{R})$ which contains the support of μ .

<u>THEOREM 5.4.</u> <u>(Central limit theorem, second form)</u> : *Let* $S_n = Y_n \ldots Y_1$ *where* $\{Y_n, n \quad 1\}$ *are i.i.d. random matrices in* $Gl(d,\mathbb{R})$ *with common distribution* μ . *Suppose that the following hold*

(a) For some $\tau > 0$, $\int \exp \tau\,\ell(M)\,d\mu(M)$ *is finite* .

(b) G_μ *is strongly irreducible* .

(c) The set $\{|\det M|^{-1/d} M\, ,\ M \in G_\mu\}$ *is not contained in a compact subgroup of* $Gl(d,\mathbb{R})$.

Then there exists $a > 0$ *such that for any* $x \neq 0$ *in* $\mathbb{R}^d$ $\frac{1}{\sqrt{n}}\{\mathrm{Log}\,\|S_n x\| - n\gamma\}$ *and* $\frac{1}{\sqrt{n}}(\mathrm{Log}\,\|S_n\| - n\gamma)$ *converge in distribution to* $N(0,a^2)$.

We shall be needing the following algebraic lemma which is proved in Chevalley, (§5, ch. IV of [13]).

<u>LEMMA 5.5.</u> *Let* G *be an irreducible subgroup of* $Gl(d,\mathbb{R})$. *For any* p , $1 \leq p \leq d$, *we can find a direct-sum decomposition of* $\Lambda^p \mathbb{R}^d$:

$$\Lambda^p \mathbb{R}^d = W_1 \oplus W_2 \oplus \ldots \oplus W_k$$

such that for any $i \in \{1,\ldots,k\}$:

(a) if $M \in G$ *then* $(\Lambda^p M)(W_i) = W_i$,

(b) there is no proper subspace W *of* W_i *such that* $(\Lambda^p M)(W) = W$ *for all* M *in* G .

This amounts saying that the action of G on $\Lambda^p \mathbb{R}^d$ is completely reducible.

Proof of Theorem 5.4. : First recall (see Proposition III.3.2) that since G_μ is strongly irreducible, the sequence $\{|\mathrm{Log}\,\|S_n\| - \mathrm{Log}\,\|S_n x\|\,|\,,\ n \geq 1\}$ is almost surely bounded for each $x \neq 0$ in $\mathbb{R}^d$. Whence $\frac{1}{\sqrt{n}}\{\mathrm{Log}\,\|S_n x\| - n\gamma\}$ and $\frac{1}{\sqrt{n}}\{\mathrm{Log}\,\|S_n\| - n\gamma\}$ converge to the same distribution (if one of them does converge). Now let p denote the index of T_μ . By Lemmas III.1.4 and III.5.3 the sequence $\{|\mathrm{Log}\,\|S_n\|^p - \mathrm{Log}\,\|\Lambda^p S_n\|\,|\,,\ n \geq 1\}$ is also bounded so that $\frac{1}{\sqrt{n}}\{\mathrm{Log}\,\|S_n\| - n\gamma\}$ and $\frac{1}{p\sqrt{n}}\{\mathrm{Log}\,\|\Lambda^p S_n\| - n\gamma p\}$ have the same limit distribution.

It is therefore enough to show that $\frac{1}{\sqrt{n}}\{\mathrm{Log}\,\|\Lambda^p S_n\| - np\gamma\}$ converges in distribution to some gaussian law $N(0,b^2)$, $b > 0$, whenever $\{|\det M|^{-1/d} M\ ;\ M \in G_\mu\}$ is not in a compact group, i.e. when $p < d$ (see Proposition III.7.1).

Let $\gamma = \gamma_1 \geq \gamma_2 \geq \ldots \geq \gamma_d$ be the Lyapunov exponents associated with μ . We know (see Proposition III.6.2) that $\gamma_i = \gamma_p$ for $1 \leq i \leq p$ but that $\gamma_{p+1} < \gamma_p$. The two top Lyapunov exponents λ_1 and λ_2 associated with $(\Lambda^p S_n)$ are

$$\lambda_1 = \sum_{i=1}^{p} \gamma_i \ , \qquad \lambda_2 = \gamma_{p+1} + \sum_{i=1}^{p-1} \gamma_i$$

(see the proof of Proposition IV.1.2). Therefore, here,

$$\lambda_1 = p\gamma \quad \text{and} \quad \lambda_1 > \lambda_2 \ .$$

Consider now the direct-sum decomposition

$$\Lambda^p \mathbb{R}^d = W_1 \oplus W_2 \oplus \ldots \oplus W_k$$

given by the preceding lemma, applied to $G = G_\mu$. It is easy to see that the set of all Lyapunov exponents associated with $(\Lambda^p S_n)$ is the union of the sets of all Lyapunov exponents associated with the restriction of $(\Lambda^p S_n)$ to each W_i , $1 \leq i \leq k$. Therefore we may choose W_1 among the W_i's in such a way that, if A_n is the restriction of $\Lambda^p S_n$ to W_1 and B_n the restriction of $\Lambda^p S_n$ to $W_2 \oplus \ldots \oplus W_k$, then

$$\lim_{n\to\infty} \frac{1}{n} \operatorname{Log} \|A_n\| = \lambda_1$$

and
$$\lim_{n\to\infty} \frac{1}{n} \operatorname{Log} \|B_n\| \leq \lambda_2 .$$

Since $\lambda_2 < \lambda_1$ and $\|\Lambda^p S_n\| = \operatorname{Sup}(\|A_n\|, \|B_n\|)$, we are done if we prove that $\frac{1}{\sqrt{n}} \{\operatorname{Log} \|A_n\| - n\lambda_1\}$ converges in distribution to some $N(0,b^2)$. Let us check that we may apply Theorem 5.1 to the sequence (A_n). If $r = \dim W_1$ we may view A_n as a product of n i.i.d. random matrices on $Gl(r, \mathbb{R})$ with a common distribution, say ρ. We first show that $r > 1$. If $r = 1$ each limit point of $\|\Lambda^p S_n\|^{-1} \Lambda^p S_n$ sends all of $\Lambda^p \mathbb{R}^d$ into the line W_1. Therefore there is a decomposable p-vector $v_1 \wedge \ldots \wedge v_p$ which spans W_1. But $(\Lambda^p M)(W_1) = W_1$ for any M in G_μ, hence if V is the subspace of $\mathbb{R}^d$ spanned by $v_1,\ldots,v_p$, $M(V) = V$ for all M in G_μ, in contradiction with the irreducibility of G_μ. Knowing that $r \geq 2$ there are at least two Lyapunov exponents associated with ρ. The largest one is λ_1 and the next one, being also an exponent associated with $(\Lambda^p S_n)$, is not bigger than λ_2. Therefore the two top exponents associated with ρ are distinct. Since T_ρ is clearly irreducible on W_1 by construction (see (a) of the above lemma) we know, by the converse statement in Theorem III.6.1, that T_ρ is actually a contracting and strongly irreducible subsemigroup of $Gl(r, \mathbb{R})$. We may thus apply Theorem 5.1 to the probability measure ρ and $\frac{1}{\sqrt{n}} (\operatorname{Log} \|A_n\| - n p \gamma)$ converges in distribution to some $N(0,b^2)$ with $b \neq 0$. This completes the proof of the theorem.

Remark 5.6. Suppose that $\{|\det M|^{-1/d} M,\ M \in G_\mu\}$ is contained in a compact subgroup of $Gl(d, \mathbb{R})$. We know (see Remark III.1.8) that there exists a scalar product on $\mathbb{R}^d$ for which all these matrices are orthogonal. If we work with the associated norm, then for any unit vector x in $\mathbb{R}^d$,

$$\operatorname{Log} \|S_n\| = \operatorname{Log} \|S_n x\| = \frac{1}{d} \operatorname{Log} |\det S_n| = \frac{1}{d} \sum_{i=1}^{n} \operatorname{Log} |\det Y_i|$$

is a sum of i.i.d. real random variables. Therefore, if $\gamma = \mathbb{E}\{\operatorname{Log} |\det Y_i|\}$, then $\{\frac{1}{\sqrt{n}} \operatorname{Log} \|S_n\| - n\gamma\}$ and $\frac{1}{\sqrt{n}} \{\operatorname{Log} \|S_n x\| - n\gamma\}$ converge to $N(0,a^2)$ where a^2 is the variance of $\operatorname{Log} |\det Y_i|$. Note that $a = 0$ if andonly if $e^{-\gamma} Y_1$ is contained in a compact subgroup.

In these theorems the assumption that $\int \exp \tau\, \ell(M)\, d\mu(M)$ is finite is unpleasant. The condition "$\int \ell(M)^2\, d\mu(M) < \infty$" seems more natural. In this direction we have the following result of Tutubalin [66], [68].

PROPOSITION 5.7. *Suppose that the probability measure* μ *has a density with respect to the Haar measure of* $Gl(d, \mathbb{R})$ *(or* $Sl(d, \mathbb{R})$*, or* $Sp(d, \mathbb{R})$*). Then, if* $\int \ell(M)^2\, d\mu(M) < \infty$ *, the distribution of* $(\frac{1}{\sqrt{n}}(\mathrm{Log}\, \|S_n x\| - n\gamma),\ S_n \cdot \bar{x})$ *converges to* $N(0,a^2) \otimes \nu$ *, for some* $a > 0$.

Sketch of proof : We define by (2) a family of operators $\{T(it),\ t \in \mathbb{R}\}$ acting on the space $\mathcal{C}$ of continuous functions on $P(\mathbb{R}^d)$. By Proposition 2.6 (see also Exercise 2.9) we have a decomposition analogous to the one given in Theorem 4.3 . For t small enough,

$$T(it) = \lambda(it)N(it) + Q(it)$$

where λ , N and Q have the same properties as in this theorem except that they are only of class C^2 . To verify this, note that if $\int \ell^2(M)\, d\mu(M)$ is finite, then $t \longrightarrow T(it)$ is of class C^2 and use remark 3.3 . The proposition is thus proved as was (i) of Theorem 5.1 .

Making use of the arguments developed for proving Lemma III.7.1 it is not difficult to show the following.

PROPOSITION 5.8. *Assume that the common distribution* μ *of the i.i.d. random matrices* (Y_n) *satisfies the assumptions (a) and (c) of Theorem 5.4 . If* G_μ *is merely irreducible then there exist a direct sum decomposition*

$$\mathbb{R}^d = V_1 \oplus V_2 \oplus \ldots \oplus V_k$$

and a real number $a > 0$ *such that for any* $1 \le i \le k$ *and any nonzero* x *in* V_i , $\frac{1}{\sqrt{n}}\,\mathrm{Log}\,\{\|S_n x\| - n\gamma\}$ *converges in distribution to* $N(0,a^2)$.

But it is worthwhile to remark that it may happen that for some nonzero x in $\mathbb{R}^d$, $\frac{1}{\sqrt{n}}\{\mathrm{Log}\,\|S_n x\| - n\gamma\}$ does not converge to $N(0,a^2)$. Consider for instance the case where for some sequence (a_n, b_n) of i.i.d. random vectors of $\mathbb{R}^2 - (0,0)$,

$$Y_n = \begin{bmatrix} a_n & 0 \\ 0 & a_n^{-1} \end{bmatrix} \quad \text{or} \quad \begin{bmatrix} 0 & b_n^{-1} \\ b_n & 0 \end{bmatrix}$$

If $\mathbb{P}(Y_1 = \begin{pmatrix} 0 & b_1^{-1} \\ b_1 & 0 \end{pmatrix}) \neq 0$ then G_μ is irreducible. It follows from Proposition III.7.4 that $\gamma = 0$. For $x = \binom{1}{1}$, $\|S_n x\| \geq 1$ and $\frac{1}{\sqrt{n}} \operatorname{Log} \|S_n x\| \geq 0$, so that this latter sequence cannot converge to a centered gaussian distribution. Notice that the same remark holds for $\frac{1}{\sqrt{n}} \operatorname{Log} \|S_n\|$. In this particular case $\frac{1}{\sqrt{n}} \operatorname{Log} \|S_n x\|$ does converge to some $N(0,a^2)$, if $x = \binom{1}{0}$ or $\binom{0}{1}$. And $a \neq 0$ when Y_1 is not a.s. orthogonal.

In the following exercise we propose to prove, using the arguments developed in this chapter, the central limit theorem for positive matrices due to Furstenberg and Kesten [25].

Exercise 5.9. Let $c > 0$ and μ be a probability measure on the set of matrices $M \in M(d, \mathbb{R})$ whose entries satisfy, for $1 \leq i,j,k,\ell \leq d$

$$\frac{1}{c} M_{i,j} \leq M_{k,\ell} \leq c\, M_{i,j} .$$

Using Exercise 2.11 and the same route as above show that if $\int \operatorname{Log}^2 \|M\| \, d\mu(M)$ is finite, then the distribution of $(\frac{1}{\sqrt{n}} \{\operatorname{Log} \|S_n x\| - n\gamma\},\ S_n \cdot \bar{x})$ converges to some product measure $N(0,a^2) \otimes \nu$ on $\mathbb{R} \times P(\mathbb{R}^d)$, for each nonzero x with nonnegative coordinates. (This is rather long but requires only the former arguments, working on $\mathbb{L}_\alpha(B_c)$ instead of $\mathbb{L}(\alpha)$).

V.6. Large deviations

Consider a probability measure μ on $Gl(d, \mathbb{R})$ such that T_μ is strongly irreducible and contracting and such that for some $\tau > 0$, $\int \exp \tau\, \ell(M)\, d\mu(M)$ is finite. By Theorem 4.3 we know that if we consider the operators $T(z)$ as acting on $\mathbb{L}(\alpha)$, then for $|z|$ small enough

$$T(z) = \lambda(z) N(z) + Q(z) .$$

In particular for all real small enough t and unit vectors x

$$\mathbb{E}\{e^{t \operatorname{Log} \|S_n x\|}\} = T(t)^n \mathbb{1}(\bar{x}) = \lambda(t)^n N(t) \mathbb{1}(\bar{x}) + Q^n(t) \mathbb{1}(\bar{x}) .$$

Since the left hand side is positive and since $\|Q^n(t)\|$ goes to zero faster than $|\lambda(t)|^n$, $\lambda(t)$ is positive and

$$\lim_{n\to\infty} \frac{1}{n} \operatorname{Log} \mathbb{E}\{e^{t \operatorname{Log} \|S_n x\|}\} = \operatorname{Log} \lambda(t) .$$

But $\lambda(t)$ is analytic , $\lambda'(0) = \gamma$ and $\lambda''(0) = a^2 + \gamma^2$ with $a^2 > 0$ (see Lemmas 5.2 and 5.3). This implies the following result due to Le Page [49] , (a similar result holds of course also for $\mathbb{P}(\operatorname{Log} \|S_n x\| - n\gamma < -n\varepsilon))$.

<u>THEOREM 6.1</u>. *Let* $S_n = Y_n \ldots Y_1$ *where* $\{Y_n,\ n \geq 1\}$ *satisfies the hypotheses of Theorem 5.1 . Then there exist* $A, B > 0$ *such that, for any unit vector* x *and* $0 < \varepsilon < B$

$$\lim_{n\to\infty} \frac{1}{n} \operatorname{Log} \mathbb{P}(\operatorname{Log} \|S_n x\| - n\gamma > n\varepsilon) = \phi(\varepsilon)$$

where $\phi(\varepsilon) = - \sup_{0<t<A} \{t\varepsilon - \operatorname{Log} \lambda(t) + t\gamma\} < 0$.

This is a direct consequence of the following lemma applied to $Z_n = \operatorname{Log} \|S_n x\| - n\gamma$. This lemma is now well known (see Freidlin and Ventsel [19], Theorem 1.1 and 1.2, Ch. V), we just sketch its proof.

<u>LEMMA 6.2</u>. *Let* $\{Z_n,\ n \geq 1\}$ *be a sequence of real random variables such that for some* $\eta > 0$,

$$\lim_{n\to\infty} \frac{1}{n} \operatorname{Log} \mathbb{E}(\exp t Z_n) = c(t)$$

exists on an open set containing $[-\eta, \eta]$. *If* c *is continuously differentiable and strictly convex on* $[-\eta, \eta]$ *and if* $c'(0) = 0$, *then for* $0 < \varepsilon < \frac{c(\eta)}{\eta}$

$$\lim_{n\to\infty} \frac{1}{n} \operatorname{Log} \mathbb{P}(Z_n > n\varepsilon) = - \sup_{0<t<\eta} [t\varepsilon - c(t)] < 0 .$$

<u>Proof</u> : a. We have for any $t \geq 0$

$$\mathbb{P}(Z_n > n\varepsilon) \leq \mathbb{E}(e^{tZ_n - n\varepsilon}) \leq e^{-t\varepsilon n} \mathbb{E}(e^{tZ_n})$$

so that

$$\overline{\lim_{n\to\infty}} \frac{1}{n} \operatorname{Log} \mathbb{P}(Z_n > n\varepsilon) \leq - \sup_{0<t<\eta} [t\varepsilon - c(t)] .$$

It is clear that the right hand side is negative since $c'(0) = 0$.

b. We define, for $0 \leq t \leq \eta$, new probability measures $\mathbb{P}_{n,t}$ by

$$\mathbb{P}_{n,t}(A) = \mathbb{E}\{e^{tZ_n}\}^{-1}\, \mathbb{E}\{e^{tZ_n} 1_A\} .$$

For $0 < \varepsilon < z < \frac{c(\eta)}{\eta}$ and $0 < a < z - \varepsilon$, the inequality

$$\begin{aligned}\mathbb{P}(Z_n > n\varepsilon) &\geq \mathbb{P}\{|\tfrac{Z_n}{n} - z| < a\} \\ &\geq \mathbb{E}(e^{tZ_n})\, \mathbb{P}_{n,t}(e^{-tZ_n} 1_{|Z_n - nz| < na})\end{aligned}$$

yields that

$$\frac{1}{n}\operatorname{Log}\mathbb{P}(Z_n > n\varepsilon) \geq \frac{1}{n}\operatorname{Log}\mathbb{E}(e^{tZ_n}) - tz - na + \frac{1}{n}\operatorname{Log}\mathbb{P}_{n,t}(|Z_n - nz| < na).$$

One chooses t_z in $[0,\eta]$ such that $c'(t_z) = z$. Making use of the relation $\lim \frac{1}{n}\operatorname{Log}\mathbb{E}_{n,t_z}(e^{sZ_n}) = c(t_z + s) - c(s)$ and of part a we have

$$\lim_{n\to\infty} \frac{1}{n}\operatorname{Log}\mathbb{P}_{n,t_z}(|Z_n - nz| < na) = 0 .$$

Therefore

$$\lim_{n\to\infty} \frac{1}{n}\operatorname{Log}\mathbb{P}(Z_n > n\varepsilon) \geq c(t_z) - t_z z = c(t_z) - t_z c'(t_z) .$$

When $z \longrightarrow \varepsilon$ then $c(t_z) - t\,c'(t_z) \longrightarrow c(t_\varepsilon) - t_\varepsilon c'(t_\varepsilon) = \inf_{0<t<\eta} c(t) - t\varepsilon$, which yields

$$\lim_{n\to\infty} \frac{1}{n}\operatorname{Log}\mathbb{P}(Z_n > n\varepsilon) \geq -\sup_{0<t<\eta}\left[t\varepsilon - c(t)\right] .$$

It is easy to check that the convergence stated in the theorem holds uniformly in x if $\|x\| = 1$.

The following result is maybe more useful since it does not require any information about T_μ.

<u>THEOREM 6.2</u>. *Assume only that* G_μ *is strongly irreducible and that for some* $\tau > 0$, $\int \exp\{\tau\, \ell(M)\}\, d\mu(M)$ *is finite. Then there exists* $\alpha > 0$ *such that for any* $\varepsilon > 0$ *and* $x \neq 0$,

$$\overline{\lim_{n\to\infty}} \frac{1}{n}\operatorname{Log}\mathbb{P}(|\operatorname{Log}\|S_n x\| - n\gamma| > n\varepsilon) < -\alpha$$

and

$$\overline{\lim_{n\to\infty}} \frac{1}{n}\operatorname{Log}\mathbb{P}(|\operatorname{Log}\|S_n\| - n\gamma| > n\varepsilon) < -\alpha .$$

Proof : First, replacing each Y_n by $e^{-\gamma} Y_n$, we may assume without loss of generality that $\gamma = 0$.

Let p be the index of T_μ. If $p = d$ then the theorem follows from the usual estimate on large deviations for sums of i.i.d. real valued random variables (see Proposition III.1.7 and Remark 5.6). Henceforth we shall assume that $p < d$. Recall some facts we have seen in the proof of Theorem 5.4. There exists a direct-sum decomposition

$$\Lambda^p \mathbb{R}^d = W \oplus V$$

such that for any M in G_μ, $(\Lambda^p M)(W) = W$, $(\Lambda^p M)(V) = V$ and, if A_n (resp. B_n) is a matrix which represents the action of $\Lambda^p S_n$ on W (resp. V) then the following holds.

a) If ρ is the distribution of A_1, then T_ρ is a strongly irreducible and contracting semigroup of $Gl(r,\mathbb{R})$, where $r = \dim W$.

Moreover $$\lim_{n\to\infty} \frac{1}{n} \operatorname{Log} \|A_n\| = \gamma = 0 .$$

b) The upper Lyapunov exponent associated with (B_n) is negative.

If we apply Theorem 6.1 to (A_n) we know that for some $\alpha > 0$ and each w in W, $w \neq 0$,

$$\overline{\lim_{n\to\infty}} \frac{1}{n} \operatorname{Log} \mathbb{P}(|\operatorname{Log} \|A_n w\|| > n\varepsilon) < -\alpha . \tag{7}$$

If we write this relation for a basis of W we readily see that

$$\overline{\lim_{n\to\infty}} \frac{1}{n} \operatorname{Log} \mathbb{P}(\operatorname{Log} \|A_n\| > n\varepsilon) < -\alpha . \tag{8}$$

Since the upper Lyapunov exponent associated with B_n is negative, for m large enough $\mathbb{E}(\operatorname{Log} \|B_m\|) < 0$. Making use of the relation

$$\operatorname{Log} \|B_n\| \leq \operatorname{Log} \|B_n B_{km}^{-1}\| + \operatorname{Log} \|B_{km} B_{(k-1)m}^{-1}\| + \dots + \operatorname{Log} \|B_m\|$$

for $km \leq n$, we deduce from the usual estimate on large deviations that, for α small enough,

$$\overline{\lim_{n\to\infty}} \frac{1}{n} \operatorname{Log} \mathbb{P}(\operatorname{Log} \|B_n\| > n\varepsilon) < -\alpha . \tag{9}$$

Since we may choose $\|\Lambda^p S_n\| = \operatorname{Sup}(\|A_n\|, \|B_n\|)$, (8) and (9) yield

$$\overline{\lim} \frac{1}{n} \operatorname{Log} \mathbb{P}(\operatorname{Log} \|\Lambda^p S_n\| > n\varepsilon) < -\alpha$$

and since (see e.g. Lemma III.1.4) the index of T_μ being p, $(\operatorname{Log} \|\Lambda^p S_n\| - p \operatorname{Log} \|S_n\|)$ is uniformly bounded we have

$$\overline{\lim} \frac{1}{n} \operatorname{Log} \mathbb{P}(\operatorname{Log} \|S_n\| > n\varepsilon) < -\alpha \tag{10}$$

This implies that for each $x \neq 0$ in $\mathbb{R}^d$,

$$\overline{\lim} \frac{1}{n} \operatorname{Log} \mathbb{P}(\operatorname{Log} \|S_n x\| > n\varepsilon) < -\alpha \quad . \tag{11}$$

The study of $\mathbb{P}(\operatorname{Log} \|S_n x\| < -n\varepsilon)$ is more intricate. We fix an $x \neq 0$ in $\mathbb{R}^d$ and first prove that there exist $x_2,\ldots,x_p$ such that $x \wedge x_2 \wedge \ldots \wedge x_p$ is not in V. It follows from Mostow [54] that we may suppose that the scalar product on $\mathbb{R}^d$ is such that W and V are orthogonal in $\Lambda^p \mathbb{R}^d$ (sketch of proof : the algebraic hull G of G_μ is irreducible. Therefore, for some scalar product, G is self adjoint. Since W is G_μ-invariant, W is also G-invariant. Thus its orthogonal is G-invariant and we may choose V as the orthogonal of W. It is G_μ-invariant). We have seen in the proof of Theorem 5.4 that W contains a decomposable p-vector, say $y_1 \wedge \ldots \wedge y_p$. If for any $x_2,\ldots,x_p$, $x \wedge x_2 \wedge \ldots \wedge x_p$ is in V, then $y_1 \wedge \ldots \wedge y_p$ is orthogonal to $x \wedge x_2 \wedge \ldots \wedge x_p$ for any $x_2,\ldots,x_p$. It is easily seen that this implies that y_1 is orthogonal to x. But for any M in G_μ, $My_1 \wedge \ldots \wedge My_p$ is also in W and we obtain, in the same way, that My_1 is orthogonal to x. This contradicts the fact that G_μ is irreducible.

We thus can choose $x_2,\ldots,x_p$ such that the component w of $x \wedge x_2 \wedge \ldots \wedge x_p$ on W is not zero. By (7)

$$\overline{\lim} \frac{1}{n} \operatorname{Log} \mathbb{P}(\operatorname{Log} \|A_n w\| < -n\varepsilon) < -\alpha$$

and, since $\|\Lambda^p S_n (x \wedge x_2 \wedge \ldots \wedge x_p)\| \geq \|A_n w\|$,

$$\overline{\lim} \frac{1}{n} \operatorname{Log} \mathbb{P}(\operatorname{Log} \|\Lambda^p S_n (x \wedge x_2 \wedge \ldots \wedge x_p)\| < -n\varepsilon) < -\alpha \quad . \tag{12}$$

Now, using the inequality

$$\|\Lambda^p S_n (x \wedge \ldots \wedge x_p)\| \leq \|S_n x\| \, \|S_n x_2\| \ldots \|S_n x_p\|$$

we have

$$\operatorname{Log} \|S_n x\| \geq \operatorname{Log} \|\Lambda^p S_n (x \wedge \ldots \wedge x_p)\| - \sum_{i=2}^{p} \operatorname{Log} \|S_n x_i\| \; .$$

Therefore $\mathbb{P}(\operatorname{Log} \|S_n x\| < -n\varepsilon) \leq \mathbb{P}(\operatorname{Log} \|\Lambda^p S_n (x \wedge \ldots \wedge x_p)\| \leq -n \frac{\varepsilon}{p})$

$$+ \sum_{i=2}^{p} \mathbb{P}(\operatorname{Log} \|S_n x_i\| \geq n \frac{\varepsilon}{p})$$

and, by (11) and (12)

$$\overline{\lim_{n\to\infty}} \frac{1}{n} \operatorname{Log} \mathbb{P}(\operatorname{Log} \|S_n x\| < -n\varepsilon) < -\alpha \; .$$

Since this implies that

$$\overline{\lim}\ \frac{1}{n}\ \mathrm{Log}\ \mathbb{P}(\mathrm{Log}\ \|S_n\| < -n\,\varepsilon) < -\alpha\,,$$

the proof of the theorem is complete.

V.7. Convergence to γ_p

We shall see, on a typical example, how the arguments developed for studying $\sigma(S_n,\bar{x}) = \mathrm{Log}\ \dfrac{\|S_n\,x\|}{\|x\|}$ can be generalized in order to study other cocycles.

PROPOSITION 7.1. *Let* $S_n = Y_n \ldots Y_1$ *where* $\{Y_n,\ n \geq 1\}$ *are i.i.d. random elements of* $Gl(d,\mathbb{R})$ *with common distribution* μ *. Assume that, for some* $1 < p \leq d$, T_μ *is* p *and* $(p-1)$ *contracting and strongly irreducible and that for some* $\tau > 0$, $\int \exp \tau\, \ell(M)\, d\mu(M)$ *is finite.*

Then for each nonzero v *in* $\Lambda^{p-1}\mathbb{R}^d$ *and each nonzero* w *in* $\Lambda^p\mathbb{R}^d$ $\frac{1}{\sqrt{n}}\ \{\mathrm{Log}\ \|\Lambda^p S_n\,w\| - \mathrm{Log}\ \|\Lambda^{p-1} S_n\,v\| - n\,\gamma_p\}$ *converges in distribution to a gaussian law* $N(0,a^2)$.

Proof : Let δ_1 (resp. δ_2) be the natural angular metric on $P(\Lambda^{p-1}\mathbb{R}^d)$ (resp. $P(\Lambda^p\mathbb{R}^d)$) and put on $B = P(\Lambda^{p-1}\mathbb{R}^d) \times P(\Lambda^d\mathbb{R}^d)$ the metric δ defined by, if $(\bar{v}_1,\bar{w}_1) \in B$, $(\bar{v}_2,\bar{w}_2) \in B$

$$\delta((\bar{v}_1,\bar{w}_1),(\bar{v}_2,\bar{w}_2)) = \delta_1(\bar{v}_1,\bar{v}_2) + \delta_2(\bar{w}_1,\bar{w}_2)\ .$$

If we apply Proposition 2.3 to the random products $\Lambda^{p-1} S_n$ and $\Lambda^p S_n$ we see that for α small enough

$$\lim_{n\to\infty}\ \sup_{\bar{v}_1,\bar{v}_2 \in P(\Lambda^{p-1}\mathbb{R}^d)} \left[\mathbb{E}\left\{\frac{\delta_1(\Lambda^{p-1}S_n\cdot\bar{v}_1,\Lambda^{p-1}S_n\cdot\bar{v}_2)}{\delta_1(\bar{v}_1,\bar{v}_2)}\right\}^\alpha\right]^{1/n} < 1\ ,$$

and

$$\lim_{n\to\infty}\ \sup_{\bar{w}_1,\bar{w}_2 \in P(\Lambda^{p}\mathbb{R}^d)} \left[\mathbb{E}\left\{\frac{\delta_2(\Lambda^{p}S_n\cdot\bar{w}_1,\Lambda^{p}S_n\cdot\bar{w}_2)}{\delta_2(\bar{w}_1,\bar{w}_2)}\right\}^\alpha\right]^{1/n} < 1\ .$$

So that if we set for M in $Gl(d,\mathbb{R})$ and $b = (\bar{v},\bar{w})$ in B

$$M\cdot b = (M\cdot\bar{v}\ ,\ M\cdot\bar{w})$$

then, for α small enough

$$\lim_{n\to\infty} \sup_{b_1,b_2\in B} \left[\mathbb{E}\left\{ \frac{\delta(S_n\cdot b_1, S_n\cdot b_2)}{\delta(b_1,b_2)} \right\}^{\alpha} \right]^{1/n} < 1 .$$

Therefore if $\mathbb{L}_\alpha(B)$ is the space of Hölder α functions on B and if s is the additive cocycle on $Gl(d,\mathbb{R})\times B$ defined by

$$s(M,b) = \mathrm{Log}\,\frac{\|\Lambda^p\, M w\|}{\|w\|} - \mathrm{Log}\,\frac{\|\Lambda^{p-1}\, M\, v\|}{\|v\|}$$

(where $M \in Gl(d,\mathbb{R})$ and $b = (\bar{v},\bar{w})$), then the formulas

$$T(z)f(b) = \int e^{z\, s(M,b)}\, f(M\cdot b)\, d\mu(M)$$

define, for $|z|$ small enough, operators on $\mathbb{L}_\alpha(B)$ which admit a decomposition similar to the one given in Theorem 4.3.
Working with this decomposition one easily shows the proposition (as in section 5).

In the same way one proves a large deviation result.

Remark 7.2. Let $B = P(\mathbb{R}^d)\times P(\Lambda^2\,\mathbb{R}^d) \times \ldots \times P(\Lambda^{d-1}\,\mathbb{R}^d)$, (we may also work with the flag manifold). Under suitable hypotheses on T_μ one can in fact study $\rho(S_n,b)$ as soon as

$$\rho(M,b) = (\sigma_1(M,b),\ldots,\sigma_{d-1}(M,b))$$

where each σ_i is an additive cocycle on $Gl(d,\mathbb{R})\times B$ which satisfies $\sigma_i(K,b) = 0$ for all K in $O(d)$ and b in B.

V.8. Convergence in distribution without normalization

We shall prove, under weak hypotheses, that S_n or $S_n x$ cannot converge in distribution unless it is to 0. By the same token we shall show that the variance of the limit law which appears in the central limit theorem is not zero. A fact we have taken for granted above.

We first study invariant probability measures on $\mathbb{R}^d$. It is clear that $\mathbb{R}^d$ is a $Gl(d,\mathbb{R})$-space for the natural action $M\cdot x = Mx$, $M \in Gl(d,\mathbb{R})$, $x \in \mathbb{R}^d$. Therefore we may define, given a probability measure μ on $Gl(d,\mathbb{R})$, a μ-invariant probability measure ν on $\mathbb{R}^d$ as a distribution which satisfies

$$\iint f(Mx)\, d\mu(M)\, d\nu(x) = \int f(x)\, d\nu(x)$$

for any bounded Borel function on $\mathbb{R}^d$. Notice that the Dirac measure at zero is always μ-invariant. Actually,

PROPOSITION 8.1. *Let* μ *be a probability measure on* $Gl(d,\mathbb{R})$ *such that* G_μ *is irreducible and not compact. Then the Dirac measure at zero is the only* μ*-invariant probability measure on* $\mathbb{R}^d$.

Proof : Suppose that there exists a μ-invariant probability measure ν on $\mathbb{R}^d$ which is not the Dirac measure at zero. Let us prove that in this case there is a real $c > 0$ such that for each M in T_μ,

$$1 \leq \|M\| \leq c \quad . \tag{12}$$

Let V be the linear subspace of $\mathbb{R}^d$ spanned by the support of ν. Then

$$\nu(V) = \iint 1_V(Mx)\, d\mu(M)\, d\nu(x) = 1$$

so that $\nu(M^{-1}V) = \nu(V)$ for μ-almost all M. This entails that the support of μ is contained in $\{M \in Gl(d,\mathbb{R})\ ;\ MV = V\}$. But this set is a closed subgroup. It must therefore contain G_μ. Taking into account the fact that G_μ is irreducible we see that $V = \mathbb{R}^d$.

Let $\{X_n, n \geq 1\}$ be a sequence of independent random matrices with distribution μ and $M_n = X_1 \ldots X_n$. By Lemma II.2.1 there exists for almost all ω a probability measure ν_ω on $\mathbb{R}^d$ such that $M_n(\omega)M\nu$ converges weakly to ν_ω for $\sum_{n=o}^{\infty} 2^{-n-1} \mu^n$-almost all M. Furthermore

$$\int f\, d\nu = \mathbb{E}\{\int f\, d\nu_\omega\} \tag{13}$$

for each bounded Borel function f on $\mathbb{R}^d$.

We now show that $\{M_n(\omega), n \geq 1\}$ is almost surely bounded. If for some ω $\operatorname{Sup}_n \|M_n(\omega)\| = \infty$, we can find a sequence $n(i)$ such that $\|M_{n(i)}(\omega)\|^{-1} M_{n(i)}(\omega)$ converges to a nonzero matrix $A(\omega)$ and such that $\lim_{i\to\infty} \|M_{n(i)}(\omega)\| = \infty$. But, if $M_n(\omega)\nu$ converges to ν_ω, then for any continuous f on $\mathbb{R}^d$ with compact support we have

$$\int f\, d\nu_\omega = \lim_{i\to\infty} \int f(M_{n(i)}(\omega)x)\, d\nu(x)$$

$$= \lim_{i\to\infty} \int 1_{\mathrm{Ker}\,A(\omega)}(x)\, f(M_{n(i)}(\omega)x)\, d\nu(x) ,$$

so that $\nu(\mathrm{Ker}\,A(\omega)) = 1$, a contradiction. Let now $M(\omega)$ be a limit point of the bounded sequence $\{M_n(\omega),\ n \geqq 1\}$.

For $\sum_{n=o}^{\infty} 2^{-n-1} \mu^n$-almost all M , we have

$$M(\omega) M \nu = \nu_\omega . \tag{14}$$

This equality holds also for each M in the closure T of T_μ in $M(d,\mathbb{R})$, (because T is the support of $\sum_{n=o}^{\infty} 2^{-n-1} \mu^n$ if we view this distribution as a probability measure on $M(d,\mathbb{R})$) . If there were some M in T_μ such that $\|M\| < 1$, then the zero matrix would be in T and (14) would imply that for each ω , ν_ω is the Dirac mass at zero. In view of (13), this would contradict our assumption. We have thus proved the L.H.S. of (12).

It remains to show that T_μ is bounded. Suppose that there exists a sequence $\{A_n,\ n \geqq 1\}$ in T_μ such that $\lim \|A_n\| = \infty$. We may assume that $\|A_n\|^{-1} A_n$ converges to some nonzero matrix, say A . By (14), for almost all ω and every M in T , the fact that MA_n is in T_μ implies that $M(\omega) M A_n \nu = \nu_\omega$.

Fix such an ω and set $W = \{x \in \mathbb{R}^d ;\ M(\omega) M A x = 0\}$. If x is not in W , $\|M(\omega) M A_n x\| \longrightarrow \infty$ as $n \longrightarrow \infty$. Therefore, for any continuous function f on $\mathbb{R}^d$ with compact support

$$\int f(x)\, d\nu_\omega(x) = \int f(M(\omega) M A_n x) d\nu(x)$$

$$= \lim_{n\to\infty} \int 1_W(x)\, f(M(\omega) M A_n x) d\nu(x) ,$$

but this implies that $\nu(W) = 1$, so that $W = \mathbb{R}^d$.

Choose some x such that Ax is not zero. For all M in T_μ ,

$$M(\omega)\, M A x = 0$$

and the subspace H of $\mathbb{R}^d$ spanned by $\{MAx ,\ M \in T_\mu\}$ is contained in the null space of $M(\omega)$. Since H is invariant under G_μ , $H = \mathbb{R}^d$ and $M(\omega) = 0$ a.s. , which implies that ν is the Dirac measure at 0.

We have thus proved that if ν is not the Dirac measure at 0 then (12) holds. The proof of the proposition is completed by the following lemma.

LEMMA 8.2. *Let* T *be an irreducible semigroup contained in* $Gl(d,\mathbb{R})$ *such that for some* $c > 0$

$$1 \leq \|M\| \leq c$$

for each M *in* T . *Then* T *is contained in a compact subgroup of* $Gl(d,\mathbb{R})$.

Proof : Let S be the closure of T in $M(d,\mathbb{R})$ and

$$\rho = \inf \{\text{rank}(M) \; ; \; M \in S\}.$$

(i) We first suppose that $p = 1$. It is easily seen using a Jordan decomposition that S contains a rank-one projection P . Without loss of generality we may assume that P is orthogonal. Choose a unit vector y in the range of P . For any x

$$Px = \langle x,y\rangle y$$

and if M_1, M_2 and M are in S

$$PM_1MM_2Px = \langle M_1 M M_2 y, y\rangle \langle x,y\rangle y \; .$$

But the spectral radius of each matrix in S is one. Therefore this relation leads to

$$\langle M_1 M M_2 y \, , \, y\rangle = \pm 1 \; . \tag{15}$$

Since T is irreducible we can find $M_1, M_2, \ldots, M_{2d}$ in T such that $\{u_1 = M_1^* y \, , \ldots, u_d = M_d^* y\}$ and $\{v_1 = M_{d+1} y \, , \ldots, v_d = M_{2d} y\}$ are two basis of $\mathbb{R}^d$. By (15) we obtain that the set $\{\langle Mu_i \, , v_j\rangle \; ; \; 1 \leq i,j \leq d, M \in T\}$ is finite, so that T itself is finite. But in that case T is closed and so $S = T$. In particular the rank-one projection P is invertible and d must be equal to one. When $d = 1$ the lemma is obvious.

(ii) We now consider the general case. First remark that

$$1 \leq \|\Lambda^p M\| \leq c^p \tag{16}$$

for each M in S . The right hand side inequality is obvious and if there exists M in S with $\|\Lambda^p M\| < 1$ then any limit point N of $\{M^n, n \geq 1\}$ is in S and satisfies $\Lambda^p N = 0$. But this latter relation entails that the rank of N is smaller than p , in contradiction with the definition of p .

Let Q be a rank $-p$ matrix in S and $\{u_1, u_2, \ldots, u_p\}$ be a basis of the range of Q . The range of $\Lambda^p Q$ is spanned by

$w_o = u_1 \wedge u_2 \wedge \ldots \wedge u_p$. By (16) the spectral radius of $\Lambda^p Q$ is one so that, for $P = Q^2$, $\Lambda^p P$ is a projection on $\mathbb{R} w_o$. Consider now the direct sum decomposition

$$\Lambda^p \mathbb{R}^d = W_1 \oplus \ldots \oplus W_k$$

given by Lemma 5.5 (where we take G as the group generated by T_μ) . Since for any i , $(\Lambda^p P)(W_i)$ is contained in W_i we may assume that w_o is in W_1 . As in (15), for some scalar product on $\Lambda^p \mathbb{R}^d$, and for any M_1, M_2, M in S we then have

$$<(\Lambda^p M_1)(\Lambda^p M)(\Lambda^p M_2) w_o , w_o> = \pm 1 .$$

Since the action of S on W_1 is irreducible this implies, as above, that the set whose elements are the restrictions of $\Lambda^p M$, $M \in T$, to W_1 is finite. Since P is in the closure of T it must exist some M in T such that

$$\Lambda^p P w = \Lambda^p M w$$

for all $w \in W_1$. But $\Lambda^p M$ is invertible, whence this relation leads to the fact that W_1 is one-dimensional. This implies that the linear span of $u_1,\ldots,u_p$ is invariant under T . Taking into account the irreducibility of T , we see that $p = d$.

Finally when $p = d$, S is a compact semigroup contained in $Gl(d,\mathbb{R})$. By (9.16) of Hewitt and Ross [36] or Exercise II.2.4, S is contained in a compact subgroup of $Gl(d,\mathbb{R})$.

Remark 8.3. The above proof is the only one which makes explicit use of the fact that the considered matrices are real. But the lemma holds also for matrices in $Gl(d,\mathbb{C})$. (To verify this write any matrix M in $Gl(d,\mathbb{C})$ as $M = A + iB$ with A, B in $Gl(d,\mathbb{R})$ and work with the matrices

$$\begin{bmatrix} A & -B \\ B & A \end{bmatrix}$$

of $Gl(2d,\mathbb{R})$. See the details in Remark 2.4 of [8]).

COROLLARY 8.4. *Let* $S_n = Y_n \ldots Y_1$, *where* $Y_1, Y_2, \ldots$ *are i.i.d. matrices in* $Gl(d,\mathbb{R})$ *with common distribution* μ . *Assume that* G_μ *is irreducible and not compact.*

If for some x *in* $\mathbb{R}^d$, $S_n x$ *converges in distribution, its limit is zero.*

Proof : It is clear that the limit distribution of $S_n x$ is μ-invariant. Therefore the claim follows from the proposition.

If we consider the distribution μ^n of S_n as a probability measure on $M(d,\mathbb{R})$, we see as above that when μ^n converges weakly then its limit distribution is the Dirac measure at the zero matrix. One may find in [8] generalizations of this result.

COROLLARY 8.5. *Under the hypotheses of Corollary 8.4, for any non zero* x *in* $\mathbb{R}^d$ *the sequence of the distributions of* $\mathrm{Log}\,\|S_n x\|$, $n=1,2,\ldots$ *is not tight on* $\mathbb{R}$.

Proof : Suppose that this sequence is tight. Then, if μ_x^i denotes the distribution of $S_i x$, the sequence $\{\frac{1}{n}\sum_{i=1}^{n}\mu_x^i\ ,\ n\geq 1\}$ is tight on $\mathbb{R}^d$. But it is easily seen (look at the proof of Lemma I.3.5) that each limit point of this sequence is a μ-invariant probability measure on $\mathbb{R}^d$. The proposition yields that $\frac{1}{n}\sum_{i=1}^{n}\mu_x^i$ converges to δ_o . Therefore for any $c > 0$,

$$\lim_{n\to\infty}\frac{1}{n}\sum_{i=1}^{n}\mathbb{P}(\mathrm{Log}\,\|S_i x\| < -c) = 1$$

which entails that the sequence $\mathrm{Log}\,\|S_n x\|$ is not tight.

In particular, applying the above corollary to the sequence $\{e^{-\gamma}Y_n,\ n\geq 1\}$ we obtain the following which proves Lemma 5.3 .

COROLLARY 8.6. *Suppose that* G_μ *is irreducible and that, for some* $x\neq 0$, *the sequence* $\mathbb{E}\{(\mathrm{Log}\,\|S_n x\| - n\gamma)^2\}$, $n = 1,2,\ldots$ *is bounded. Then there exists a compact subgroup* K *of* $Gl(d,\mathbb{R})$ *such that* $e^{-\gamma}Y_1$ *is a.s. in* K .

V.9. Complements : linear stochastic differential equations

We now consider briefly the continuous time analogue of the random products $S_n = Y_n \ldots Y_1$. Namely

DEFINITION 9.1. *A random process* $\{S_t, t \geq 0\}$ *with values in* $Gl(d,\mathbb{R})$ *is called a Process with Stationary Multiplicative Independent Increments (P.S.M.I.I) if the following hold*

(i) (S_t) *is continuous in probability.*

(ii) For any $0 \leq t_1 < t_2 < \dots < t_n$, *the random matrices* S_{t_1}, $S_{t_2} S_{t_1}^{-1}, \dots, S_{t_n} S_{t_{n-1}}^{-1}$ *are independent.*

(iii) The distribution of $S_{t+s} S_t^{-1}$ *depends only on* $s \geq 0$.

In particular, if μ_t is the distribution of S_t, then for any $t \geq 0$, $s \geq 0$, $\mu_{t+s} = \mu_t * \mu_s$. We denote by μ the collection $\{\mu_t, t \geq 0\}$ and by T_μ (resp. G_μ) the smallest closed subsemigroup (resp. subgroup) of $Gl(d,\mathbb{R})$ which contains the support of μ_t for all $t \geq 0$.

It is straightforward to check that all the previous results, which were proved for the discrete time model, are in fact valid for P.S.M.I.I. with these definitions of T_μ and G_μ. For instance if this G_μ is strongly irreducible, and if some S_t has an exponential moment, then for any $x \neq 0$, $\{\text{Log} \|S_t x\|, t > 0\}$ satisfies a central limit theorem, and a large deviation estimate (one may use Exercise 9.4 in order to prove this).

An important application (see e.g. Arnold and Kliemann [1]) is given by the so-called Linear Stochastic Differential Equations on $\mathbb{R}^d$. Namely let $A_o, A_1, \dots, A_r$ be arbitrary matrices of order d and $\{(B_t^1,\dots,B_t^r), t \geq 0\}$ be the usual r-dimensional Brownian motion. Consider the Stratonovitch stochastic differential equation on $\mathbb{R}^d$

$$dx_t = A_o x_t \, dt + \sum_{i=1}^{r} A_i x_t \circ d B_t^i \tag{17}$$

or, equivalently, the Ito equation

$$dx_t = A'_o x_t \, dt + \sum_{i=1}^{r} A_i x_t \, d B_t^i \, , \quad A'_o = A_o + \frac{1}{2} \sum_{i=1}^{r} A_i^2 \, .$$

Let $\{x_t(x), t \geq 0\}$ denote the solution of (17) which starts from x (i.e. $x_o(x) = x$). It is clear that $x_t(x)$ is a linear function of x. In other words we can write $x_t(x) = S_t x$ where S_t is a $d \times d$ matrix which satisfies the following S.D.E. on $M(d,\mathbb{R})$

$$d S_t = A_o S_t \, dt + \sum_{i=1}^{r} A_i S_t \circ d B_t^i \, , \quad S_o = Id \, . \tag{18}$$

In fact $\{S_t, t \geq 0\}$ is the stochastic flow of diffeomorphisms associated with (17). As such (S_t) is a P.S.M.I.I. with values in $Gl(d,\mathbb{R})$ (see e.g. Kunita [44] or Ikeda and Watanabe [37]). Let G be the Lie subgroup of $Gl(d,\mathbb{R})$ whose Lie algebra is the one generated by $A_o, A_1, \ldots, A_r$. The S.D.E. (18) can be considered as a S.D.E. on the manifold G. Therefore S_t takes its values in G. Moreover if $\mu = \{\mu_t, t \geq 0\}$, where μ_t is the distribution of S_t, then it is easy to see that T_μ contains an open set in G (see Lemma 7.1 of [8]). This implies that G_μ is the closure $\overline{G}$ of G.

Since G is connected it is easy to see that G_μ is strongly irreducible if and only if G is irreducible, which amounts to say that $\{A_o, A_1, \ldots, A_r\}$ acts irreducibly on $\mathbb{R}^d$.

Besides, note that $M_t = (\det S_t)^{-1/d} S_t$ satisfies the equation

$$dM_t = \overline{A}_o M_t \, dt + \sum_{i=1}^{r} \overline{A}_i M_t \circ dB_t^i$$

where $\overline{A}_i = A_i - \frac{1}{d}(\text{trace } A_i) \, Id$, and that

$$d(\det M_t) = (\text{trace } A_o) \det M_t \, dt + \sum_{i=1}^{r} (\text{trace } A_i) \det M_t \circ dB_t^i$$

so that $\det M_t = \text{Exp}\{(\text{trace } A_o)t + \sum_{i=1}^{r} (\text{trace } A_i) B_t^i\}$.

This gives that $\mathbb{E}(\text{Log} \det M_t) = t \, \text{trace } A_o$ and $\sum_{i=1}^{d} \gamma_i = \text{trace } A_o$.

Finally using well known estimates on solutions of S.D.E. (see e.g. Theorem IV.2.4 of [ ]) one obtains that $\mathbb{E}(\exp 2\ell(S_t))$ is finite for each t.

Getting these results together we may apply the theorems III.6.3, V.5.4 and V.6.2 to the process (S_t). We obtain

<u>PROPOSITION 9.2</u>. *Let* $x_t(x)$ *be the solution of*

$$dx_t = A_o x_t \, dt + \sum_{i=1}^{r} A_i x_t \circ dB_t^i$$

starting from x *at time* 0. *Suppose that the following hold.*

(i) There is no proper linear subspace V *of* $\mathbb{R}^d$ *such that* $A_i(V)$ *is contained in* V *for all* $i = 0, 1, \ldots, r$.

(ii) There is no matrix Q *such that all the matrices* $Q(A_i - \frac{1}{d}(\text{trace } A_i) \, Id)Q^{-1}$, $i = 0, \ldots, r$ *are skew-symmetric.*

Then (a) For all $x \neq 0$, $\lim_{t\to\infty} \frac{1}{t} \operatorname{Log} \|x_t(x)\| = \gamma$ *and* $\gamma > \frac{1}{d}$ trace A_o .

(b) For some $\sigma^2 > 0$ *and all* $x \neq 0$, $\{\frac{1}{\sqrt{t}} (\operatorname{Log} \|x_t(x)\| - t\gamma) , t > 0\}$ *converges in distribution to the gaussian measure* $N(0, \sigma^2)$.

(c) For some $\alpha > 0$ *and all* $x \neq 0$, $\varepsilon > 0$,

$$\overline{\lim_{t\to\infty}} \frac{1}{t} \operatorname{Log} \mathbb{P}(|\operatorname{Log} \|x_t(x)\| - t\gamma| > t\varepsilon) < -\alpha .$$

Note that when (ii) is not satisfied then all the matrices $(\det S_t)^{-1/d} S_t$ lie in a compact subgroup of $Gl(d,\mathbb{R})$. Therefore we may find a norm on $\mathbb{R}^d$ such that $\|x_t(x)\| = (\det S_t)^{1/d} \|x\|$ for any $x \in \mathbb{R}^d$ and $t \geq 0$.
In this case (a) holds but $\gamma = \frac{1}{d}$ trace A , (b) holds but $\sigma = 0$ if and only if trace $A_i = 0$, $i = 1,\ldots,r$, and (c) always holds.

We have of course much better results when T_μ is contracting. To check this condition on examples it is worth noting that we have a concrete description of T_μ . Namely, T_μ is the smallest closed subsemigroup of $Gl(d,\mathbb{R})$ which contains all the matrices $\{\operatorname{Exp} t S_o, t > 0\}$ and $\{\operatorname{Exp} t S_i , t \in \mathbb{R}, i = 1,\ldots,r\}$. This follows either from Hörmander's hypoellipticity theorem (see Siebert [65] and especially Mc Crudden and Wood [53] , Proposition 3) or from the support theorem of Stroock and Varadhan (see e.g. Arnold and Kliemann [1]) .

<u>Example 9.3</u>. Suppose that $A_o, A_1, \ldots, A_r$ are $d \times d$ matrices such that trace $(A_i) = 0$, $0 \leq i \leq r$, and such that the Lie algebra generated by $A_o, A_1, \ldots, A_r$ is the set of all trace-zero matrices. Then the d Lyapunov exponents associated with the solution of (17) are distinct $(\gamma_1 > \gamma_2 > \ldots > \gamma_d)$, the full flag manifold is a μ_t-boundary for each $t > 0$ (see IV.4) and we have a complete description of the behaviour of the solutions of (17) at infinity. This follows readily from the fact that in this case $G_\mu = S\ell(d,\mathbb{R})$ and from Proposition IV.2.3 since T_μ contains an open subset of G_μ .

<u>Exercise 9.4</u>. Let $\{S_t, t > 0\}$ be a P.S.M.I.I. with values in $G\ell(d,\mathbb{R})$ such that $\mathbb{E}(\|S_1\|^\tau)$ is finite for some $\tau > 0$.

a. Show that there exists $c > 0$ such that for any $0 \leqq t \leqq 1$, $\mathbb{P}(\|S_t^{-1}\| \leqq c) \geqq 3/4$.

b. Deduce that for each $0 \leqq s \leqq 1$ there exists a matrix A in $Gl(d,\mathbb{R})$ such that $\|A^{-1}\| \leqq c$ and

$$\mathbb{E}(\|S_s A\|^\tau) \leqq 3\, \mathbb{E}(\|S_1\|^\tau)$$

so that
$$\mathbb{E}(\|S_s\|^\tau) \leqq 3\, c^\tau\, \mathbb{E}(\|S_1\|^\tau).$$

c. Prove that for any $r > 0$, $\sup_{0 \leqq t \leqq r} \mathbb{E}(\|S_t\|^\tau)$ is finite.

CHAPTER VI

PROPERTIES OF THE INVARIANT MEASURE AND APPLICATIONS

The main result to be proved in this chapter is that under suitable hypotheses, the μ-invariant distribution ν on $P(\mathbb{R}^d)$ satisfies for some $\alpha > 0$

$$\int \left|\frac{\|x\|}{\langle x,y\rangle}\right|^{\alpha} d\nu(\bar{x}) < \infty . \qquad (1)$$

Since for any nonzero x in $\mathbb{R}^d$ the distribution of the direction $S_n \cdot \bar{x}$ is close to ν , this relation indicates that in somme sense $\frac{\|S_n x\|}{\langle S_n x,y\rangle}$ is not too big. More precisely we shall show that for any $r > 0$

$$\lim_{n\to\infty} \frac{1}{n^r} \operatorname{Log} \frac{\|S_n x\|}{|\langle S_n x,y\rangle|} = 0 \qquad \text{a.s.} \ .$$

This will imply that the previous limit theorems about $\|S_n x\|$ are actually valid for the coefficients of S_n.

All the main results of this chapter are due to Guivarc'h and Raugi.

To get an idea of the way we shall prove (1) we first consider a degenerate situation. The i.i.d. matrices $Y_1, Y_2, \ldots$ are in a compact subset of $Gl(2,\mathbb{R})$ and of the form

$$Y_i = \begin{bmatrix} a_i & 0 \\ b_i & c_i \end{bmatrix} .$$

In that case T_μ is of course not irreducible. We suppose that $\gamma_1 > \gamma_2$ where $\gamma_1 = \mathbb{E}(\mathrm{Log}|a_1|)$ and $\gamma_2 = \mathbb{E}(\mathrm{Log}|c_1|)$.
Setting $T_n = Y_n \dots Y_1$ it is readily seen that

$$\frac{T_n e_1}{\langle T_n e_1, e_1\rangle} = e_1 + u_n e_2$$

where

$$u_n = a_1^{-1} b_1 + \frac{c_1}{a_1}(a_2^{-1} b_2) + \dots + \frac{c_1 \dots c_{n-1}}{a_1 \dots a_{n-1}}(a_n^{-1} b_n) \quad . \tag{2}$$

Since $\limsup \{|a_n^{-1} b_n \frac{c_1 \dots c_{n-1}}{a_1 \dots a_{n-1}}|\}^{1/n} \leq e^{\gamma_2 - \gamma_1} < 1$, the series u_n converges a.s. to a random variable u_∞.
Moreover for any $0 < \alpha < 1$

$$\begin{aligned} \mathbb{E}(|u_\infty|^\alpha) &\leq \sum_{n=o}^{\infty} \mathbb{E}\{|\frac{c_1 \dots c_{n-1}}{a_1 \dots a_{n-1}}|^\alpha\} \, \mathbb{E}\{|a_n^{-1} b_n|^\alpha\} \\ &\leq \sum_{n=o}^{\infty} \{\mathbb{E}|\frac{c_1}{a_1}|^\alpha\}^{n-1} \, \mathbb{E}\{|a_1^{-1} b_1|^\alpha\} \, . \end{aligned}$$

The negativity of $\mathbb{E}(\mathrm{Log}|\frac{c_1}{a_1}|)$ implies that for $\alpha > 0$ small enough

$$\mathbb{E}\{|\frac{c_1}{a_1}|^\alpha\} < 1 \, ,$$

so that $\mathbb{E}(|u_\infty|^\alpha)$ is finite.

Now notice that $T_n \cdot \bar{e}_1$ converges a.s. to $\overline{e_1 + u_\infty e_2}$, which implies that the distribution ν of $\overline{e_1 + u_\infty e_2}$ is μ-invariant (in fact every μ-invariant distribution can be written as $\lambda\nu + (1-\lambda)\delta_{\bar{e}_2}$, $0 \leq \lambda \leq 1$, as is easily deduced from Proposition II.3.3).

Therefore

$$\int \left|\frac{\|x\|}{|\langle x, e_1\rangle|}\right|^\alpha d\nu(x) = \mathbb{E}\{(1+u_\infty^2)^{\alpha/2}\} < \infty$$

and, in this non-irreducible situation, (1) holds for $y = e_1$.

Consider now a sequence $Y_1, Y_2, \dots$ of independent random matrices in $Gl(d,\mathbb{R})$ whose common distribution μ is such that T_μ is strongly irreducible and contracting. To get an analogy with the preceding example we write $M_n = T_n K_n$, where K_n is orthogonal and

$$T_n = \begin{bmatrix} a_n & 0 \\ B_n & C_n \end{bmatrix}$$

(where a_n is a real number, B_n a (d-1)-dimensional column vector and

C_n a matrix of order $d-1$).

We shall see that $\frac{T_n e_1}{\langle T_n e_1, e_1\rangle}$ can be written as a series similar to (2) and that $T_n \cdot \bar{e}_1$ converges to a random direction $\bar{Z}$ with distribution ν. By the same type of argument as above we shall deduce that if μ has an exponential moment, then

$$\int \left|\frac{\|x\|}{\langle x, e_1\rangle}\right|^\alpha d\nu(\bar{x}) < \infty$$

for some $\alpha > 0$. In this irreducible situation e_1 has no particular role and (1) will follow.

The first four sections of this chapter are devoted to the proof of these results and to their applications to limit theorems. In the following we study the regularity properties of the invariant measure ν and give some examples.

6.1. Convergence in the Iwasawa decomposition

As said above we shall write $M_n = Y_1 \dots Y_n$ as $M_n = T_n K_n$ where K_n is orthogonal and T_n satisfies $\langle T_n e_i, e_1\rangle = 0$, $i = 2,\dots,d$. As before, $\{e_1,\dots,e_d\}$ stands for the canonical basis of $\mathbb{R}^d$. They are many ways to express M_n as such a product. As far as we are only concerned with the behaviour of $T_n e_1$, any of these will do. Just for convenience we choose a particular one, sometimes called the Iwasawa decomposition, given by

LEMMA 1.1. *Any matrix* M *in* $Gl(d,\mathbb{R})$ *can be written as*

$$M = s(M)\, k(M)$$

where $k(M)$ *is orthogonal and* $s(M)$ *is lower triangular with positive diagonal entries. Such a decomposition is unique.*

Proof : We apply the Gram-Schmidt orthogonalization procedure to the vectors $M^* e_1, \dots, M^* e_d$. We can write

$$M^* e_1 = \alpha_{11} u_1$$

$$M^* e_2 = \alpha_{21} u_1 + \alpha_{22} u_2$$

....

$$M^* e_d = \alpha_{d1}\, u_1 + \ldots + \alpha_{dd}\, u_d$$

where $\{u_1, \ldots, u_d\}$ is an orthonormal basis of $\mathbb{R}^d$. We choose $\alpha_{ii} > 0$, $i = 1,\ldots,d$. This decomposition is then unique. Define the orthogonal matrix $k(M)$ by the relations $k(M)u_i = e_i$ and the lower triangular matrix $s(M)$ by

$$\langle s(M)e_i, e_j\rangle = \begin{cases} \alpha_{ji} & \text{if} \quad i \geq j \\ 0 & \text{if} \quad i < j \end{cases}$$

Then $M^* = k(M)^*\, s(M)^*$ so that $M = s(M)\, k(M)$.

Using Theorem III.3.1 we may introduce the following

Notation 1.2. If $Y_1, Y_2, \ldots$ are independent random matrices distributed according to μ we set $M_n = Y_1 \ldots Y_n$ and $S_n = Y_n \ldots Y_1$. When T_μ is strongly irreducible and contracting, ν denotes the unique μ-invariant distribution on $P(\mathbb{R}^d)$ and Z any non zero random vector such that a.s. $\lim_{n\to\infty} M_n \nu = \delta_{\bar{Z}}$.

Recall that for any $x \neq 0$, $\mathbb{P}(\langle Z,x\rangle = 0) = 0$ since ν is proper.

PROPOSITION 1.3. *Suppose that* T_μ *is strongly irreducible and contracting and write* $M_n = s(M_n)\, k(M_n)$ *according to the Iwasawa decomposition. Then, a.s.*

$$\lim_{n\to\infty} \frac{s(M_n)e_1}{\langle s(M_n)e_1, e_1\rangle} = \frac{Z}{\langle Z, e_1\rangle}.$$

Proof : First notice that

$$s(M_n)^* e_1 = \langle s(M_n)e_1, e_1\rangle e_1$$

so that $|\langle s(M_n)e_1, e_1\rangle| = \|s(M_n)^*\, e_1\| = \|k(M_n)^*\, s(M_n)^*\, e_1\| = \|M_n^* e_1\|$ and

$$\frac{M_n\, M_n^*\, e_1}{\|M_n^*\, e_1\|^2} = \frac{s(M_n)\; s(M_n)^*\; e_1}{\langle s(M_n)e_1, e_1\rangle^2} = \frac{s(M_n)e_1}{\langle s(M_n)e_1, e_1\rangle}.$$

Now write a polar decomposition $M_n^* = K_n A_n U_n$ of M_n^* (K_n and U_n are orthogonal matrices and $A_n = \mathrm{diag}(a_1(n), \ldots, a_d(n))$, with

$a_1(n) \geqq a_2(n) \geqq \ldots \geqq a_d(n) > 0$). Recall that the index of T_μ is one since T_μ is contracting. It thus follows from Proposition III.3.2 (applied to M_n^*) that

$$\lim_{n\to\infty} \frac{a_2(n)}{a_1(n)} = 0 \qquad \text{a.s. ,}$$

and that $U_n^* \cdot \bar{e}_1$ converges a.s. . It is clear, at least from the proof of this proposition, that the limit of $U_n^* \cdot \bar{e}_1$ is actually $\bar{Z}$.

Since for any y in $\mathbb{R}^d$

$$\begin{aligned} \langle M_n M_n^* e_1 , y\rangle &= \langle U_n^* A_n^2 U_n e_1 , y\rangle \\ &= a_1(n)^2 \langle U_n e_1, e_1\rangle\langle e_1, U_n y\rangle + O(a_1(n)a_2(n)) \end{aligned}$$

it follows that

$$\begin{aligned} \lim_{n\to\infty} \frac{\langle s(M_n)e_1, y\rangle}{\langle s(M_n)e_1, e_1\rangle} &= \lim_{n\to\infty} \frac{\langle M_n M_n^* e_1 , y\rangle}{\langle M_n M_n^* e_1, e_1\rangle} \\ &= \lim_{n\to\infty} \frac{\langle U_n^* e_1 , y\rangle}{\langle U_n^* e_1, e_1\rangle} \\ &= \frac{\langle Z , y\rangle}{\langle Z , e_1\rangle} \quad . \end{aligned}$$

This proves the proposition.

For any M in $Gl(d,\mathbb{R})$ let $a(M)$ denote the diagonal matrix satisfying $a(M)_{i,i} = s(M)_{i,i}$, $i = 1,\ldots,d$, and $\eta(M) = s(M)\, a(M)^{-1}$. The matrix $\eta(M)$ is lower triangular with 1 on the diagonal.

<u>COROLLARY 1.4</u>. *Suppose that for any* $p = 1,2,\ldots,d-1$, T_μ *is* p-*contracting and* p-*strongly irreducible. Then if*

$$Y_1 \ldots Y_n = M_n = \eta(M_n)\, a(M_n)\, k(M_n)$$

$\eta(M_n)$ *converges a.s. to a random matrix* N_∞ .

If, moreover, $\mathbb{E}\{\ell(Y_1)\}$ *is finite then* $a(M_n)^{1/n}$ *converges a.s. to* $\mathrm{diag}(e^{\gamma_1},\ldots,e^{\gamma_d})$.

<u>Proof</u> : Since $s(M_n) = \eta(M_n)\, a(M_n)$ we have

$$\frac{s(M_n)e_1}{\langle s(M_n)e_1, e_1\rangle} = \eta(M_n)e_1 \quad .$$

Therefore the proposition yields that $\eta(M_n)e_1$ converges a.s. .

In the same way, for any $p = 1,\dots,d-1$

$$\Lambda^p M_n = \{\Lambda^p \eta(M_n)\}\{\Lambda^p a(M_n)\}\{\Lambda^p k(M_n)\}$$

is an Iwasawa decomposition of $\Lambda^p M_n$ (when we consider on $\Lambda^p \mathbb{R}^d$ the basis $\{e_{i_1} \wedge \dots \wedge e_{i_d} \,;\, i_1 < i_2 < \dots < i_d\}$ in the lexicographic order). The proposition applied to these matrices gives that, for any p, $\Lambda^p \eta(M_n)(e_1 \wedge e_2 \wedge \dots \wedge e_p)$ converges a.s. . This implies that $\eta(M_n)$ converges since for $i > j$,

$$\langle \eta(M_n)e_j, e_i\rangle = \langle \Lambda^j \eta(M_n)(e_1 \wedge e_2 \wedge \dots \wedge e_{j-1} \wedge e_i), e_1 \wedge e_2 \wedge \dots \wedge e_{j-1} \wedge e_j\rangle$$

On the other hand if $a(M_n) = \mathrm{diag}(\alpha_1(n),\dots,\alpha_d(n))$ then

$$\alpha_1(n)\alpha_2(n)\dots\alpha_p(n) = \|\Lambda^p M_n^*(e_1 \wedge e_2 \wedge \dots \wedge e_p)\| .$$

Therefore when $\mathbb{E}(\ell(Y_1))$ is finite and T_μ p-contracting and strongly irreducible, then

$$\lim \frac{1}{n} \mathrm{Log}\ \alpha_1(n)\dots\alpha_p(n) = \lim \frac{1}{n} \mathrm{Log}\ \|\Lambda^p M_n^*(e_1 \wedge\dots\wedge e_p)\| = \gamma_1 + \dots + \gamma_p$$

which implies that $a(M_n)^{1/n}$ converges a.s. to $\mathrm{diag}(e^{\gamma_1},\dots,e^{\gamma_d})$.

<u>Remark 1.5</u>. The process $k(M_n)$ is a Markov chain on $O(d)$. If $\bar{\nu}$ is the unique distribution on the flag manifold F which is invariant under the image μ^* of μ by transposition (see Proposition IV.4.1) then the distribution of

$$k(M_n)^* \cdot (\bar{e}_1 , \overline{e_1 \wedge e_2}, \dots, \overline{e_1 \wedge \dots \wedge e_{d-1}})$$

converges to $\bar{\nu}$. In order to show this, write $S_n = Y_n \dots Y_1$ as $S_n = \eta(S_n)\, a(S_n)\, k(S_n)$. Since S_n and M_n have the same distribution, it's also the case for $k(S_n)$ and $k(M_n)$. Thus it is enough to notice that

$$k(S_n)^* \cdot (\bar{e}_1,\dots,\overline{e_1 \wedge \dots \wedge e_{d-1}}) = S_n^* \cdot (\bar{e}_1,\dots,\overline{e_1 \wedge \dots \wedge e_{d-1}})$$

converges in probability to a random variable with law $\bar{\nu}$ (this follows readily from (ii) of Theorem III.4.3).

We now suppose that μ has a first order moment. In this case the two upper Lyapunov exponents γ_1 and γ_2 are distinct when T_μ is strongly irreducible and contracting (see Theorem III.6.1) .

THEOREM 1.6. *If* T_μ *is strongly irreducible and contracting and if* $\int \ell(M)\, d\mu(M) < \infty$ *, then a.s.*

$$\overline{\lim_{n\to\infty}} \frac{1}{n} \operatorname{Log} \left\| \frac{s(M_n)e_1}{\langle s(M_n)e_1, e_1\rangle} - \frac{Z}{\langle Z, e_1\rangle} \right\| \leq \gamma_2 - \gamma_1 .$$

Proof : For any M in $Gl(d,\mathbb{R})$ set

$$t(M) = \frac{s(M)}{\langle s(M)e_1, e_1\rangle} = \frac{s(M)}{\|M^* e_1\|} \quad \text{and} \quad x(M) = t(M)e_1 - e_1 . \tag{3}$$

We have, for M and M' in $Gl(d,\mathbb{R})$,

$$MM' = s(M)\ k(M)M' = s(M)\ s(k(M)M')\ k(k(M)M')$$

and

$$MM' = s(MM')\ k(MM') .$$

Therefore by the uniqueness of the Iwasawa decomposition

$$s(MM') = s(M)\ s(k(M)M') .$$

Since $\langle s(MM')e_1, e_1\rangle = \langle s(M)e_1, e_1\rangle \langle s(k(M)M')e_1, e_1\rangle$ this yields immediately

$$x(MM') = x(M) + t(M)\ x(k(M)M') .$$

We thus get the following formula, analogous to (2) ,

$$x(M_n) = x(Y_1 \dots Y_n) = x(Y_1) + t(M_1)x(k(M_1)Y_2) + \dots + t(M_{n-1})x(k(M_{n-1})Y_n) .$$

To study this series we shall use the relations

$$\|x(M)\| \leq \exp(2\ell(M)) \tag{4}$$

and

$$\sup \left\{\frac{\|s(M)x\|}{\|x\|} ;\ x \text{ orthogonal to } e_1\right\} \leq \frac{\|\Lambda^2 M^*\|}{\|M^* e_1\|} \tag{5}$$

Since $x(M)$ is always orthogonal to e_1 these imply that for all positive integers m and n ,

$$\begin{aligned} \|t(M_{n+m})e_1 - t(M_n)e_1\| &= \|x(M_{n+m})e_1 - x(M_n)e_1\| \\ &\leq \sum_{j=n}^{n+m-1} \|t(M_j)\ x(k(M_j)Y_{j+1})\| \\ &\leq \sum_{j=n}^{\infty} \frac{\|\Lambda^2 M_j^*\|}{\|M_j^* e_1\|^2} e^{2\ell(Y_{j+1})} . \end{aligned} \tag{6}$$

Now, the exponents associated with (Y_n) and (Y_n^*) being the same, we know that a.s.

$$\lim_{j\to\infty} \frac{1}{j} \operatorname{Log} \frac{\|\Lambda^2 M_j^*\|}{\|M_j^* e_1\|^2} = \lim_{j\to\infty} \frac{1}{j} \operatorname{Log} \|\Lambda^2 Y_j^* \ldots Y_1^*\| - 2 \lim \frac{1}{j} \operatorname{Log} \|Y_j^* \ldots Y_1^* e_1\|$$
$$= (\gamma_1 + \gamma_2) - 2\gamma_1$$
$$= \gamma_2 - \gamma_1 .$$

Therefore for each $\varepsilon > 0$ we can find N_1 such that, if $j > N_1$,

$$\frac{\|\Lambda^2 M_j^*\|}{\|M_j^* e_1\|^2} \leqq e^{j(\gamma_2 - \gamma_1 + \varepsilon)} .$$

On the other hand since $\mathbb{E}(\ell(Y_1))$ is finite,

$$\sum_{j=o}^{\infty} \mathbb{P}(\operatorname{Log} \ell(Y_{j+1}) > j\varepsilon) < \infty$$

so that, using the Borel-Cantelli lemma, there exists a.s. $N_2 > 0$ such that for $j > N_2$

$$\exp 2\ell(Y_{j+1}) \leqq \exp 2j\varepsilon .$$

Hence we obtain for $n > \operatorname{Sup}(N_1, N_2)$ and any $m > 0$

$$\|t(M_{n+m})e_1 - t(M_n)e_1\| \leqq \sum_{j=n}^{\infty} e^{j(\gamma_2 - \gamma_1 + 3\varepsilon)}$$

and, since $\frac{Z}{\langle Z, e_1\rangle}$ is the limit of $t(M_{m+n})e_1$ as $m \longrightarrow \infty$, (see Proposition 1.3)

$$\|t(M_n)e_1 - \frac{Z}{\langle Z, e_1\rangle}\| \leqq \frac{e^{n(\gamma_2 - \gamma_1 + 3\varepsilon)}}{1 - e^{(\gamma_2 - \gamma_1 + 3\varepsilon)}} .$$

This yields that for each $\varepsilon > 0$, a.s.

$$\overline{\lim_{n\to\infty}} \frac{1}{n} \operatorname{Log} \|\frac{s(M_n)e_1}{\langle s(M_n)e_1, e_1\rangle} - \frac{Z}{\langle Z, e_1\rangle}\| \leqq \gamma_2 - \gamma_1 + 3\varepsilon .$$

The theorem will follow once we have proved the relations (4) and (5).

On the one hand we have (see the beginning of the proof of Proposition 1.3)

$$\|x(M)\| \leqq \frac{\|s(M)e_1\|}{|\langle s(M)e_1, e_1\rangle|} \leqq \frac{\|MM^* e_1\|}{\|M^* e_1\|^2} \leqq \|M\| \, \|M^{*-1}\| \leqq e^{2\ell(M)} .$$

On the other hand, since $s(M)$ leaves invariant the subspace orthogonal to e_1,

$$\sup\{\frac{\|s(M)x\|}{\|x\|} ; x \text{ orthogonal to } e_1\} = \sup_{y\in\mathbb{R}^d} \frac{\|e_1 \wedge s(M)^* y\|}{\|y\|}$$
$$= \sup_y \frac{\|s(M)^* e_1 \wedge s(M)^* y\|}{|\langle s(M)e_1, e_1\rangle| \, \|y\|}$$

$$= \sup_y \frac{\|M^*e_1 \wedge M^*y\|}{\|M^*e_1\| \, \|y\|}$$

$$\leq \frac{\|\Lambda^2 M^*\|}{\|M^*e_1\|} .$$

The proof is now complete.

We have already noticed that it is in general not true that for almost all ω, $\frac{1}{n} \operatorname{Log} \|S_n(\omega)x\|$ converges to $\gamma = \gamma_1$ for every nonzero x (see in particular the end of section II.3). In fact we have, working for convenience with transpose matrices,

COROLLARY 1.7. *If* T_μ *is strongly irreducible and contracting and if* $\mathbb{E}(\ell(Y_1))$ *is finite, then for almost all* ω

(i) If x *is not orthogonal to* $Z(\omega)$ *then*

$$\lim_{n\to\infty} \frac{1}{n} \operatorname{Log} \|Y_n^*(\omega) \ldots Y_1^*(\omega)x\| = \gamma_1 .$$

(ii) If x *is orthogonal to* $Z(\omega)$ *then*

$$\overline{\lim_{n\to\infty}} \frac{1}{n} \operatorname{Log} \|Y_n^*(\omega) \ldots Y_1^*(\omega)x\| \leq \gamma_2 < \gamma_1 .$$

Proof : Making use of the relation

$$\|u \wedge v\|^2 = \|u\|^2 \|v\|^2 - \langle u,v\rangle^2$$

we have, if $M_n = Y_1 \ldots Y_n$,

$$\|M_n^* x\|^2 = \|M_n^* e_1\|^2 \left\langle x , \frac{M_n M_n^* e_1}{\|M_n^* e_1\|^2} \right\rangle^2 + \frac{\|M_n^* e_1 \wedge M_n^* x\|^2}{\|M_n^* e_1\|^2} .$$

Since, a.s. ,

$$\lim_{n\to\infty} \frac{1}{n} \operatorname{Log} \|M_n^* e_1\| = \gamma_1$$

and

$$\overline{\lim_{n\to\infty}} \frac{1}{n} \operatorname{Log} \frac{\|M_n^* e_1 \wedge M_n^* x\|}{\|M_n^* e_1\|} \leq \gamma_2$$

the claim follows from the fact that, because $\dfrac{M_n M_n^* e_1}{\|M_n^* e_1\|^2} = \dfrac{s(M_n)e_1}{\langle s(M_n)e_1, e_1\rangle}$,

$$\overline{\lim_{n\to\infty}} \frac{1}{n} \operatorname{Log} \|M_n^* e_1\| \left|\left\langle x , \frac{M_n M_n^* e_1}{\|M_n^* e_1\|^2} - \frac{Z}{\langle Z, e_1\rangle} \right\rangle\right| \leq \gamma_1 + \gamma_2 - \gamma_1 = \gamma_2 .$$

For 2×2 matrices, this corollary is Osseledec's theorem. To see this, note that since the upper exponent associated with $(Y_n^*)^{-1}$ is

$-\gamma_2$ we always have in this case

$$\liminf_{n\to\infty} \frac{1}{n} \operatorname{Log} \|Y_n^*\dots Y_1^* x\| \geq \liminf -\frac{1}{n}\operatorname{Log}\|(Y_n^*)^{-1}\dots(Y_1^*)^{-1}\| = \gamma_2 .$$

So that, if x is a nonzero vector orthogonal to $Z(\omega)$,

$$\lim_{n\to\infty} \frac{1}{n} \operatorname{Log} \|Y_n^*\dots Y_1^* x\| = \gamma_2 .$$

We also have

COROLLARY 1.8. *Let* $Y_1, Y_2, \dots$ *be independent random matrices distributed according to* μ *and* $S_n = Y_n \dots Y_1$. *Suppose that* $T_\mu^{-1} = \{M ; M^{-1} \in T_\mu\}$ *is strongly irreducible and contracting, and that* $\mathbb{E}(\ell(Y_1))$ *is finite. Then the two lower Lyapunov exponents* γ_{d-1} *and* γ_d *are distinct. Moreover let* ν' *be the distribution on* $P(\mathbb{R}^d)$ *which is invariant under the law of* Y_1^{-1} *and* $\delta_{\bar{Z}}$ *be the limit of* $S_n^{-1}\nu'$. *Then, for almost all* ω *and every vector* x *not colinear with* $Z(\omega)$,

$$\liminf_{n\to\infty} \frac{1}{n} \operatorname{Log} \|S_n(\omega)x\| \geq \gamma_{d-1} .$$

Proof : We know, applying the preceding corollary to the sequence $Y_1^{-1}, Y_2^{-1}, \dots$ that a.s., for any y orthogonal to $Z(\omega)$

$$\limsup_{n\to\infty} \frac{1}{n} \operatorname{Log} \|(Y_n^{-1})^*\dots(Y_1^{-1})^* y\| \geq -\gamma_{d-1} .$$

(We have used the fact that $-\gamma_d \geq -\gamma_{d-1} \geq \dots \geq -\gamma_1$ are the exponents associated with (Y_n^{-1})). Let now x be a vector such that $\bar{x} \neq \bar{Z}(\omega)$. We can find a vector y such that $\langle x,y\rangle \neq 0$ and $\langle y,Z\rangle = 0$. Since

$$|\langle x,y\rangle| = |\langle (S_n^{-1})^* y , S_n x\rangle| \leq \|(S_n^{-1})^* y\| \, \|S_n x\|$$

we have

$$\liminf_{n\to\infty} \frac{1}{n} \operatorname{Log} \|S_n x\| \geq -\limsup_{n\to\infty} \frac{1}{n} \operatorname{Log} \|(S_n^{-1})^* y\| \geq \gamma_{d-1} .$$

Remark 1.9. Since Osseledec's theorem states that there exists a vector $z(\omega)$ such that $\lim \frac{1}{n} \operatorname{Log} \|S_n(\omega) z(\omega)\| = \gamma_d$ (see e.g. Ledrappier [46]) we see that actually

$$\lim_{n\to\infty} \frac{1}{n} \operatorname{Log} \|S_n(\omega) Z(\omega)\| = \gamma_d .$$

Suppose now that for any $p = 1,..,d-1$, T_μ is p-contracting and p-strongly irreducible. Then one can show that with the notation of Corolarry 1.4, a.s. ,

$$\lim_{n\to\infty} \frac{1}{n} \operatorname{Log} \| M_n^*(N_\infty^*)^{-1} e_i \| = \gamma_i$$

$i = 1,...d$ (see Guivarc'h and Raugi [34] for a direct proof). The Osseledec's subspaces are thus described by the random matrix N_∞ .

Exercise 1.10. Show that the first part of Corolarry 1.4 is, for 2×2 matrices, a direct consequence of Corollary II.7.1 .

Exercise 1.11. Prove that if T_μ is strongly irreducible and contracting then there exists a random vector Z such that for any nonzero x in $\mathbb{R}^d$,

$$\lim_{n\to\infty} \frac{S_n^* S_n}{\|S_n x\|^2} = \frac{Z^* Z}{\langle Z,x\rangle^2} \qquad \text{a.s.}$$

Find the limit of $\dfrac{S_n^* S_n}{\|S_n\|^2}$.

VI.2. Limit theorems for the coefficients

We first prove the following important result of Guivarc'h and Raugi.

THEOREM 2.1. *Let* μ *be a probability measure on* $Gl(d,\mathbb{R})$ *such that* T_μ *is strongly irreducible and contracting. If for some* $\tau > 0$, $\int \exp \tau\, \ell(M)\, d\mu(M)$ *is finite then there exists* $\alpha > 0$ *such that the* μ-*invariant distribution* ν *on* $P(\mathbb{R}^d)$ *satisfies*

$$\sup_{\|y\|=1} \int \left\{\frac{\|x\|}{|\langle x,y\rangle|}\right\}^{\alpha} d\nu(\bar{x}) < \infty .$$

Proof : Let $B = P(\mathbb{R}^d) \times P(\Lambda^2 \mathbb{R}^d)$ and $\rho : Gl(d,\mathbb{R}) \times B \longrightarrow \mathbb{R}$ be the cocycle defined by

$$\rho(M,(\bar{x},\bar{w})) = \operatorname{Log} \frac{\|\Lambda^2 M w\| \, \|x\|^2}{\|w\| \, \|Mx\|^2}$$

for M in $Gl(d,\mathbb{R})$, x in $\mathbb{R}^d$ and w in $\Lambda^2\mathbb{R}^d$. We have

$$\underset{x,w}{\underline{\mathrm{Sup}}}\ |\rho(M,(\bar{x},\bar{w}))| \leq 2\ell(M) \ .$$

Moreover, for $M_n = Y_1 Y_2 \dots Y_n$,

$$\underset{x,w}{\underline{\mathrm{Sup}}}\ \mathbb{E}\{\rho(M_n^*,(\bar{x},\bar{w}))\} \leq \mathbb{E}\{\mathrm{Log}\ \|\Lambda^2 M_n^*\|\} - 2\ \inf_x \mathbb{E}(\mathrm{Log}\ \frac{\|M_n^* x\|}{\|x\|})$$

is negative for n large enough since $\gamma_1 > \gamma_2$ (see Corollary III.3.4). It thus follows from Lemma V.2.3 that there exists $\alpha > 0$ such that

$$\overline{\lim_{n\to\infty}}\ \frac{1}{n}\ \mathrm{Log}\ \underset{x,w}{\mathrm{Sup}}\ \mathbb{E}\{\frac{\|\Lambda^2\ M_n^* w\|\,\|x\|^2}{\|M_n^* x\|^2\ \|w\|}\}^\alpha < 0 \ .$$

Hence there exist $C > 0$ and β , $0 < \beta < 1$, such that for any unit vector x

$$\mathbb{E}\{\frac{\|\Lambda^2\ M_n^*\|}{\|M_n^* x\|^2}\}^\alpha \leq C\ \beta^n \ . \tag{7}$$

We may of course suppose that $\alpha < 1$. In this case we have, taking into account Proposition 1.3 and (6) ,

$$\begin{aligned}\mathbb{E}\{\frac{\|Z\|}{|\langle Z,e_1\rangle|}\}^\alpha &\leq \lim_{n\to\infty} \mathbb{E}\ \|\frac{s(M_n)e_1}{\langle s(M_n)e_1,e_1\rangle}\|^\alpha \\ &\leq \sum_{j=o}^{\infty} \mathbb{E}\{\frac{\|\Lambda^2\ M_j^*\|}{\|M_j^*\ e_1\|^2}\}^\alpha\ \mathbb{E}\{e^{2\alpha\ \ell(Y_{j+1})}\} \\ &\leq \mathbb{E}\ \{e^{2\alpha\ \ell(Y_1)}\}\ \frac{C}{1-\beta} \ .\end{aligned}$$

This is finite for α small enough. Now for any unit vector y in $\mathbb{R}^d$ choose an orthogonal matrix K such that $Ke_1 = y$. If we replace each Y_i by $K^{-1}Y_i K$, then M_n is replaced by $K^{-1}M_n K$ and Z by $K^{-1}Z$. Since the formula (7) is valid for any x we obtain as above that

$$\mathbb{E}\{\frac{\|K^{-1}Z\|}{|\langle K^{-1}Z,e_1\rangle|}\}^\alpha \leq \mathbb{E}\ \{e^{2\alpha\ \ell(K^{-1}Y_1K)}\}\ \frac{C}{1-\beta} \ .$$

But $\ell(Y_1) = \ell(K^{-1}Y_1K)$, $\|K^{-1}Z\| = \|Z\|$ and $\langle K^{-1}Z,e_1\rangle = \langle Z,y\rangle$. Therefore

$$\int \frac{\|x\|^\alpha}{|\langle x,y\rangle|^\alpha}\, d\nu(\bar{x}) = \mathbb{E}\{\frac{\|Z\|}{|\langle Z,y\rangle|}\}^\alpha \leq \mathbb{E}\ \{e^{2\alpha\ \ell(Y_1)}\}\ \frac{C}{1-\beta} \ .$$

The proof is complete.

As a consequence let us show

PROPOSITION 2.2. *Let* $S_n = Y_n \dots Y_1$ *where* Y_1, Y_2 *are independent matrices whose distribution* μ *satisfies the hypotheses of the above theorem. Then for any* $r > 0$ *and any nonzero* x, y *in* $\mathbb{R}^d$

$$\lim_{n\to\infty} \frac{1}{n^r} \operatorname{Log} \frac{|\langle S_n x, y\rangle|}{\|S_n x\|} = 0 \qquad \text{a.s.}$$

Proof : We shall show that for β large enough

$$\lim_{n\to\infty} \frac{1}{n^\beta} \frac{\|S_n x\|}{\langle S_n x, y\rangle} = 0 , \qquad \text{a.s.} , \qquad (8)$$

which implies the proposition. By the Borel-Cantelli lemma it suffices to prove that for any $\varepsilon > 0$

$$\sum_{n=1}^{\infty} \mathbb{P}\left(\left|\left\langle \frac{S_n x}{\|S_n x\|} , y\right\rangle\right| < \varepsilon\, n^{-\beta}\right) < \infty .$$

We shall first use the exponential convergence of the law of $\frac{S_n x}{\|S_n x\|}$ to ν. Define $f_n : [0,1] \longrightarrow \mathbb{R}$ by

$$f_n(t) = \begin{cases} 1 & \text{if } 0 \leq t \leq \varepsilon\, n^{-\beta} \\ 2 - t\, n^\beta \varepsilon^{-1} & \text{if } \varepsilon\, n^{-\beta} \leq t \leq 2\varepsilon\, n^{-\beta} \\ 0 & \text{if } 2\varepsilon\, n^{-\beta} \leq t \leq 1 . \end{cases}$$

It is easy to see that for all t, t' in $[0,1]$,

$$|f_n(t) - f_n(t')| \leq |t-t'|\, n^\beta \varepsilon^{-1} .$$

Therefore if we define $\phi_n : P(\mathbb{R}^d) \longrightarrow \mathbb{R}$ by

$$\phi_n(\bar{x}) = f_n\left(\left|\left\langle \frac{x}{\|x\|} , y\right\rangle\right|\right)$$

we have for $\|x\| = \|x'\| = 1$

$$|\phi_n(\bar{x}) - \phi_n(\bar{x}')| \leq \frac{(|\langle x,y\rangle| - |\langle x',y\rangle|)n^\beta}{\varepsilon} \leq \frac{2n^\beta}{\varepsilon}\, \delta(\bar{x},\bar{x}')$$

(see the beginning of the proof of Lemma V.4.2). This entails that, with the notations of Chapter V,

$$\|\phi_n\|_\alpha = \|\phi_n\|_\infty + m_\alpha(\phi_n) \leq \frac{2n^\beta}{\varepsilon} + 1$$

for any α , $0 < \alpha < 1$. Taking into account Theorem V.2.5 we obtain, for α small enough,

$$\sum_{n=1}^{\infty} \mathbb{P}(|<\frac{S_n x}{\|S_n x\|}, y>| < \varepsilon n^{-\beta}) \leq \sum_n \mathbb{E}\{\phi_n(S_n \cdot \bar{x})\}$$

$$\leq \sum_n |\mathbb{E}\{\phi_n(S_n \cdot \bar{x})\} - \int \phi_n \, d\nu| + \sum_n \int \phi_n \, d\nu$$

$$\leq \sum_n \|P^n - N\|_\alpha \, \|\phi_n\|_\alpha + \sum_n \int \phi_n \, d\nu$$

$$\leq C + \sum_n \int \phi_n \, d\nu$$

for some $C > 0$.

But $$\sum_{n=1}^{\infty} \int \phi_n \, d\nu \leq \sum_n \nu\{\bar{x} ; |<\frac{x}{\|x\|}, y>| \leq 2\varepsilon \, n^{-\beta}\}$$

$$\leq \sum_n \nu\{\bar{x} ; \{\frac{\|x\|}{|<x,y>|}\}^{1/\beta} \geq \frac{n}{(2\varepsilon)^{1/\beta}}\}$$

is finite as soon as

$$\int \{\frac{\|x\|}{|<x,y>|}\}^{1/\beta} \, d\nu(\bar{x}) < \infty .$$

This holds for β sufficiently large by Theorem 2.1 . For such β (8) holds.

<u>COROLLARY 2.3</u>. *Let* $S_n = Y_n \ldots Y_1$ *where* $Y_1, Y_2, \ldots$ *are independent random matrices with common distribution* μ . *Suppose that* T_μ *is strongly irreducible and contracting and that for some* $\tau > 0$, $\mathbb{E}\{\exp \tau \, \ell(Y_1)\}$ *is finite. Then there exists* $a > 0$ *such that*

(i) For any nonzero vectors x, y ,

$$\lim_{n \to +\infty} \frac{1}{n} \operatorname{Log} |<S_n x, y>| = \gamma$$

and $\frac{1}{\sqrt{n}} \{\operatorname{Log}|<S_n x, y>| - n\gamma\}$ *converges in distribution to the gaussian law* $N(0,a^2)$.

(ii) If $(S_n)_{i,j}$, $1 \leq i, j \leq d$ *are the coefficients of the matrix* S_n, then the $\mathbb{R}^{d^2}$*-valued random vector*

$$\{\frac{1}{\sqrt{n}} (\operatorname{Log} |(S_n)_{i,j}| - n\gamma) , \ 1 \leq i, j \leq d\}$$

converges in distribution to a random vector X *which satisfies* $X_{i,j} = X_{k,\ell}$ *for any* $1 \leq i,j,k,\ell \leq d$ *and such that* X_{11} *is distributed according to* $N(0,a^2)$.

Proof : The statements in (i) are direct consequences of Proposition 2.2, the law of large numbers for $\frac{1}{n}$ Log $\|S_n x\|$ (see Corollary III.3.4) and the central limit theorem (Theorem V.5.1). In order to prove (ii), taking into account the convergence in distribution of $\frac{1}{\sqrt{n}}(\mathrm{Log}|(S_n)_{1,1}| - n\gamma)$ to $N(0,a^2)$, it suffices to show that, for any $1 \leq i, j \leq d$

$$\frac{1}{\sqrt{n}} \mathrm{Log}\ |(S_n)_{1,1}| - \frac{1}{\sqrt{n}} \mathrm{Log}\ |(S_n)_{i,j}|$$

converges to zero in probability. It follows from Proposition 2.2 that a.s. ,

$$\lim \frac{1}{\sqrt{n}} (\mathrm{Log}|(S_n)_{1,1}| - \mathrm{Log}|(S_n)_{1,j}|) =$$
$$\lim \frac{1}{\sqrt{n}} (\mathrm{Log}\ \frac{|\langle S_n e_1, e_1\rangle|}{\|S_n\ e_1\|} - \mathrm{Log}\ \frac{|\langle S_n e_1, e_j\rangle|}{\|S_n\ e_1\|} = 0 \quad .$$

On the other hand

$$\mathrm{Log}|(S_n)_{1,j}| - \mathrm{Log}|(S_n)_{i,j}| = \mathrm{Log}\ \frac{|\langle S_n e_1, e_j\rangle|}{\|S_n^*\ e_j\|} - \mathrm{Log}\ \frac{|\langle S_n e_i, e_j\rangle|}{\|S_n^*\ e_j\|}$$
$$= \mathrm{Log}\ |\langle e_1 , \frac{S_n^* e_j}{\|S_n^* e_j\|}\rangle| - \mathrm{Log}|\langle e_i, \frac{S_n^* e_j}{\|S_n^* e_j\|}\rangle|$$

converges in probability to an a.s. finite limit (this follows from the fact, seen in Theorem III.4.3, that $S_n^* \cdot \bar{e}_j$ converges in probability to a random direction $\bar{Z}$ satisfying $\mathbb{P}(\langle Z,x\rangle = 0) = 0$ for any nonzero x). Therefore

$$\frac{1}{\sqrt{n}} (\mathrm{Log}\ |(S_n)_{1,1}| - \mathrm{Log}\ |(S_n)_{i,j}|) \quad \text{converges to 0 in probability.}$$

Exercise 2.4. Suppose that T_μ^{-1} is strongly irreducible and contracting and that for some $\tau > 0$, $\int \exp \tau\, \ell(M)\, d\mu(M)$ is finite. Show that for any nonzero x , $\frac{1}{n}$ Log $\|S_n^{-1} x\|$ converges almost surely to $-\gamma_d$.

VI.3. Behaviour of the rows

As the preceding results, the following is due to Guivarc'h and

Raugi (see [59]). We use the notation 1.2 .

THEOREM 3.1. *Suppose that* T_μ *is strongly irreducible and contracting and that* $\int \exp \tau\, \ell(M)\, d\mu(M)$ *is finite for some* $\tau > 0$ *. Then for any* $\bar{x}$ *in* $P(\mathbb{R}^d)$, $Y_1 \ldots Y_n \cdot \bar{x}$ *converges to* $\bar{Z}$ *a.s.* .

Proof : We first show that $Y_1 \ldots Y_n \cdot \bar{e}_1$ converges to $\bar{Z}$. Using the notations of the first section we have

$$Y_1 \ldots Y_n e_1 = M_n e_1 = s(M_n)\, k(M_n) e_1 = \sum_{i=1}^{d} \langle k(M_n)e_1, e_i\rangle\, s(M_n) e_i \ .$$

Hence

$$\frac{M_n e_1}{\langle M_n e_1, e_1\rangle} = \frac{s(M_n)e_1}{\langle s(M_n)e_1, e_1\rangle} + \sum_{i=2}^{d} \frac{\langle k(M_n)e_1, e_i\rangle\, s(M_n)e_i}{\langle M_n e_1, e_1\rangle} \ .$$

Taking into account (5) and (i) of Corollary 2.3 we see that, for $i = 2,\ldots,d$,

$$\overline{\lim_{n\to\infty}} \frac{1}{n} \operatorname{Log} \frac{\|\langle k(M_n)e_1, e_i\rangle\, s(M_n)e_i\|}{|\langle M_n e_1, e_1\rangle|}$$

$$\leq \overline{\lim_{n\to\infty}} \frac{1}{n} \operatorname{Log} \frac{\|s(M_n)e_i\|}{|\langle M_n e_1, e_1\rangle|}$$

$$\leq \overline{\lim_{n\to\infty}} \frac{1}{n} \operatorname{Log} \frac{\|\Lambda^2 M_n^*\|}{\|M_n^* e_1\|\, |\langle e_1, M_n^* e_1\rangle|}$$

$$\leq (\gamma_1 + \gamma_2) - 2\gamma_1 < 0 \ .$$

Therefore a.s.

$$\lim_{n\to\infty} \frac{M_n e_1}{\langle M_n e_1, e_1\rangle} = \lim_{n\to\infty} \frac{s(M_n)e_1}{\langle s(M_n)e_1, e_1\rangle} = \frac{Z}{\langle Z, e_1\rangle}$$

(see Proposition 1.3) and $M_n \cdot \bar{e}_1$ converges to $\bar{Z}$.

Now for any $\bar{x}$ in $P(\mathbb{R}^d)$ choose an orthogonal matrix K such that $K \cdot \bar{x} = \bar{e}_1$. Applying the latter result to the matrices $K^{-1} Y_i K$ we deduce that $K^{-1} M_n K \cdot \bar{e}_1$ converges a.s. to $K^{-1} \cdot \bar{Z}$ so that $M_n \cdot \bar{x}$ converges to $\bar{Z}$.

When we apply this theorem to the sequence $Y_1^*, Y_2^*, \ldots$ we find that if $S_n = Y_n \ldots Y_1$ then $S_n^* \cdot \bar{x}$ converges a.s. to some direction $\bar{Z}$ (this improves Theorem III.4.3). Choosing $\bar{x}$ equal to some basis vector $\overline{e}_i$ we see that the directions of the rows of S_n converge a.s. to the same limit.

VI.4. Regularity of the invariant measure

We shall now see, following Guivarc'h and Raugi, that although the μ-invariant distribution ν is often singular, it always has some regularity.

PROPOSITION 4.1. *Suppose that* T_μ *is contracting, strongly irreducible and that, for some* $\tau > 0$, $\int \exp \tau \ell(M)\, d\mu(M)$ *is finite. Then there exists* $\alpha > 0$ *such that the* μ*-invariant distribution* ν *on* $P(\mathbb{R}^d)$ *satisfies*

$$\sup_{\overline{y} \in P(\mathbb{R}^d)} \int \delta(\overline{x},\overline{y})^{-\alpha}\, d\nu(\overline{x}) < \infty .$$

Proof : If y and u are two orthogonal unit vectors, we have, for any x

$$\|x \wedge y\|^2 = \|x\|^2 - \langle x,y\rangle^2 \geqq \langle x,u\rangle^2 .$$

Hence

$$\delta(\overline{x},\overline{y})^{-\alpha} = \left\{\frac{\|x\|}{\|x \wedge y\|}\right\}^\alpha \leqq \left\{\frac{\|x\|}{|\langle x,u\rangle|}\right\}^\alpha .$$

Since, by Theorem 1.2, there exists some $\alpha > 0$ such that

$$\sup_{\|u\|=1} \int \left\{\frac{\|x\|}{|\langle x,u\rangle|}\right\}^\alpha d\nu(\overline{x}) < \infty ,$$

the proposition is clear .

COROLLARY 4.2. *Under the above hypotheses, there exist* $\alpha > 0$ *and* $C > 0$ *such that for any* $\overline{x}$ *in* $P(\mathbb{R}^d)$ *and* $\varepsilon > 0$

$$\nu(\{\overline{y}\ ;\ \delta(\overline{x},\overline{y}) \leqq \varepsilon\}) \leqq C\, \varepsilon^\alpha .$$

Proof : This follows readily from the fact that if B is the ball with center $\overline{x}$ and radius ε ,

$$\nu(B) = \int_B \frac{\delta(\overline{x},\overline{y})^\alpha}{\delta(\overline{x},\overline{y})^\alpha}\, d\nu(\overline{y}) \leqq \varepsilon^\alpha \int \delta(\overline{x},\overline{y})^{-\alpha}\, d\nu(\overline{y}) .$$

As a consequence let us prove that the dimension of ν is positive. First recall the definition of the Hausdorff dimension of a subset A of the metric space $(P(\mathbb{R}^d), \delta)$. The reader is referred to Billingsley [5] for details. For any $\beta > 0$ and $\varepsilon > 0$ let

$$\ell_\beta(A,\varepsilon) = \inf \sum_i (\mathrm{diam}\, B_i)^\beta$$

where the infinimum extends over all countable coverings of A by closed balls B_i of diameter less than ε, and let

$$\ell_\beta(A) = \lim_{\varepsilon\to 0} \ell_\beta(A,\varepsilon) .$$

The Hausdorff dimension of A is defined by

$$\dim A = \sup\{\beta : \ell_\beta(A) = \infty\} = \inf\{\beta : \ell_\beta(A) = 0\} .$$

The dimension of ν is

$$\dim \nu = \inf\{\dim A \,;\, A \text{ is a Borel subset such that } \nu(A) \neq 0\}.$$

Now if $\nu(A) \neq 0$ and if (B_i) is such a covering we have

$$\nu(A) \leqq \Sigma\, \nu(B_i) \leqq C\, \Sigma\, (\mathrm{diam}\, B_i)^\alpha$$

so that $\ell_\alpha(A) \geqq \nu(A)$ and $\dim A \geqq \alpha$. This proves that $\dim \nu \geqq \alpha > 0$.

According to Ledrappier [46] one may conjecture that, for 2×2 matrices,

$$\dim \nu = -\frac{1}{2\gamma} \int \mathrm{Log}\, \frac{d(g^{-1}\nu)}{d\nu}(\bar{x})\, d\nu(\bar{x})\, d\mu(g) \qquad (9)$$

(This is actually proved in [46] but for a weaker notion of dimension).

We have seen (see II.7.2) that $Sl(2,\mathbb{R})$ acts on the upper half plane $H = \{z \in \mathbb{C} \,;\, \mathrm{Im}\, z > 0\}$ by the formula

$$g\cdot z = \frac{az + b}{cz + d}$$

when $g = \begin{bmatrix} a & b \\ c & d \end{bmatrix}$. Also that the extended action on the real axis is just the action of $Sl(2,\mathbb{R})$ on $P(\mathbb{R}^2)$ if we identify $P(\mathbb{R}^2)$ with $\mathbb{R} \cup \{\infty\}$. If G_μ is the closed subgroup generated by the support of μ, then ν is carried by the so-called limit set L of G_μ, i.e.

$$L = \{x \in \mathbb{R} \cup \{\infty\} \,;\, \exists (g_n) \in G_\mu ,\ \lim_{n\to\infty} g_n\cdot i = x\} .$$

(This follows from the fact that L is closed and G_μ-invariant or directly from Corollary II.7.2 which asserts that $Y_1 \ldots Y_n \cdot i$ converges to a random variable with law ν) .

It is known that the dimension of L may be arbitrarily small (see e.g. Beardon [3]). In particular ν may be (and is often) singular.

It is worth rephrasing the above corollary in the context of projective transformations.

COROLLARY 4.3. *Let* μ *be a probability on* $Gl(2,\mathbb{R})$ *such that* T_μ *is strongly irreducible and contracting. If for some* $\tau > 0$, $\int e^{\tau \ell(M)} d\mu(M)$ *is finite and if* ν *is the probability measure on* $\mathbb{R}$ *which satisfies*

$$\int f\left(\frac{ax+b}{cx+d}\right) d\mu(a,b,c,d)\, d\nu(x) = \int f(x)\, d\nu(x)$$

for any bounded continuous function f *on* $\mathbb{R}$, *then there exist* $\alpha > 0$ *and* $C > 0$ *such that*

(i) For any y *in* $\mathbb{R}$

$$\int |x-y|^{-\alpha} d\nu(x) \leq C .$$

(ii) $\int (1+x^2)^{\alpha/2} d\nu(x) \leq C$.

In particular the distribution function $F(x) = \nu(-\infty,x]$ *is Hölder continuous.*

Proof : The mapping $\Psi : \mathbb{R} \cup \{\infty\} \longrightarrow P(\mathbb{R}^2)$ defined by $\Psi(x) = \overline{x e_1 + e_2}$ and $\Psi(\infty) = \bar{e}_1$ sends ν onto the unique μ-invariant measure on $P(\mathbb{R}^2)$. Therefore, by Proposition 4.1 , there exist $C > 0$ and $\alpha > 0$ such that for any real y

$$\int_{\mathbb{R}} \delta(\Psi(y),\Psi(x))^{-\alpha} d\nu(x) \leq C$$

and

$$\int_{\mathbb{R}} \delta(\psi(\infty),\Psi(x))^{-\alpha} d\nu(x) \leq C$$

But $\delta(\Psi(y),\Psi(x)) = \dfrac{\|(x e_1+e_2) \wedge (y e_1 + e_2)\|}{\|x e_1 + e_2\| \; \|y e_1 + e_2\|}$

$$= \frac{|x-y|}{\sqrt{1+x^2}\,\sqrt{1+y^2}}$$

and $\delta(\Psi(\infty),\Psi(x)) = \dfrac{\|(xe_1+e_2)\wedge e_1\|}{\sqrt{1+x^2}} = \dfrac{1}{\sqrt{1+x^2}}$.

The proposition follows immediatly.

Finally we notice that the probability measure ν satisfies a law of pure type with respect to the natural measure m_d . Recall that m_d is the unique rotation-invariant distribution on $P(\mathbb{R}^d)$, (see II.5.1). On the projective line $\mathbb{R}\cup\{\infty\}$, m_2 can be identified with the Cauchy distribution and is equivalent to the Lebesgue measure.

PROPOSITION 4.4. *Let* μ *be a probability measure on* $Gl(d,\mathbb{R})$ *which has only one* μ*-invariant distribution* ν *on* $P(\mathbb{R})^d)$. *Then*

-either ν *is absolutely continuous w.r.t.* m_d ,

- or ν *and* m_d *are mutually singular* ,

- or ν *is purely discrete.*

Proof : Write ν_1 (resp. ν_2) for the discrete (resp. continuous) component of ν . It is readily seen that $\mu \overset{\bullet}{*} \nu_2$ is continuous. Therefore the equality

$$\nu = \mu \overset{\bullet}{*} \nu = \mu \overset{\bullet}{*} \nu_1 + \mu \overset{\bullet}{*} \nu_2$$

entails that $\nu_2 \geq \mu \overset{\bullet}{*} \nu_2$. But ν_2 and $\mu \overset{\bullet}{*} \nu_2$ have the same mass, thus $\nu_2 = \mu \overset{\bullet}{*} \nu_2$. By uniqueness either $\nu = \nu_2$ or $\nu = \nu_1$.

Suppose now that ν is continuous and write $\nu_{a.c}$ (resp. ν_s) for its absolutely continuous (resp. singular) component w.r.t. m_d . Since for any M in $Gl(d,\mathbb{R})$ Mm_d is equivalent to m_d (see II.5.1) ,

$$\mu \overset{\bullet}{*} \nu_{a.c} = \int M\,\nu_{a.c}\, d\mu(M)$$

is absolutely continuous, entailing as above that $\mu \overset{\bullet}{*} \nu_{a.c} = \nu_{a.c}$. By uniqueness of the μ-invariant measure either ν is absolutely continuous or ν is singular.

Example 4.5. It was already noticed by Kaijser in [38] that is it easy to exhibit singular invariant measures. Consider for instance two non negative 2×2 matrices

$$M_1 = \begin{bmatrix} a_1 & b_1 \\ c_1 & d_1 \end{bmatrix} \qquad M_2 = \begin{bmatrix} a_2 & b_2 \\ c_2 & d_2 \end{bmatrix}$$

such that $|\det M_i| = 1$, $c_i > 0$ and $d_i^2 > 2$ for $i = 1,2$. Let us show that if μ is the Bernoulli measure defined by

$$\mu(\{M_1\}) = \mu(\{M_2\}) = 1/2$$

then the μ-invariant distribution ν on $P(\mathbb{R}^2)$ is singular.
We shall consider ν as a measure on the projective line $\mathbb{R} \cup \{\infty\}$ and prove that the Lebesgue measure of the support S of ν is zero.

For $i = 1,2$ and x in $\mathbb{R} \cup \{\infty\}$ set

$$\phi_i(x) = M_i \cdot x = \frac{a_i x + b_i}{c_i x + d_i} \quad .$$

It is readily seen that if $\alpha = \mathrm{Max}(\frac{a_1}{c_1}, \frac{a_2}{c_2}, \frac{b_1}{d_1}, \frac{b_2}{d_2})$ then $\phi_i([0,\alpha])$ is contained in $[0,\alpha]$. Therefore $[0,\alpha]$ is invariant under T_μ , implying that S is contained in $[0,\alpha]$. Now, since for any $x \geqq 0$

$$|\phi_i'(x)| = \frac{1}{(c_i x + d_i)^2} \leqq \frac{1}{d_i^2}$$

we see that if $\rho = \mathrm{Min}(d_1^2, d_2^2)$ then

$$(\phi_{i_1} \circ \phi_{i_2} \circ \cdots \circ \phi_{i_n}) \; [0,\alpha]$$

is an interval of length smaller than $\rho^{-n} \alpha$. Hence if

$$A = \bigcup \{(\phi_{i_1} \circ \cdots \circ \phi_{i_n}) [0,\alpha] \; ; \; i_1, \ldots, i_n = 1,2\}$$

then the Lebesgue measure of A is smaller than $2^n \rho^{-n} \alpha$. But

$$\nu(A) = (\mu^n \overset{\bullet}{*} \nu) = \frac{1}{2^n} \sum_{i_1,\ldots,i_n} \nu(\{\phi_{i_1} \circ \cdots \circ \phi_{i_n}\}^{-1}(A)) = 1$$

so that S is contained in A . This shows that for any $n > 0$, $m(A) \leqq 2^n \rho^{-n} \alpha$. Since $\rho > 2$, $m(A) = 0$.

Exercise 4.6. Let μ be a probability measure on $Gl(d,\mathbb{R})$ which has a unique invariant distribution ν on $P(\mathbb{R}^d)$. Show that the support of ν is the smallest closed subset S of $P(\mathbb{R}^d)$ which satisfies $M \cdot (S) \subset S$ for any M in T_μ .

Exercise 4.7. (a) Suppose that μ is a probability measure on $Gl(d,\mathbb{R})$ (or $Sl(d,\mathbb{R})$) which has a bounded density with compact support w.r.t.

the Haar measure. Show that ν has a bounded density.

(b) Deduce that if μ is not singular then ν is absolutely continuous.

Exercise 4.8. Show that the Cantor distribution is μ-invariant for some non-irreducible probability measure μ on $Gl(2,\mathbb{R})$.

VI.5. An example : Random continued fractions

(5.1) Let $\{A_n, n \geq 1\}$ be a sequence of independent random positive integers with a common distribution p , defined on some $(\Omega, \mathcal{A}, \mathbb{P})$. Consider the 2×2 matrices

$$Y_n = \begin{bmatrix} 0 & 1 \\ 1 & A_n \end{bmatrix}$$

and call μ their distribution. For any x in the projective line $\mathbb{R} \cup \{\infty\}$,

$$Y_1 Y_2 \dots Y_n \cdot x = \cfrac{1}{A_1 + \cfrac{1}{A_2 + \cdot \, \cdot + \cfrac{1}{A_n + x}}}$$

Notice that $Y_1 \dots Y_n \cdot 0$ converges a.s. to a real random variable Z with values in $[0,1]$. The distribution ν of Z is the μ-invariant distribution on the projective line. This follows readily from the fact that the A_n's are positive integers and from the elementary properties of continued fractions (see e.g. Billingsley [5]). (Much less obvious is the fact that this also holds for arbitrary real i.i.d. A_n under an integrability assumption, which is an immediate consequence of Theorem 3.1).

Let us show that ν is singular as soon as the A_i's are not a.s. constant (in this case T_μ is strongly irreducible and contracting, cf. Exercise 5.4). This result is due to Chatterji [12].

First recall that any real number t in $(0,1)$ may be expanded in a continued fraction, i.e.

$$t = \lim_{n\to\infty} (a_1(t) \; ; \; a_2(t) \; ; \; \dots \; ; \; a_n(t))$$

where we define inductively

$$(a_1) = \frac{1}{a_1} , \quad (a_1;\ldots;a_n) = \frac{1}{a_1+(a_2;\ldots;a_n)} .$$

When t is irrational this expansion is unique. In particular since ν is continuous, the expansion is ν-a.s. unique. As a consequence the law of $A_1(\omega),\ldots,A_n(\omega)$ under $\mathbb{P}$ is the law of $a_1(t),\ldots,a_n(t)$ under ν.

Consider now the mapping $T : (0,1) \longrightarrow (0,1)$ defined by

$$T(t) = \frac{1}{t} - \left[\frac{1}{t}\right] .$$

Since $(a_1(T(t))\ ;\ a_2(T(t))\ ;\ a_3(T(t))\ ;\ \ldots) = (a_2(t)\ ;\ a_3(t)\ ;\ \ldots)$, T is equivalent to the shift on the sequences of integers. The 0-1 law thus implies that ν is invariant and ergodic under T. But it is well known that the measure $(1+x)^{-1}\,dx$ on $(0,1)$ is also ergodic under T (see e.g. Billingsley). Therefore either ν is singular w.r.t. the Lebesgue measure or $\nu = 1_{(0,1)}(\text{Log } 2)^{-1}\,(1+x)^{-1}\,dx$. But $a_1(t),\ a_2(t),\ \ldots$ are not independent under this latter distribution (elementary computations). This proves that ν is singular.

(5.2) Let us outline a proof of Ledrappier's formula (9) in this simple situation, which is in this case due to Kinney and Pitcher [43]. Let

$$J_n(\omega) = Y_1(\omega)\ldots Y_n(\omega)\cdot[0,1] .$$

It is readily seen that we may write

$$M_n = Y_1\ldots Y_n = \begin{bmatrix} p_{n-1} & p_n \\ q_{n-1} & q_n \end{bmatrix}$$

where $q_n = A_n q_{n-1} + q_{n-2}$. The length $|J_n(\omega)|$ of $J_n(\omega)$ satisfies

$$|J_n(\omega)| = |M_n(\omega)\cdot 0 - M_n(\omega)\cdot 1| = \{q_{n-1}(\omega)(q_n(\omega)+q_{n-1}(\omega))\}^{-1}$$

But, since $q_{n-1} \leqq q_n$,

$$q_{n-2}^2 + q_{n-1}^2 \leqq q_{n-1}(q_n + q_{n-1}) \leqq q_{n-1}^2 + q_n^2 .$$

Therefore, if $\mathbb{E}(\text{Log } A_1)$ is finite and if γ is the Lyapunov exponent associated with (Y_n), then

$$\lim_{n\to\infty} \frac{1}{n} \text{Log } |J_n(\omega)| = -\lim_{n\to\infty} \frac{1}{n} \text{Log } \|M_n^*(\omega)e_2\|^2 = -2\gamma \qquad \mathbb{P} \text{ a.s.}$$

On the other hand

$$\nu(J_n(\omega)) = \nu(\{t \in (0,1)\ ;\ a_1(t) = A_1(\omega),\dots,a_n(t) = A_n(\omega)\})$$
$$= \prod_{i=1}^{n} p(\{A_i(\omega)\}) .$$

(Recall that under ν , $a_1(t),\dots,a_n(t)$ are independent r.v. distributed according to p). Making use of the law of large numbers, if $p(i) = p_i$,

$$\lim_{n\to\infty} \frac{1}{n} \operatorname{Log} \nu(J_n(\omega)) = \mathbb{E}(\operatorname{Log} p(A_1))$$
$$= \sum_{i=1}^{\infty} p_i \operatorname{Log} p_i \qquad \mathbb{P} \text{ a.s.}$$

(Note that since $\operatorname{Log} x \leq x-1$ and since $\Sigma p_i \operatorname{Log} i = \mathbb{E}(\operatorname{Log} A_1)$ is finite

$$|\Sigma p_i \operatorname{Log} p_i| = \Sigma p_i \operatorname{Log} \frac{1}{i^2 p_i} + 2\Sigma p_i \operatorname{Log} i \leq (\Sigma \frac{1}{i^2})-1 + 2\,\mathbb{E}(\operatorname{Log} A_1)$$

is also finite).

Therefore

$$\lim_{n\to\infty} \frac{\operatorname{Log} \nu(J_n(\omega))}{\operatorname{Log} |J_n(\omega)|} = -\frac{\Sigma p_i \operatorname{Log} p_i}{2\gamma} \qquad \mathbb{P} \quad \text{a.s.} .$$

In other words if for any t in $(0,1)$ we set

$$I_n(t) = \{s \in (0,1)\ ;\ a_1(s) = a_1(t),\dots,a_n(t) = a_n(s)\}$$

then

$$\lim_{n\to\infty} \frac{\operatorname{Log} \nu(I_n(t))}{\operatorname{Log} |I_n(t)|} = -\frac{\Sigma p_i \operatorname{Log} p_i}{2\gamma} \qquad \nu\text{-a.s.} .$$

Onecan deduce from this formula (see Kenney and Pitcher [43]) that

$$\dim \nu = -\frac{\Sigma p_i \operatorname{Log} p_i}{2\gamma} .$$

In order to recover Ledrappier's formula (9) it suffices to note that if $M = \begin{pmatrix} 0 & 1 \\ 1 & a \end{pmatrix}$ then

$$\nu(M\cdot I_n(t)) = p(a) \prod_{i=1}^{n} p(a_i(t)) = p(a)\, \nu(I_n(t))$$

(use the formula $\nu(M\cdot I_n(t)) = \nu(\{s\ ;\ a_1(s) = a,\ a_2(s) = a_1(t),\ \dots,$ $a_{n+1}(s) = a_n(t)\}))$.

Thus $\frac{d(M^{-1}\nu)}{d\nu} = p(a)$ on $(0,1)$ and

$$\int \operatorname{Log} \frac{d(M^{-1}\nu)}{d\nu}\, d\nu\, d\mu(M) = \Sigma p_i \operatorname{Log} p_i .$$

(5.3) An interesting example of such a measure is the Minkowski measure denoted by ? on (0,1). We may define it as the μ-invariant distribution on $\mathbb{R} \cup \{\infty\}$ where μ is defined by

$$\mu\begin{pmatrix}0 & 1\\ 1 & n\end{pmatrix} = \frac{1}{2^n} \quad , \qquad n \geq 1 \ .$$

In order to describe it let us notice that if $\tilde{\mu}$ is the Bernoulli measure which verifies $\tilde{\mu}(A) = \tilde{\mu}(B) = 1/2$ for

$$B = \begin{pmatrix}0 & 1\\ 1 & 1\end{pmatrix} \quad , \qquad B = \begin{pmatrix}1 & 0\\ 1 & 1\end{pmatrix}$$

then ? is also $\tilde{\mu}$-invariant. (This is proved as follows : if ν is $\tilde{\mu}$-invariant then

$$\frac{1}{2}(\delta_A + \delta_B) \dot{*} \nu = \nu$$

and
$$\nu = \frac{1}{2}\,\delta_A \dot{*} \nu + \frac{1}{2}\,\delta_B * (\frac{1}{2}\,\delta_A + \frac{1}{2}\,\delta_B) \dot{*} \nu$$
$$= \frac{1}{2}\,\delta_A \dot{*} \nu + \frac{1}{4}\,\delta_{BA} \dot{*} \nu + \frac{1}{4}\,\delta_{B^2} \dot{*} \nu \ .$$

By induction we have for any $n > 0$,

$$\nu = \sum_{m=1}^{n} \frac{1}{2^m}\,\delta_{B^{m-1}A} \dot{*} \nu + \frac{1}{2^n}\,\delta_{B^n} \dot{*} \nu$$

and, going to the limit ,

$$\nu = \sum_{m=1}^{\infty} \frac{1}{2^m}\,\delta_{B^{m-1}A} \dot{*} \nu \ .$$

But $\mu = \sum_{m=1}^{\infty} \frac{1}{2^m}\,\delta_{B^{m-1}A}$, hence $\nu = ?$ by uniqueness of the invariant measure).

We now define inductively the Brocot sequence of order n , $\frac{p_o(n)}{q_o(n)} \leq \frac{p_1(n)}{q_1(n)} \leq \ldots \leq \frac{p_{2^n}(n)}{q_{2^n}(n)}$ as follows.

For $n = 0$, $\frac{p_o(0)}{q_o(0)} = \frac{0}{1}$ and $\frac{p_1(0)}{q_1(0)} = \frac{1}{1}$. The sequence of order $n + 1$ is obtained by adding to the sequence of order n the fractions

$$\frac{p_i(n) + p_{i+1}(n)}{q_i(n) + q_{i+1}(n)} \quad , \qquad i = 0, \ldots, 2^n-1 \ .$$

For instance the first three Brocot sequences are

order 0 $\quad \frac{0}{1} \qquad\qquad\qquad \frac{1}{1}$

order 1 $\quad \frac{0}{1} \qquad \frac{0+1}{1+1} = \frac{1}{2} \qquad \frac{1}{1}$

order 2 $\quad \frac{0}{1} \quad \frac{0+1}{1+2} = \frac{1}{3} \quad \frac{1}{2} \quad \frac{1+1}{2+1} = \frac{1}{3} \quad \frac{1}{1}$.

If M is a 2×2 matrix such that $|\det M| = 1$, $M \cdot 0 = \frac{p}{q}$ and $M \cdot 1 = \frac{p'}{q'}$ it follows from Bezout's theorem that

$$M = \begin{pmatrix} p'-p & p \\ q'-q & q \end{pmatrix}.$$

Whence, since

$$A \cdot x = \frac{1}{1+x} \quad \text{and} \quad B \cdot x = \frac{x}{x+1},$$

$MA \cdot 0 = \frac{p'}{q'}$, $MB \cdot 0 = \frac{p}{q}$, $MA \cdot 1 = MB \cdot 1 = \frac{p+q}{p'+q'}$. Making use of this result one sees immediately that if $Y_1, \ldots, Y_n$ are in $\{A,B\}$ then

$$Y_1 \ldots Y_n \cdot [0,1] = \left[\frac{p_i(n)}{q_i(n)}, \frac{p_{i+1}(n)}{q_{i+1}(n)}\right]$$

for some $0 \leqq i < 2^n$. Since there exists only one sequence $Y_1, \ldots, Y_n$ (with $Y_i = A$ or B) which satisfies this relation we conclude that

$$?\left(\left[\frac{p_i(n)}{q_i(n)}, \frac{p_{i+1}(n)}{q_{i+1}(n)}\right]\right) = (\tilde{\mu}^n * ?)\left(\left[\frac{p_i(n)}{q_i(n)}, \frac{p_{i+1}(n)}{q_{i+1}(n)}\right]\right) = \frac{1}{2^n}.$$

Obviously this formula defines $?$ uniquely.

Related examples may be found in Chassaing, Letac and Mora [11]. Exercise 5.5 is also taken from this paper (see also Letac and Seshadri [51]). It gives a non-trivial example where the Lyapunov exponent is explicitely computed.

<u>Exercise 5.4</u>. Let A be a not a.s. constant real random variable, and μ be the distribution of

$$Y = \begin{pmatrix} 0 & 1 \\ 1 & A \end{pmatrix}$$

a. Show that T_μ is strongly irreducible and contracting.

b. Let ν be the μ-invariant distribution on the projective line $\mathbb{R} \cup \{\infty\}$. Prove that the Lyapunov exponent γ associated with μ is equal to $\int \operatorname{Log} \frac{1}{x}\, d\nu(x)$, when there exists some $\tau > 0$ such that $\mathbb{E}\{|A|^\tau\}$ is finite. Show that this is also true when A is posicive-integer-valued, as soon as $\mathbb{E}(\operatorname{Log}|A|)$ is finite.

<u>Exercise 5.5</u>. [11] [51]. For $a > 0$, let λ_a and ν_a be the two distributions on $\mathbb{R}^+$ defined by

$$\lambda_a(dx) = \frac{a}{2} \exp - \frac{ax}{2} \; 1_{(0,\infty)}\, dx$$

$$\nu_a(dx) = \frac{1}{2K_1(a)} \exp - \frac{a}{2}\left(x + \frac{1}{x}\right) 1_{(0,\infty)}\, dx$$

where $K_r(a) = \frac{1}{2}\int_0^\infty u^{r-1} \exp - \frac{a}{2}\left(u + \frac{1}{u}\right) du$.

a. Consider two independent random variables X and A such that the law of X (resp. A) is ν_a (resp. λ_a) . Show that the law of $(X + A)^{-1}$ is ν_a . (Hint : compute the Laplace transforms).

b. Show that the Lyapunov exponent γ associated with the distribution of $\begin{pmatrix} 0 & 1 \\ 1 & A \end{pmatrix}$ is equal to $\frac{2K_o(a)}{aK_1(a)}$.

SUGGESTIONS FOR FURTHER READINGS

We provide some recent references on limit theorems for random matrices and related topics. We make no claim for completeness but the quoted papers often contain a large bibliography.

(1) Lyapunov exponent for stationary sequences

The main properties of Lyapunov exponents in the stationary case can be found in Ledrappier [46] . Guivarc'h [32] (see also Royer [62] , Virtser [69], [70]) gives a criterion ensuring that two given exponents are distinct for markovian products. See Ledrappier [48] for an assumption implying that the exponents are not all equal, in the general stationary setting.

(2) Boundary theory

After the fundamental work of Furstenberg (see [20], [22], [23]) the set of bounded harmonic functions was determined

- for absolutely continuous distributions on connected groups by Raugi [59] (see also Guivarc'h [31]),

- for distributions on discrete groups of matrices by Ledrappier [47] .

(3) Limit theorems

(3.1) Onecan find a proof, under our usual irreducibility assumptions, of

- the functional central limit theorem,
- the law of iterated logarithm,
- the renewal theorem,

- the local limit theorem,

for the sequence $\operatorname{Log} \|S_n x\|$ in Le Page [49], [50] . Recurrence properties are studied in Guivarc'h [33] (see also Bougerol [9]) .

The central limit theorem for S_n written in the polar and the Iwasawa decomposition is proved in Raugi [59] . References to earlier proofs and applications can be found in Tutubalin [68].

(3.2) Properties of the solutions of the difference equation on $\mathbb{R}^d$

$$X_{n+1} = Y_n X_n + B_n$$

(where B_n is in $\mathbb{R}^d$ and Y_n in $Gl(d,\mathbb{R})$) are studied in Kesten [41] and in Le Page [50] . Stationary solutions are given in [8] . For $d = 1$, Grincevicius [34] proves a central limit theorem.

(3.3) Without irreducibility assumptions, the central limit theorem is not yet fully understood. The latest reference is Raugi [60].

(4) Positive matrices

The reader will find in Cohen [14] a nice account of the applications of products of positive random matrices to demography and an extensive bibliography. Kesten and Spitzer [42] study the convergence in distribution of such products.

(5) Linear stochastic differential equation

A goog introduction to this subject is the survey of Arnold and Kliemann [1] . A nice application is given in Pardoux and Pignol [58].

<u>BIBLIOGRAPHY</u>

[1] ARNOLD, L. and KLIEMANN, W. (1983). Qualitative theory of stochastic systems. In "Probabilistic Analysis and Related topics", A.T. Bharucha-Reid (ed.), Vol. 3, 1-79, Academic Press, New York.

[2] ARNOLD, L., CRAUEL, H. and WIHSTUTZ, V. (1983). Stabilization of linear systems by noise. SIAM J. Control Optim. (21), 451-461.

[3] BEARDON A.F. (1966). The Hausdorff dimension of singular sets of properly discontinuous subgroups. Amer. Journ. of Math. (88), 722-736.

[4] BELLMAN, R. (1954). Limit theorem for non-commutative operations. I. Duke Math. J. (21), 491-500.

[5] BILLINGSLEY, P. (1965). <u>Ergodic theory and information</u>. Wiley and Sons. New York.

[6] BIRKHOFF, G. (1957). Extensions of Jentzsch's theorem. Trans. Amer. Math. Soc. (85), 219-227.

[7] BOUGEROL, P. (1984). Stabilité en probabilité des équations différentielles stochastiques linéaires et convergence de produits de matrices aléatoires. C.R. Acad. Sc. Paris, (299), Série 1, 631-634.

[8] BOUGEROL, P. (1984). Tightness of products of Random matrices and stability of linear stochastic systems. To appear in Ann. Probab.

[9] BOUGEROL, P. (1985). Oscillation des produits de matrices aléatoires dont l'exposant de Lyapounov est nul. To appear.

[10] BREIMAN, L. (1968). <u>Probability</u>. Addison Wesley. Reading Massachusetts.

[11] CHASSAING, P., LETAC, G. and MORA, M. (1984). Brocot sequences and random walks on Sl(2,ℝ). In "Probability measures on groups 7", ed. H. Heyer. Lecture Notes in Math. n° 1064. Springer Verlag. Berlin, Heidelberg, New York, 36-48.

[12] CHATTERJI, S.D. (1966). Masse, die von regelmässigen Kettenbrüchen induziert sind. Math. Annalen (164), 113-117.

[13] CHEVALLEY, C. (1951). Théorie des groupes de Lie, t.2 : groupes algébriques. Hermann, Paris.

[14] COHEN, J.E. (1979). Ergodicity theorems in demography. Bull. Amer. Math. Soc. (3), 275-295.

[15] COHEN, J.E. and NEWMAN, C.M. (1984). The stability of large random matrices and their products. Ann. Probab. (12), 283-310.

[16] DEKKING, F.M. (1982). On transience and recurrence of generalized random walks. Zeit. für Wahrscheinlichkeitstheorie und Verw. Gebiete. (61), 459-465.

[17] DUNFORD, N. and SCHWARTZ, J. (1958). Linear Operators, Vol. 1. Interscience.

[18] FELLER, W. (1971). An Introduction to Probability Theory and its Applications. Vol. 2. Wiley. New York, London, Sydney, Toronto.

[19] FREIDLIN, M.I. and VENTSEL, A.D. (1984). Random perturbations of dynamical systems. Springer Verlag. Berlin, Heidelberg, New York.

[20] FURSTENBERG, H. (1963). A Poisson Formula for semisimple groups. Annals of Math. (77), 335-383.

[21] FURSTENBERG, H. (1963). Non-commuting random products. Trans. Amer. Math. Soc. (108), 377-428.

[22] FURSTENBERG, H. (1971). Random walks and discrete subgroups of Lie groups. Adv. Proba. and related topics (1), 3-63.

[23] FURSTENBERG, H. (1972). Boundary theory and stochastic processes in homogeneous spaces. In "Harmonic analysis on homogeneous spaces", Symposia on Pure and Applied Math. Willamstone, Mass.

[24] FURSTENBERG, H. (1980). Random walks on Lie groups. In "Harmonic analysis and representations of semi-simple Lie groups". J.A. Wolf, M. Cahen and De Wilde (eds.). D. Reidel Publishing

Company. Dordrecht, Holland.

[25] FURSTENBERG, H. and KESTEN, H. (1960). Products of random matrices. Ann. Math. Statist. (31), 457-469.

[26] FURSTENBERG, H. and KIFER, Y. (1983). Random matrix products and measures on projective spaces. Israel J. Math. (10), 12-32.

[27] GLASNER, S. (1976). Proximal Flows. Lecture Notes 517. Springer Verlag. Berlin, Heidelberg, New York.

[28] GORDIN, M.I. and LIFSCHITZ, B.A. (1978). The central limit theorem for stationary Markov processes Soviet. Math. Dokl. (19), N° 2, 392-394.

[29] GREENLEAF, F. (1969). Invariant means on topological groups. Van Nostrand. New York.

[30] GRINCEVICIUS, A.K. (1974). A central limit theorem for the group of linear transformations of the real axis. Soviet Math. Dokl. (15), 1512-1515.

[31] GUIVARC'H, Y. (1980). Quelques propriétés asymptotiques des produits de matrices aléatoires. In "Ecole d'été de Saint-Flour 7 - 1978", ed. P.L. Hennequin. Lecture Notes in Math. n° 774. Springer Verlag. Berlin, Heidelberg, New York.

[32] GUIVARC'H, Y. (1984). Exposants caractéristiques des produits de matrices aléatoires en dépendance markovienne. In "Probability measures on groups 7", ed. H. Heyer. Lecture Notes in Math. n° 1064. Springer Verlag. Berlin, Heidelberg, New York, 161-181.

[33] GUIVARC'H, Y. (1984). Application d'un théorème limite local à la transcience et à la récurrence de marches de Markov. In "Théorie du Potentel". Lecture Notes in Math. n° 1096. Springer Verlag. Berlin, Heidelberg, New York, 301-332.

[34] GUIVARC'H, Y. and RAUGI, A. (1985). Frontière de Furstenberg. propriétés de contraction et théorèmes de convergence. Zeit. für Wahrscheinlichkeitstheorie und Verw. Gebiete. (69), 187-242.

[35] HENNION, H. (1984). Loi des grands nombres et perturbations pour des produits réductibles de matrices aléatoires. Zeit. für Wahrscheinlichkeitstheorie und Verw. Gebiete (67), 265-278.

[36] HEWITT, K. and ROSS, A. (1963). Abstract Harmonic Analysis 1. Springer Verlag. Berlin, Heidelberg, New York.

[37] IKEDA, N. and WATANABE, S. (1981). Stochastic differential equations and diffusion processes. North Holland-Kodansha . Amsterdam, Takio.

[38] KAIJSER, T. (1970). On distribution problems for random products of non-commutative matrices. Report n° 22. Uppsala University.

[39] KAIJSER, T. (1972). Some limit theorem for Markov chains with applications to learning models and products of random matrices. Report Institute Mittag-Leffler, Djursholm, Sweden.

[40] KAIJSER, T. (1978). A limit theorem for Markov chains in compact metric spaces with applications to products of random matrices. Duke Math. Journ. (45), 311-349.

[41] KESTEN, H. (1973). Random difference equations and renewal theory for products of random matrices. Acta Math. (131), 207-248.

[42] KESTEN, H. and SPITZER, F. (1984). Convergence in distribution for products of random matrices. Zeit. für Wahrscheinlichkeitstheorie und Verw. Gebiete. (67), 363-386.

[43] KINNEY, J.R. and PITCHER, T.S. (1966). The dimension of some sets defined in terms of f-expansions. Zeit. für Wahrscheinlichkeitstheorie und Verw. Gebiete. (4), 293-315.

[44] KUNITA, H. (1984). Stochactic differential equations and stochastic flow of diffeomorphisms. In " Ecole d'été de Saint-Flour 12-1982", ed. P.L. Hennequin. Lecture Notes in Math. n° 1097. Springer Verlag. Berlin, Heidelberg, New York.

[45] LANG S. (1965). Algebra. Addison-Wesley. Reading, Massachusetts.

[46] LEDRAPPIER, F. (1984). Quelques propriétés des exposants caractéristiques. In "Ecole d'été de Saint-Flour 12-1982", ed. P.L. Hennequin. Lecture Notes in Math. n° 1097. Springer Verlag. Berlin, Heidelberg, New York.

[47] LEDRAPPIER, F. (1984). Poisson formula of discrete groups of matrices. To appear in Israël Jour. Math.

[48] LEDRAPPIER, F. (1985). Positivity of the exponent for stationary sequences of matrices. To appear.

[49] LE PAGE, E. (1982). Théorèmes limites pour les produits de matrices aléatoires. In "Probability measures on groups", ed. H. Heyer. Lecture Notes in Math. n° 928. Springer Verlag. Berlin, Heidelberg, New York, 258-303.

[50] LE PAGE, E. (1983). Théorèmes de renouvellement pour les produits de matrices aléatoires. Equations aux différences aléatoires. Séminaire de Probabilités. Université de Rennes.

[51] LETAC, G. and SESHADRI, V. (1983). A characterization of the generalized inverse Gaussian distribution by continued fractions. Zeit. für Wahrscheinlichkeitstheorie und Verw. Gebiete. (62), 485-489.

[52] LOOMIS, L.H. and STERNBERG, S. (1968). Advanced Calculus. Addison Wesley. Reading, Massachusetts.

[53] Mc CRUDDEN, M. and WOOD, R.M. (1984). On the supports of absolutely continuous Gauss measures on $Sl(2,\mathbb{R})$. In "Probability measures on groups 7", ed. H. Heyer. Lecture Notes in Math. n° 1064. Springer Verlag. Berlin, Heidelberg, New York, 379-397.

[54] MOSTOW, G.D. (1955). Self-adjoint groups. Ann. of Math. (62), 44-55.

[55] NAGAEV, S.V. (1957). Some limit theorems for stationary Markov chains. Theor. Proba. Appl. (2), 378-406.

[56] NEWMAN, C.M. (1984). The distribution of Lyapunov exponents : Exact results for random matrices. To appear.

[57] OSSELEDEC, V.I. (1968). A multiplicative ergodic theorem. Trans. Moscow Math. Soc. (19), 197-231.

[58] PARDOUX, E. and PIGNOL, M. (1984). Etude de la stabilité de la solution d'une E.D.S. bilinéaire à coefficients périodiques. Application au mouvement des pales d'hélicoptère. In "Analysis and Optimization of Systems, Part 2" ; A. Bensoussan and J.L. Lions. (Eds.), Lecture Notes in Control and Info. Sci. 63. Springer Verlag.

[59] RAUGI, A. (1977). Fonctions harmoniques et théorèmes limites pour les marches aléatoires sur les groupes. Bull. Soc. Math. France, Mémoire 54, 127 p.

[60] RAUGI, A. (1980). Quelques remarques sur le théorème de la limite centrale sur un groupe de Lie. C.R. Acad. Sc. Paris, (290), 103-106.

[61] REVUZ, D. (1984). Markov chains. North Holland. Amsterdam, New York, Oxford.

[62] ROYER, G. (1980). Croissance exponentielle de produits markoviens de matrices aléatoires. Ann. I.H.P. (16), 49-62.

[63] SAZONOV, V.V. and TUTUBALIN, V.N. (1966). Probability distributions on topological groups. Th. Proba. Appl. (13), 1-45.

[64] SENETA, E. (1981). Non-negative matrices and Markov chains, 2nd ed., Springer Verlag. Berlin, Heidelberg, New York.

[65] SIEBERT, E. (1982). Absolute continuity, singularity, and supports of Gauss semigroups on a Lie group. Mh. Math. (93), 239-253.

[66] TUTUBALIN, V.N. (1965). On limit theorems for products of random matrices. Theor. Proba. Appl. (10), 15-27.

[67] TUTUBALIN, V.N. (1969). Some theorems of the type of the strong law of large numbers. Theor. Proba. Appl. (14), 313-319.

[68] TUTUBALIN, V.N. (1977). The central limit theorem for products of random matrices and some of its applications. Symposia Math. (21), 101-116.

[69] VIRTSER, A.D. (1979). On products of random matrices and operators. Theor. Proba. Appl. (24), 367-377.

[70] VIRTSER, A.D. (1984). On the simplicity of the spectrum of the Lyapunov characteristic indices of a product of random matrices. Theor. Proba. Appl. (28), 122-135.

PART B

RANDOM SCHRÖDINGER OPERATORS

INTRODUCTION

Numerous equations arising from one dimensional discrete physical systems lead to the analysis of a linear second order difference operator H, acting on a complex sequence Ψ_n , $n \in \mathbb{Z}$, by :

$$(H\Psi)_n = b_n^{-1} ((-\Delta\Psi)_n + a_n \Psi_n)$$

In this formula, Δ is the discrete Laplacian $(\Delta\Psi)_n = \Psi_{n+1} + \Psi_{n-1} - 2\Psi_n$ and a_n, b_n , are two fixed sequences of real numbers with $b_n > 0$, representing the physical properties of the medium. Generally such operators are associated to "time dependent" equations and we give some typical examples :

(i) A solution of the Schrödinger equation $i \frac{\partial\phi}{\partial t} = H\phi$ of the form $\phi(n,t) = \Psi_n e^{-i\lambda t}$ satisfies $H\Psi = \lambda\Psi$ where H is the classical Schrödinger operator, that is the operator associated to $b_n = 1$, $\forall n \in \mathbb{Z}$, and a_n is the potential at site n.

(ii) A solution of the wave equation $b_n \frac{\partial^2\phi}{\partial t^2} = (\Delta\phi)_n$ of the form $\phi(n,t) = \Psi_n e^{i\lambda t}$ satisfies $H\Psi = \lambda^2\Psi$ where H is the "Helmotz operator" that is the operator associated to $a_n = 0$, $\forall n \in \mathbb{Z}$, and b_n is the diffusion coefficient at site n.

(iii) A solution of the heat equation $b_n \frac{\partial\phi}{\partial t} = (\Delta\phi)_n$ of the form $\phi(n,t) = \Psi_n e^{-\lambda t}$ satisfies $H\Psi = \lambda\Psi$ where H is the Helmotz operator.

Similar equations and operators also appear in quasi-one dimensional systems associated to an infinite wire of finite cross section with ℓ sites. We have only to replace the sites n by (i,n) where $i \in (1, \ldots, \ell)$, (the integer ℓ is called the width of the strip) and the real sequences a_n, b_n by matrices sequences.

It is well known that the spectral properties of the operator H viewed as a self adjoint operator on an Hilbert space, govern the asymptotic behavior of the solutions of the associated time dependent equation. In quantum mechanics the number $|\phi(t,n)|^2$ (normalized in a suitable way) associated to a solution $\phi(t,n) = (e^{-itH}\Psi)_n$ of the time dependent Schrödinger equation, represents the probability of presence of a particle at the site n at time t. Roughly speaking, when Ψ is associated to the continuous spectrum of H then we have the "diffusion" behavior $\lim_{T\to+\infty} \frac{1}{2T}\int_{-T}^{T} |\phi(t,n)|^2 dt = 0$, and if Ψ is associated to the point spectrum of H then we have the "localization" property $\lim_{N\to+\infty} \sup_t \sum_{|n|\geq N} |\phi(t,n)|^2 = 0$. (See D. Ruelle [53] for more details). In deterministic systems with periodic structure, it is known that only diffusion behavior occurs. But it has been remarked that localization appears when this regular structure is perturbed by impurities or inhomogeneity in the medium. Thus a "metallic" wire suddenly becomes an insulator. We first give below a brief historical survey of the mathematical approaches to this subject.

It was first announced by P.W. Anderson [1] (1958) that for the classical multi-dimensional Schrödinger operator with an independent identically distributed family of random potentials, the spectrum has to be pure point for a "typical sample" assuming the disorder "large enough". It was later conjectured by N. Mott and W.D. Twose [46] (1961) that in the one dimensional case, this should be true at any disorder. The works of H. Furstenberg, H. Kesten [20] (1960), H. Furstenberg [19] (1963), V. Osseledec [49] (1968) provided the essential mathematical background used in the first rigorous approaches of the subject. It was first proved by H. Matsuda, K. Ishii [44] (1970), A. Casher J.L. Lebowitz [10] (1971), L.A. Pastur [50] (1973), Y. Yoshioka [64] (1973) that there does not exist an absolutely continuous component in the spectrum of H. An essential step was achieved in 1973 when I. Ja. Goldsheid, S.A. Molcanov, and L.A. Pastur [24] gave the first proof to the conjecture of Mott and Twose (they actually dealed with the continuous case). Their original proof has been later considerably simplified and extended by R. Carmona [7] (1982), [8] (1983), G. Royer [52] (1983), J. Brossard (1983). In the "discrete" case the same result has been obtained by H. Kunz, B. Souillard [37] (1980), J. Lacroix [38] (1982), F. Delyon, H. Kunz, B. Souillard [13] (1983). Moreover Goldsheid gave a similar announcement in a strip [23] (1981) and the proof can be found in J. Lacroix [39] (1983) [40] [41] (1984). All these previous proofs of localization are rather technical and at times the essential guiding principles are not easily understood. Fortunately, in the late of 1984, S. Kotani [36] clarified the situation, giving a rigorous statement to an earlier claim of R.E. Borland [5] (1963).

Our essential goal is to provide a direct and unified treatment to the foregoing problems, in the general setting of operators H introduced at the begining of this discussion here in after called Schrödinger operators. The essential tool will be the theory of products of i.i.d. random matrices developed in the first part of this book. We are mainly concerned with the independent case but a lot of definitions and properties are given in the general ergodic case. The "almost periodic case" is also of great physical and theoretical interest but most of the proofs have nothing to do with random matrices. Interested readers have to look at the survey of B. Simon [57] where they can also find an extensive bibliography. Since it seems that the theory of random matrices can hardly be used in the multidimensional case (up to now, limiting procedures in strips whose width goes to infinity have not been successful) we restrict ourselves to the one dimensional case and strips.

In chapter I the essential definitions and properties related to the spectral analysis of the deterministic operator H are given. In particular we construct a sequence of "good approximations" of the spectral measure of H and establish the existence of "slowly" increasing generalized eigenfunctions. Moreover the links between the singularity of the spectrum and the fundamental notion of hyperbolic behavior of a product of matrices are pointed out.

In chapter II we define an ergodic family of Schrödinger operators which contains as essential examples the classical Schrödinger operator and the Helmotz operator. Some weak properties of the spectrum of H considered as a subset of $\mathbb{R}$ are given, before introducing the essential concept of Lyapunov exponent. Positivity of this exponent is carefuly studied since this property is crucial in order to obtain absence of absolutely continuous spectrum. The distribution of states describing the asymptotic behavior of the eigenvalues distribution for the operator restricted to "boxes" is of great physical interest and we discuss in detail its regularity properties together with the links with the Lyapunov exponent (Thouless formula). Kotani's criterion insuring localization property is then introduced in the general ergodic case but its main application to the independent case is discussed in the following chapter. Finaly we give a straightforward application of the central limit theorem on $SL(2,\mathbb{R})$ to the asymptotic behavior of the conductance.

Chapter III is devoted to the proof of the conjecture of Mott and Twose both in classical Schrödinger and Helmotz case. In the first model, Kotani's criterion gives immediately the solution, but in the general case the proof is more involved and requires some Laplace analysis on $SL(2,\mathbb{R})$. As a consequence extra properties of the distribution of states are obtained.

All these foregoing results are generalized in the chapter IV to the case of a strip. Most of the previous proofs in the one dimensional case can be translated with some care. But some problems are much more involved, especialy positivity of Lyapunov exponents. General results given in the first part of the book are then very useful. The proof of localization in the general case requires also much more work since Laplace analysis on symplectic groups needs some knowledge about symplectic geometry.

Numerous related topics, non-stationary processes for instance, are not tackled when they don't appear as direct applications of products of i.i.d. matrices, thus we don't intend to provide a complete survey in the theory of random Schrödinger operators. Moreover we are aware of that a lot of pioneer and connected works are not cited since we have focused our attention to a precise mathematical aspect of the subject.

CHAPTER I

THE DETERMINISTIC SCHRÖDINGER OPERATOR

I.1 The difference equation. Hyperbolic structures

Let $\mathcal{L}$ be the linear space of complex sequences $\Psi = (\Psi_n)$ where n runs through the set of integers $\mathbb{Z}$. The operator H is associated to two given real sequences a and b with $b_n \neq 0 \quad \forall n \in \mathbb{Z}$, and acts on $\mathcal{L}$ by the formula :

$$(H\Psi)_n = b_n^{-1} \left[-\Psi_{n-1} - \Psi_{n+1} + a_n \Psi_n\right]$$

For a complex number λ every solution of the difference equation $H\Psi = \lambda\Psi$ lies in a two dimensional subspace of $\mathcal{L}$ spanned by the solutions $p(\lambda)$ and $q(\lambda)$ constructed from the initial values $p_o(\lambda) = q_{-1}(\lambda) = 1$, $p_{-1}(\lambda) = q_o(\lambda) = 0$, such that :

$$\Psi_n(\lambda) = p_n(\lambda)\,\Psi_o(\lambda) + q_n(\lambda)\,\Psi_{-1}(\lambda)$$

From now in order to avoid too complicated notations we don't write the variable λ in the solutions of the difference equation. A solution Ψ of the difference equation is constructed from initial values Ψ_o and Ψ_{-1} by a product of "transfer matrices" Y_n defined by :

$$Y_n = \begin{bmatrix} a_n - \lambda b_n & -1 \\ 1 & 0 \end{bmatrix}$$

$$S_n = Y_n Y_{n-1} \dots Y_o \qquad \text{if } n \geqslant 0$$

$$S_n = Y_n^{-1} Y_{n+1}^{-1} \dots Y_{-1}^{-1} \qquad \text{if } n \leqslant -1$$

Thus $\begin{bmatrix} \Psi_{n+1} \\ \Psi_n \end{bmatrix} = S_n \begin{bmatrix} \Psi_o \\ \Psi_{-1} \end{bmatrix}$ if $n \geqslant 0$, $\begin{bmatrix} \Psi_n \\ \Psi_{n-1} \end{bmatrix} = S_n \begin{bmatrix} \Psi_o \\ \Psi_{-1} \end{bmatrix}$ if $n \leqslant -1$

The transfer matrices Y_n and therefore the products S_n belong to the group $SL(2,\mathbb{C})$ of two by two matrices with complex entries and of determinant one. If λ is real then Y_n and S_n belong to the subgroup $SL(2,\mathbb{R})$ with real entries. The construction of the solutions of the difference equation by such products of matrices is the essential link between the two parts of this book.

A matrix $\mathcal{g}$ of $SL(2,\mathbb{C})$ is said "hyperbolic" if the eigenvalues of $\mathcal{g}$ have distinct moduli. If V is an eigenvector associated to the eigenvalue of modulus stricly less than one, then the sequence $||\mathcal{g}^n V||$ goes to zero when n goes to $+\infty$ and for any vector W non proportional to V the sequence $||\mathcal{g}^n W||$ goes to $+\infty$ when n goes to $+\infty$. (The same situation occurs when n goes to $-\infty$, replacing V by the eigenvector associated to the eigenvalue of modulus strictly greater than one).

Let now (s_n) $n \in \mathbb{Z}$, be a sequence of matrices of $SL(2,\mathbb{C})$. We say that s_n has an hyperbolic structure in the positive direction if there exists a non zero vector V of $\mathbb{C}^2$ (called the contractive vector) such that $\lim_{n \to +\infty} || s_n V|| = 0$. It is readily seen that for any W non proportional to V then $\lim_{n \to +\infty} || s_n W|| = +\infty$ (from the determinant property). We define in the same way an hyperbolic structure in the negative direction.

The existence (for some λ) of hyperbolic structures for the sequence S_n is the most important question discussed in the following chapters. In particular for a real λ the existence of hyperbolic structures for S_n in both directions with the same contractive vector implies the existence of a real non zero solution Ψ of the difference equation with $\lim_{|n| \to +\infty} \Psi_n = 0$. The next "Osseledec's theorem" is a basic tool to prove the existence of hyperbolic structures :

<u>PROPOSITION 1.1</u> *Let* $(\mathcal{g}_n)$, $n \in \mathbb{N}$ *be a sequence of matrices of* $SL(2,\mathbb{C})$ *such that :*

(i) $\lim_{n \to +\infty} \frac{1}{n} \operatorname{Log} ||\mathcal{g}_n \mathcal{g}_{n-1} \dots \mathcal{g}_1|| = \gamma$

(ii) $\lim_{n \to +\infty} \frac{1}{n} \operatorname{Log} \|\mathcal{g}_n\| = 0$

Then there exists a non zero vector V *such that :*

$$\lim_{n \to +\infty} \frac{1}{n} \operatorname{Log} \| \mathcal{g}_n \cdots \mathcal{g}_1 V\| = -\gamma$$

and for any vector W *non proportional to* V

$$\lim_{n \to +\infty} \frac{1}{n} \operatorname{Log} \| \mathcal{g}_n \cdots \mathcal{g}_1 W\| = \gamma$$

The proof of this proposition can be found in [49]. The euclidian norm of a matrix of SL(2,$\mathbb{C}$) is no less than one, thus γ is non negative. When γ is strictly positive Osseledec's theorem asserts the existence of an hyperbolic structure for the sequence $s_n = \mathcal{g}_n \mathcal{g}_{n-1} \cdots \mathcal{g}_1$ and moreover we see that the sequences $\|s_n V\|$ or $\|s_n W\|$ converge to 0 or $+\infty$ with an exponential rate.

We now give two obvious (but useful) lemmas related to the solutions of the difference equation. The Wronskian of ϕ and Ψ in $\mathcal{L}$ is the sequence

$$W_n(\Psi,\phi) = \Psi_{n+1}\phi_n - \phi_{n+1}\Psi_n$$

LEMMA 1.2 *Let* ϕ *and* Ψ *be two solutions of the difference equation* $Hu = \lambda u$, *then* $W(\Psi,\phi)$ *is constant and this constant is zero if and only if* ϕ *and* Ψ *are proportional.*

Proof : $\begin{bmatrix} \Psi_n & \phi_n \\ \Psi_{n-1} & \phi_{n-1} \end{bmatrix} = S_n \begin{bmatrix} \Psi_o & \phi_o \\ \Psi_{-1} & \phi_{-1} \end{bmatrix}$ for $n \leqslant -1$ and

$$\begin{bmatrix} \Psi_{n+1} & \phi_{n+1} \\ \Psi_n & \phi_n \end{bmatrix} = S_n \begin{bmatrix} \Psi_o & \phi_o \\ \Psi_{-1} & \phi_{-1} \end{bmatrix} \quad \text{for } n \geqslant 0$$

Since the determinant of S_n is equal to one the result is obvious.

LEMMA 1.3 *Let* ϕ *and* Ψ *be in* $\mathcal{L}$, m *and* n *two integers with* $m \leqslant n$, *then we have the "Green formula" :*

$$\sum_{k=m}^{n} b_k \left((H\Psi)_k \phi_k - \Psi_k (H\phi)_k\right) = W_{m-1}(\Psi,\phi) - W_n(\Psi,\phi)$$

Proof : Straightforward computation.

Exercise 1.4 : Let Ψ be a solution of $H\Psi = \lambda\Psi$, m and n two integers with $m \leq -1 \leq 0 \leq n$. Prove that if b_k is positive for $m \leq k \leq n$ then :

$$|\Psi_n \Psi_{n+1}| + |\Psi_m \Psi_{m-1}| \geq |\mathfrak{J}m\,\lambda| \, (b_o|\Psi_o|^2 + b_{-1}|\Psi_{-1}|^2)$$

I.2 Self-adjointness of H. Spectral properties

The Green formula of the lemma 1.3 suggests that H should be a symmetric operator on a suitable Hilbert subspace of $\mathcal{L}$. From now we assume in all this chapter that the sequence b is strictly positive and bounded away from zero (that is the sequence b^{-1} is bounded)

We denote by $\mathcal{H}$ the subspace of $\mathcal{L}$ defined by

$$\mathcal{H} = \{\Psi \in \mathcal{L} / \Sigma\, b_n\, |\Psi_n|^2 < +\infty\}$$

(when the symbol Σ has no index of summation it's understood that this index should run through $\mathbb{Z}$). Endowed with the scalar product $\langle\phi,\Psi\rangle = \Sigma\, b_n\, \phi_n\, \overline{\Psi}_n$, $\mathcal{H}$ is an Hilbert space. If D is a dense linear subset of $\mathcal{H}$ such that $H(D) \subset \mathcal{H}$ we denote by (H,D) the linear operator with domain D. Let D_o be the dense linear subset of $\mathcal{H}$ of sequences with a finite number of terms different from zero. From the Green formula we know that (H,D_o) is a symmetric operator and it's not difficult to see that its adjoint $(H,D_o)^*$ is equal to (H,D_1) where

$$D_1 = \{\Psi \in \mathcal{H} / H\Psi \in \mathcal{H}\} = \{\Psi \in \mathcal{H} / \Sigma\, a_n^2\, b_n^{-1}|\Psi_n|^2 < +\infty\}$$

LEMMA 2.1 *If the sequence* a *is bounded then* H *is a bounded self adjoint operator.*

Proof : Let α, β be the bounds of the sequences a, b^{-1} and $\Psi \in \mathcal{H}$:

$$||H\Psi|| = (\Sigma\, b_n^{-1}\, |\Psi_{n-1} + \Psi_{n-1} - a_n\Psi_n|^2)^{1/2}$$

then by the ℓ^2 triangle inequality :

$$||H\Psi|| \leq (\Sigma\, b_n^{-1}\, |\Psi_{n+1}|^2)^{1/2} + (\Sigma\, b_n^{-1}|\Psi_{n-1}|^2)^{1/2} + (\Sigma\, b_n^{-1}\, a_n^2\, |\Psi_n|^2)^{1/2}$$

and :

$$||H\Psi|| \leq ||\Psi|| \, (2\alpha + \alpha\beta)$$

Thus H is a bounded operator on $\mathcal{H}$. The assumption on the sequence b implies that $\mathcal{H} \subset \ell^2(\mathbb{Z})$ and hence $\lim_{|n| \to +\infty} \Psi_n = 0$ for $\Psi \in \mathcal{H}$. By the Green formula we see that H is a symmetric operator on $\mathcal{H}$ and hence self adjoint.

In many applications the sequence a is not bounded, so we prove :

LEMMA 2.2 *The operator* (H,D_1) *is self adjoint.*

Proof : As remarked in the lemma 2.1, H is a symmetric operator on D_1, hence (H,D_1) is a symmetric extension of (H,D_o) with $(H,D_1) = (H,D_o)^*$, the result follows from the theory of self adjoint extensions of symmetric operators [9].

From now, when we speak of the operator H, it will be always understood that we actually deal with the self adjoint operator (H,D_1). As we shall see in the exercise 2.3 below, the sequence b has not to be bounded away from zero in order that (H,D_o) possesses a self adjoint extension, but such an assumption seems reasonable since a solution of $H\Psi = \lambda\Psi$ which is non zero at infinity does not have any physical meaning in general.

Exercise 2.3 For $n \in \mathbb{N}$ let C_n be the sequence $C_n = \min((b_n b_{n+1})^{1/2}, (b_{-n} b_{-n-1})^{1/2})$. If we assume that $\sum_{n=0}^{\infty} C_n = +\infty$ prove that the operator (H,D_o) is essentially self adjoint (and (H,D_1) is its self adjoint extension).
Hint : Use the result of 1.4 to prove that when $\mathcal{I}m\,\lambda \neq 0$ the equation $H\Psi = \lambda\Psi$ has no non zero solution in $\mathcal{H}$ (see [4] VII theorem 1.3).

We can apply to H the general theory of self adjoint operators which we may find in [9]. There exists a resolution of the identity E which associates to each borel subset Δ of $\mathbb{R}$ the projection E_Δ of $\mathcal{H}$ and for ϕ, Ψ in $\mathcal{H}$, $\sigma_{\phi,\Psi}(\Delta) = \langle E_\Delta\phi, \Psi\rangle$ is a bounded complex radon measure on $\mathbb{R}$ (We denote the positive measure $\sigma_{\phi,\phi}$ by σ_ϕ). If λ is a complex number with $\mathcal{I}m\,\lambda \neq 0$, $R_\lambda = (H - \lambda I)^{-1}$ is a bounded operator on $\mathcal{H}$ and the spectral theorem asserts that $\langle R_\lambda\phi, \Psi\rangle = \int (t-\lambda)^{-1}\, d\sigma_{\phi,\Psi}(t)$. It's also well known that $\mathcal{H}$ can be written as the orthogonal sum of three closed subspaces invariant under H (Lebesgue decomposition of $\mathcal{H}$) defined by :

$\mathcal{H}^a = \{\phi \in \mathcal{H} / \sigma_\phi$ is absolutely continuous$\}$

$\mathcal{H}^p = \{\phi \in \mathcal{H} / \sigma_\phi$ is pure point$\}$

$\mathcal{H}^s = \{\phi \in \mathcal{H} / \sigma_\phi$ is singular continuous$\}$

The spectrum $\mathfrak{S}$ of H is the union of $\mathfrak{S}^a$, $\mathfrak{S}^p$, $\mathfrak{S}^s$ which are the spectra of the restrictions of H to the invariant subspaces $\mathcal{H}^a, \mathcal{H}^p, \mathcal{H}^s$. We say that the spectrum of H is pure point if we have $\mathcal{H} = \mathcal{H}^p$ in this case $\mathcal{H}$ has an orthogonal basis of eigenvectors. In our case this Lebesgue decomposition of $\mathcal{H}$ can be obtained by the decomposition of a single measure on $\mathbb{R}$ as it is shown below. Let e^k be the orthogonal basis of $\mathcal{H}$ defined by $(e^k)_n = b_n^{-1} \delta_n^k$ (we remark that $\|e^k\|^2 = b_k^{-1}$ and for $\Psi \in \mathcal{H}, \langle\Psi, e^n\rangle = \Psi_n$). We denote by $\sigma_{m,n}$ the measure σ_{e^m, e^n}, by σ_n the measure $\sigma_{n,n}$ and σ the "spectral measure" $\sigma = \sigma_o + \sigma_{-1}$. For each n in $\mathbb{Z}$, $p_n(\lambda)$ and $q_n(\lambda)$ defined in the first section are polynomial functions in λ and thus the operators $p_n(H)$ and $q_n(H)$ are well defined.

<u>PROPOSITION 2.4</u> $p_n(H)\, e^o + q_n(H)\, e^{-1} = e^n$

Proof : The equality is obvious for n = 0 and n = -1 and is proved by induction for all n from the identity $b_k\, H e^k = -e^{k+1} - e^{k-1} + a_k e^k$

<u>COROLLARY 2.5</u> *For each Ψ in $\mathcal{H}$ the positive bounded measure σ_ϕ is absolutely continuous witn respect to σ.*

Proof : Let Δ be a Borel subset of $\mathbb{R}$ with $\sigma(\Delta) = 0$. Then $\|E_\Delta e^o\|^2 = \sigma_o(\Delta) = 0$ and $\|E_\Delta e^{-1}\|^2 = \sigma_{-1}(\Delta) = 0$, thus by the proposition 2.4 :
$E_\Delta e^n = p_n(H)\, E_\Delta e^o + q_n(H)\, E_\Delta e^{-1} = 0$. This implies that $E_\Delta = 0$ and hence $\sigma_\Psi(\Delta) = \|E_\Delta \Psi\|^2 = 0$.

This corollary has a direct consequence for the Lebesgue decomposition of $\mathcal{H}$. Let σ^a, σ^p, σ^s be the three parts of the Lebesgue decomposition of σ.

<u>PROPOSITION 2.6</u> *With the above notations, $\mathfrak{S}$, $\mathfrak{S}^a$, $\mathfrak{S}^p$, $\mathfrak{S}^s$ are the topological supports of the positive bounded measures σ, σ^a, σ^p, σ^s, moreover λ is an eigenvalue of H if and only if $\sigma(\lambda) > 0$.*

Proof : By Weyl's criterion we know that $\lambda \in \mathfrak{S}$ if and only if there exists a sequence $(\Psi_n)_{n\geq 0}$ in D_1 with $\lim_{n\to+\infty} \|(H-\lambda I)\Psi_n\| = 0$ and $\|\Psi_n\| = 1$. Taking in account that for $\lambda \in \mathbb{R}$ and $\Psi \in D_1$ we have the relation $\|(H-\lambda I)\Psi\|^2 = \int (t-\lambda)^2 \, d\sigma_\Psi(t)$ the result follows for $\mathcal{T}$ and σ. For the other parts of the spectrum we remark that if $\Psi = \Psi^a + \Psi^p + \Psi^s$ is the "Lebesgue decomposition" of $\Psi \in \mathcal{H}$ then $\sigma_\Psi^a = \sigma_{\Psi^a}$, $\sigma_\Psi^s = \sigma_{\Psi^s}$, $\sigma_\Psi^p = \sigma_{\Psi^p}$.

We associate to H the measure valued matrix $\$$ given by :

$$\$ = \begin{bmatrix} \sigma_0 & \sigma_{-1,0} \\ \sigma_{0,-1} & \sigma_{-1} \end{bmatrix}$$

<u>LEMMA 2.7</u> *The spectral measures* $\sigma_{m,n}$ *can be constructed from the spectral matrix* $\$$ *by the formula :*

$$\sigma_{m,n} = (p_m, q_m) \; \$ \begin{pmatrix} p_n \\ q_n \end{pmatrix}$$

Proof : Using the expressions of e^n and e^m given by the proposition 2.4 it's enough to compute $\sigma_{m,n}(\Delta) = \langle E_\Delta e^m, e^n \rangle$.

<u>Exercise 2.8</u> Prove that $\{E_\Delta e^0, E_\Delta e^{-1} \,/\, \Delta$ is a Borel subset of $\mathbb{R}\}$ spans a dense subspace of $\mathcal{H}$.

<u>Exercice 2.9</u> Prove that $(\sigma_n + \sigma_{n+1})$ and σ have the same nul sets.

Some authors work with the operator H restricted to the half axis $\mathbb{N}$ denoted by ^+H . This operator is defined on the sequences $\Psi = (\Psi_n)_{n\geq 0}$ by

$(^+H\Psi)_0 = b_0^{-1} (-\Psi_1 + a_0 \Psi_0)$

$(^+H\Psi)_n = (H\Psi)_n$ if $n \geq 1$

The Hilbert space $^+\mathcal{H}$, the spectral measures $^+\sigma_{m,n}$, the resolvent $^+R_\lambda$ are defined in the same way than for H and we set $^+\sigma = {}^+\sigma_0$

<u>Exercise 2.10</u> Prove that

(i) $p_n(^+H)\, e^0 = e^n \qquad (n \in \mathbb{N})$

(ii) For each Ψ in $^+\mathcal{H}$ the measure $^+\sigma_\Psi$ is absolutely continuous with respect to $^+\sigma$

(iii) $^+\sigma_{m,n} = p_n \, p_m \, {}^+\sigma \quad (m,n \in \mathbb{N})$

(iv) $\{^+E_\Delta e^0 \,/\, \Delta$ is a Borel subset of $\mathbb{R}\}$ spans a dense subspace of $^+\mathcal{H}$

Exercise 2.11 For $m,n \in \mathbb{N}$ prove the following relations :

(i) $\int p_n\, p_m\, d^+\sigma = b_n^{-1}\, \delta_n^m$

(ii) $\int \lambda\, p_n\, p_m\, d^+\sigma = b_n^{-2}\, a_n\, \delta_m^n - b_n^{-1}\, b_m^{-1}\, (\delta_m^{n+1} + \delta_m^{n-1})$

The next exercise shows that when $\Im m\, \lambda \neq 0$ then there exists an hyperbolic structure in the two directions for S_n, with distinct contractive vectors.

Exercise 2.12 Let ${}^+m(\lambda) = \int (t-\lambda)^{-1}\, d^+\sigma(t)$ for $\Im m\,\lambda \neq 0$

(i) We define the sequence $({}^+\Psi_n)$, $n \in \mathbb{Z}$, by

${}^+\Psi_n = \langle {}^+R_\lambda\, e^o, e^n\rangle$ for $n \geqslant 0$

${}^+\Psi_n = {}^+m(\lambda)\, p_n + q_n$ for $n \leqslant -1$

Prove that $H\, {}^+\Psi = \lambda\, {}^+\Psi$, ${}^+\Psi_{-1} = 1$

(ii) Deduce from (i) that

$\langle {}^+R_\lambda\, e^o, e^n\rangle = {}^+m(\lambda)\, p_n + q_n$ for $n \geqslant 0$

(iii) Prove that

$$\sum_{n=0}^{+\infty} b_n |{}^+\Psi_n|^2 = \frac{\Im m({}^+m(\lambda))}{\Im m\,\lambda}$$

Hint : ${}^+R_\lambda - {}^+R_{\bar\lambda} = (\lambda-\bar\lambda)\, {}^+R_{\bar\lambda}\, {}^+R_\lambda$ and compute

$\langle {}^+R_{\bar\lambda}\, {}^+R_\lambda\, e^o, e^o\rangle$

(iv) The same property holds for the negative half axis and a complex number λ with $\Im m\,\lambda \neq 0$ is not an eigenvalue of H. This proves the above assertion about hyperbolic structures.

Exercise 2.13 Prove that it's possible to recover the sequences $(a_n)_{n\geqslant 0}$ and $(b_n)_{n\geqslant 0}$ from ${}^+\sigma$.

Hint : use 2.11.

Exercise 2.14 The sequences ${}^+\Psi(\lambda)$, ${}^-\Psi(\lambda)$ are constructed for the two half axis in the same way than in 2.12. Prove that the symmetric matrix $\langle R_\lambda e^n , e^m\rangle$ $(\Im m\,\lambda \neq 0)$ is given for $n \leqslant m$ by

$$\langle R_\lambda e^n, e^m\rangle = \frac{{}^+\Psi(m)\ {}^-\Psi(n)}{W({}^-\Psi, {}^+\Psi)}$$

(where W is the wronskian)

Exercise 2.15 Let λ be a real number and n a positive integer. Prove that

$$\sum_{k=0}^{n} b_k\, p_k^2 = p_n \frac{d}{d\lambda} p_{n+1} - p_{n+1} \frac{d}{d\lambda} p_n$$

I.3 Slowly increasing generalized eigenfunctions

We say that $\Psi \in \mathcal{L}$ is a generalized eigenfunction if $H\Psi = \lambda\Psi$ for a real λ and Ψ is non zero. (When $\Psi \in \mathcal{H}$, Ψ is an eigenfunction). A fairly general theorem about Carleman operators asserts that such sequences are slowly increasing [4]. Actually it's possible in the simpler case of difference operators to give a direct and easy proof of this result. As we have seen in 2.5, $\sigma_{m,n}$ is absolutely continuous with respect to σ (thank's to the inequality $|\sigma_{m,n}| \leqslant \frac{1}{2}(\sigma_m + \sigma_n)$). Let S be a matrix of density of $\mathbb{S}$ with respect to σ.

LEMMA 3.1 *For σ almost all real λ, the matrix $S(\lambda)$ is symmetric and positive.*

Proof : For each borel subset Δ of $\mathbb{R}$, the real matrix $\mathbb{S}(\Delta)$ is symmetric, positive since $\sigma_{o,-1} = \sigma_{-1,o}$, and for real x,y :

$$(x,y)\ \mathbb{S}\ (\Delta)\begin{pmatrix} x \\ y \end{pmatrix} = \sigma_{x\,e^{o} + y\,e^{-1}}(\Delta) \geqslant 0.$$

The conclusion follows easily.

PROPOSITION 3.2 *For σ almost all real λ there exists a generalized eigenfunction Ψ such that for each $\varepsilon > 0$*

$$\lim_{|n| \to \infty} \frac{\Psi(n)}{n^{\frac{1}{2}+\varepsilon}} = 0$$

Proof : For σ almost all λ we can write $S = K \begin{bmatrix} \alpha & 0 \\ 0 & \beta \end{bmatrix} K^*$ where K is an orthogonal matrix, α and β are two positive numbers with $\alpha \geqslant \beta$, $\alpha + \beta = 1$. By virtue of the lemma 2.7

$$\sigma_n = (p_n, q_n)\ K \begin{bmatrix} \alpha & 0 \\ 0 & \beta \end{bmatrix} K^* \begin{pmatrix} p_n \\ q_n \end{pmatrix} \sigma$$

If we denote by $\begin{pmatrix} \Psi_o \\ \Psi_{-1} \end{pmatrix}$ the first column of K and $\Psi_n = p_n \Psi_o + q_n \Psi_{-1}$ we see that $\sigma_n \geqslant \alpha\, \Psi_n^2\, \sigma \geqslant \frac{1}{2}\Psi_n^2\, \sigma$. We know that $\sigma_n(\mathbb{R}) = b_n^{-1}$ and that the sequence b_n^{-1} is bounded then for any $\varepsilon > 0$ we have $\int \sum_{n \neq 0} \frac{\Psi_n^2}{n^{1+\varepsilon}}\, d\sigma < +\infty$

thus $\sum_{n \neq 0} \frac{\Psi_n^2}{n^{1+\varepsilon}} < +\infty \quad \sigma$ a.e.

This implies that $\lim_{|n| \to \infty} \frac{\Psi_n^2}{n^{1+\varepsilon}} = 0 \quad \sigma$ a.e.

Exercise 3.3 The same result for the operator ^{+}H on the half positive axis : for $^{+}\sigma$ almost all λ, $\lim_{n\to+\infty} \frac{p_n^2}{n^{1+\varepsilon}} = 0$.

Exercise 3.4 Prove that the Proposition 3.2 could be : For σ almost all real λ there exists a generalized eigenfunction Ψ such that for every $\phi \in \ell^2(\mathbb{Z})$ we have $\phi\cdot\Psi \in \ell^2(\mathbb{Z})$.

I.4 Approximations of the spectral measures

The spectral measures $\sigma_{m,n}$ are generally obtained as weak limits of spectral measures of the operator H restricted to "boxes". These limits are independent from the boundary conditions on the boxes and this allows us the construct new approximations which are absolutely continuous with respect to the Lebesgue measure on $\mathbb{R}$.

A box Λ is a finite subset $[M , N]$ of $\mathbb{Z}$, and we say that a sequence of boxes is going to $\mathbb{Z}$ if the associated sequences of M and N are going to $-\infty$ and $+\infty$ respectively.

For a box Λ and a real number x we define the operator $^{\wedge}H^x$ on the Hilbert space $^{\wedge}\mathcal{H}$ of sequences (Ψ_n), $n \in [M , N]$, (with the scalar product inherited from $\mathcal{H}$) by :

$$(^{\wedge}H^x\Psi)_M = b_M^{-1}(-\Psi_{M+1} + a_M\Psi_M)$$

$$(^{\wedge}H^x\Psi)_n = (H\Psi)_n \qquad \text{for } M+1 \leqslant n \leqslant N-1$$

$$(^{\wedge}H^x\Psi)_N = b_N^{-1}(-\Psi_{N-1} + (a_N-x)\ \Psi_N)$$

Let $^{\wedge}p$ and $^{\wedge}q$ be the solutions of $H\Psi = \lambda\Psi$ with initial values $^{\wedge}p_M = {}^{\wedge}q_{M-1} = 1$, $^{\wedge}p_{M-1} = {}^{\wedge}q_M = 0$. It's easy to see that $^{\wedge}H^x$ is a self adjoint matrix on $^{\wedge}\mathcal{H}$, its eigenvalues are the roots of the polynom $^{\wedge}p_{N+1}$, they are simple and if λ is an eigenvalue we have

$$^{\wedge}\sigma^x_{m,n}\ (\lambda) = {}^{\wedge}p_n\ {}^{\wedge}p_m\Big(\sum_{k=M}^{N} b_k\ {}^{\wedge}p_k^2\Big)^{-1}$$

PROPOSITION 4.1 *Let* $x \in \mathbb{R}$, $m,n \in \mathbb{Z}$ *be fixed, then the sequence* $^{\wedge}\sigma^x_{m,n}$ *converges weakly to* $\sigma_{m,n}$ *when* Λ *goes to* $\mathbb{Z}$.

Proof : Fix λ with $\Im m\,\lambda \neq 0$, $x \in \mathbb{R}$, $n \in \mathbb{Z}$. A function $^{\wedge}\Psi$ in $^{\wedge}\mathcal{H}$ can be extended to $\mathcal{H}$ by $^{\wedge}\Psi_n = 0$ if $n \notin \Lambda$; with this convention :

$$((\hat{}H-\lambda I)\hat{}\Psi) = (H-\lambda I)\hat{}\Psi + \hat{}\Psi_M\, e^{M-1} + \hat{}\Psi_N\, e^{N+1} - x\hat{}\Psi_N\, e^{N}.$$

When $|k| \to +\infty$ the sequence e^k converges weakly to zero, thus if the sequence $\hat{}\Psi$ is bounded when Λ goes to $\mathbb{Z}$ (and hence weakly compact) the sequence $\hat{}\phi = \hat{}\Psi_M\, e^{M-1} + \hat{}\Psi_N e^{N+1} - x\,\hat{}\Psi_N\, e^N$ converges weakly to zero when Λ goes to $\mathbb{Z}$. Taking $\hat{}\Psi = \hat{}R_\lambda^x\, e^n$ the sequence $\hat{}\Psi$ is bounded by $b_n^{-1}|\Im m\,\lambda|^{-1}$. With this choice we have $e^n = (H-\lambda)\hat{}\Psi + \hat{}\phi$ and $R_\lambda\, e^n = \hat{}\Psi + R_\lambda\,\hat{}\phi$. If Ψ is any weak cluster point of $\hat{}\Psi$ we have $\Psi = R_\lambda\, e^n$ and thus $\hat{}R_\lambda^x\, e^n$ converges weakly to $R_\lambda\, e^n$ when Λ goes to $\mathbb{Z}$. Taking in account that

$$<\hat{}R_\lambda^x\, e^n\,,\, e^m> = \int (t-\lambda)^{-1}\, d\,\hat{}\sigma^x_{m,n}(t)$$

$$< R_\lambda\, e^n\,,\, e^m > = \int (t-\lambda)^{-1}\, d\,\sigma_{m,n}(t)$$

we obtain the expected result since it is well known that convergence of integrals of the continuous bounded functions $t \to \frac{1}{t-\lambda}$ for $\Im m\,\lambda \neq 0$ implies weak convergence.

LEMMA 4.2 : *For* $\Im m\,\lambda \neq 0$ *we have*

$$<\hat{}R_\lambda^x\, e^M,\; e^M> = -\,\frac{\hat{}q_{N+1} - x\hat{}q_N}{\hat{}p_{N+1} - x\hat{}p_N}$$

Proof : Let $\Psi \in \hat{}\mathcal{H}$ with $(\hat{}H^x - \lambda I)\Psi = e^M$ this equation can be written :

$$\begin{pmatrix}\Psi_N\\ \Psi_{N+1}\end{pmatrix} = Y_{N-1}\,\cdots\, Y_{M+1}\begin{pmatrix}\Psi_{M+1}\\ \Psi_M\end{pmatrix}$$

$$\Psi_{M+1} = (a_M - \lambda b_M)\,\Psi_M - 1$$

$$\Psi_{N-1} = (a_N - \lambda b_N)\,\Psi_N - x\,\Psi_N$$

and we obtain the relation :

$$\begin{bmatrix}x\,\Psi_N\\ \Psi_{N^{\cdot}}\end{bmatrix} = Y_N\, Y_{N-1}\,\cdots\, Y_M\begin{bmatrix}\Psi_M\\ 1\end{bmatrix}$$

taking in account that $Y_N\,\ldots\,Y_M = \begin{bmatrix}p_{N+1} & q_{N+1}\\ p_N & q_N\end{bmatrix}$ the statement follows from $<\hat{}R_\lambda^x\, e^M,\, e^M> = \Psi_M$.

LEMMA 4.3 *If* $\Im m\,\lambda > 0$, $\Im m\, z \geqslant 0$ *then* $\hat{}p_{n+1} - z\hat{}p_N \neq 0$

Proof : If $\hat{}p_{N+1} - z\hat{}\,p_N = 0$, z is an eigenvalue of the operator $\hat{}H^z$ (defined in the same way as $\hat{}H^x$). For $z = x + iy$ the adjoint operator $(\hat{}H^z)^*$ is given by :

$((\hat{H}^z)^* \Psi)_n = (\hat{H}^z \Psi)_n \quad M \leq n \leq N-1$

$((\hat{H}^z)^* \Psi)_N = (\hat{H}^z \Psi)_N + 2iyb_N^{-1}\, \Psi_N$

For an eigenfunction Ψ associated to z

$<\hat{H}^z_\lambda \Psi, \Psi> = \lambda ||\Psi||^2 = <\Psi, (\hat{H}^z)^* \Psi> = \bar{\lambda} ||\Psi||^2 - 2iy|\Psi_N|^2$

thus $(\Im m\, \lambda)\ ||\Psi||^2 = -y|\Psi_N|^2$ and this is impossible, the right member of this equation being strictly positive.

We recall without proof some basic facts about the Poisson kernel of the upper half plane. These results are proven for the unit disc in [32]. Let C_z be the Cauchy distribution with density $\frac{1}{\Pi} \frac{\beta}{(t-\alpha)^2+\beta^2}\, dt$, $z = \alpha + i\beta$ with $\beta > 0$. Let f(z) be an analytic function on $\Im m\, z > 0$ such that f can be continuously extended to $\mathbb{R} \cup \{\infty\}$. Then this extension satisfies the Poisson formula :

$$f(z) = \int f(t)\, dC_z(t) \quad (\Im m\, z > 0)$$

Furthermore let F(z) be the function associated on $\Im m\, z > 0$ to a positive bounded measure μ by : $F(z) = \frac{1}{\Pi} \int (t-z)^{-1}\, d\mu(t)$. If $\mu = \mu^a + \mu^s + \mu^p$ is the Lebesgue decomposition of μ, then :

(i) $\lim_{\varepsilon \downarrow 0} \Im m\, F(x+i\varepsilon)$ exists for Lebesgue almost all x of $\mathbb{R}$ and is equal to the density of μ^a.

(ii) $(\mu^s+\mu^p)$ is supported by the subset $\{x \in \mathbb{R} / \lim_{\varepsilon \downarrow 0} \Im m\, F(x+i\varepsilon) = +\infty\}$

In particular if $\lim_{\varepsilon \downarrow 0} \Im m\, F(x+i\varepsilon)$ exists and is finite for all real x, then μ is absolutely continuous with respect to the Lebesgue measure on $\mathbb{R}$ and its density is given by this limit. It's also useful to remark that $\Im m\, F(z) = \int \frac{dC_z}{dt}\, d\mu(t)$, that is, $\Im m\, F(z)$ is the Poisson integral of μ and the above statements are nothing else than Fatou's theorems.

We define $\hat{\tilde{\sigma}}_{m,n}$ as the average of $\hat{\sigma}^x_{m,n}$ with respect to the Cauchy measure by the formula

$$\hat{\tilde{\sigma}}_{m,n}(\Delta) = \int \hat{\sigma}^x_{m,n}(\Delta)\, dC_i(x)$$

<u>PROPOSITION 4.4</u> *The measure* $\hat{\tilde{\sigma}}_{m,n}$ *has a density with respect to the Lebesgue measure on* $\mathbb{R}$ *given by* $\frac{1}{\Pi} \frac{\hat{p}_n \hat{p}_m}{\hat{p}^2_{N+1} + \hat{p}^2_N}$

In particular we have the formula $\frac{1}{\Pi}\int \frac{d\lambda}{{}^\wedge p_N^2 + {}^\wedge p_{N+1}^2} = 1$

Proof : Thank's to the relation ${}^\wedge\sigma^x_{m,n} = {}^\wedge p_n\, {}^\wedge p_m\, {}^\wedge\sigma^x_M$, it's enough to prove 4.4 for $m = n = M$. By the lemma 4.3, for $\Im m\,\lambda > 0$ the function $z \to -\dfrac{{}^\wedge q_{N+1} - z\,{}^\wedge q_N}{{}^\wedge p_{N+1} - z\,{}^\wedge p_N}$ is continuous on $\Im m\, z \geqslant 0$ and has obviously a limit when $|z| \to +\infty$. From the lemma 4.2 and the properties of the Poisson integral :

$$F(\lambda) = \frac{1}{\Pi}\int (t-\lambda)^{-1}\, d\,{}^\wedge\tilde{\sigma}_M(t) = -\frac{{}^\wedge q_{N+1} - i\,{}^\wedge q_N}{{}^\wedge p_{N+1} - i\,{}^\wedge p_N}$$

As it is readily seen that for λ real $\lim_{\varepsilon\downarrow 0} \Im m\, F(\lambda + i\varepsilon) = \dfrac{1}{{}^\wedge p_{N+1}^2 + {}^\wedge p_N^2}$

the result follows from the above remarks.

The sequence ${}^\wedge\tilde{\sigma}_{m,n}$ converges weakly to $\sigma_{m,n}$ by the proposition 4.1 when Λ goes to $\mathbb{Z}$ (this is true for ${}^\wedge\sigma^x_{m,n}$ and hence for the averaged measures). One of our essential goals is to obtain (under some hypothesis) that $\sigma_{m,n}$ is pure point, and it seems surprising to work with ${}^\wedge\tilde{\sigma}_{m,n}$ which is absolutely continuous (rather than with ${}^\wedge\sigma^x_{m,n}$ which is pure point). But we shall see in the chapter III that this choice is well adapted to the proof of localization in the random case.

Exercise 4.5 Compute ${}^\wedge\tilde{\sigma}_{m,n}$ and $\sigma_{m,n}$ for the Laplace operator $(\mathring{H}\Psi)_n = -\Psi_{n-1} - \Psi_{n+1}$

Hint : $p_n(\lambda) = \dfrac{\mathrm{sh}(n-M+1)\omega}{\mathrm{sh}\,\omega}$ (where $\mathrm{ch}\,\omega = -\frac{\lambda}{2}$)

and for $\lambda \in\,]-2,+2[$, $\dfrac{d\,\sigma_{m,n}}{d\lambda} = \dfrac{1}{2\Pi}\cos\left[(m-n)\ \mathrm{Arcos}\, -\frac{\lambda}{2}\right] \cdot (1-\lambda^2)^{-1/2}$

Exercise 4.6 Prescribing boundary values at each endpoint of the box Λ, we define for $x,y \in \mathbb{R}$ the operator ${}^\wedge H^{y,x}$ by

$$({}^\wedge H^{y,x}\Psi)_M = b_M^{-1}(-\Psi_{M+1} + (a_M - y)\Psi_M)$$

$$({}^\wedge H^{y,x}\Psi)_n = (H\Psi)_n \quad \text{if } M+1 \leqslant n \leqslant N-1$$

$$({}^\wedge H^{y,x}\Psi)_n = b_N^{-1}(-\Psi_{N-1} + (a_N - x)\Psi_N)$$

Prove that for $\Im m\,\lambda \neq 0$

$$\langle {}^\wedge R_\lambda^{y,x}\, e^M,\ e^M\rangle = -\frac{{}^\wedge q_{N+1} - x\,{}^\wedge q_N}{{}^\wedge p_{N+1} - x\,{}^\wedge p_N + y({}^\wedge q_{N+1} - x\,{}^\wedge q_N)}$$

Exercise 4.7 Let ${}^{+}H^{y}$ be the operator on the "box" $[0, +\infty[$ with the boundary condition y at the endpoint 0 (see 4.6). When y = 0 this is just the operator ${}^{+}H$ defined in 2.10. Let ${}^{+}m(\lambda) = \int \frac{d^{+}\sigma(t)}{t-\lambda}$ defined in 2.12.

(i) Prove that $\langle {}^{+}R_{\lambda}^{y} e^{o}, e^{o} \rangle = \dfrac{{}^{+}m(\lambda)}{1 - y\,{}^{+}m(\lambda)}$

Hint : use 4.6 with N = 0, x = 0

(ii) Prove that the measure ${}^{+}\tilde{\sigma} = \int {}^{+}\sigma^{y}\, dC_{i}(y)$ is absolutely continuous with respect to the Lebesgue measure on $\mathbb{R}$.

Hint : $\mathfrak{Im} \dfrac{{}^{+}m(\lambda)}{1 - i\,{}^{+}m(\lambda)}$ is bounded on $\mathfrak{Im}\,\lambda > 0$

(Remark that $\mathfrak{Im}\,\lambda > 0 \Longrightarrow \mathfrak{Im}\,{}^{+}m(\lambda) > 0$).

Exercise 4.8 Compute ${}^{+}m(\lambda)$ for the Laplace operator (exercises 2.12 and 4.5).

I.5 The pure point spectrum. A criterion.

Let λ be an eigenvalue of H associated to a normalized eigenfunction Ψ (the eigenspaces are one dimensional !), then

$$\sigma_{m,n}(\lambda) = \langle \Psi, e^{n} \rangle \langle \Psi, e^{m} \rangle$$

and the density matrix $S(\lambda)$ defined in I.3 has the form :

$$\frac{1}{\Psi_{o}^{2} + \Psi_{-1}^{2}} \begin{bmatrix} \Psi_{o}^{2} & \Psi_{o}\Psi_{-1} \\ \Psi_{o}\Psi_{-1} & \Psi_{-1}^{2} \end{bmatrix}$$

In this case $S(\lambda)$ is of rank one and each column vector is proportional to $\begin{bmatrix} \Psi_{o} \\ \Psi_{-1} \end{bmatrix}$

Therefore, in order to prove that λ is an eigenvalue it seems natural to prove that each column vector of $S(\lambda)$ can be taken as initial values of a solution of $H\Psi = \lambda\Psi$ which is in $\mathcal{H}$. ($S(\lambda)$ is defined σ almost everywhere so this is only an heuristic argument ...).

If I is an open subset of $\mathbb{R}$, we say that the spectrum is pure point on I if $\sigma^{a}(I) = \sigma^{s}(I) = 0$ (or $\mathfrak{S}^{a} \cap I = \mathfrak{S}^{s} \cap I = \emptyset$).

THEOREM 5.1 *Let* I *be an open subset of* $\mathbb{R}$ *for which there exists two positive constants* C *and* ρ *with* $\rho < 1$ *such that for each box* Λ *and* $m,n \in \Lambda$, $|{}^{\Lambda}\tilde{\sigma}_{m,n}|(I) \leq C\rho^{|m-n|}$.
Then if $\Sigma\, b_n^{1/2} \rho^{|n|} < +\infty$, *the spectrum of* H *is pure point on* I *and the associated eigenfunctions fall off exponentially at infinity.*

Proof : On the set of bounded measures on $\mathbb{R}$ endowed with the weak topology (that is convergence of integrals of bounded continuous functions), for each positive lowersemicontinuous function f, the map $\mu \longrightarrow |\mu|(f)$ is lowersemicontinuous. Together with proposition 4.1 this implies $|\sigma_{m,n}|(I) \leq C\rho^{(m-n)}$. Choosing m = 0 and m = -1 in the lemma 2.7 this can be written :

$$\int_I \left| p_n \frac{d\sigma_o}{d\sigma} + q_n \frac{d\sigma_{o,-1}}{d\sigma} \right| d\sigma \leq C\rho^{|n|}$$

$$\int_I \left| p_n \frac{d\sigma_{o-1}}{d\sigma} + q_n \frac{d\sigma_{-1}}{d\sigma} \right| d\sigma \leq C\rho^{|n+1|}$$

Observing that a sequence in $\ell^1(\mathbb{Z})$ is necessarily in $\ell^2(\mathbb{Z})$ the hypothesis $\Sigma\, b_n^{1/2} \rho^{|n|} < +\infty$ implies that for σ almost all λ of I the two column vectors of the matrix $S(\lambda)$ can be taken as initial values of a solution of $H\Psi = \lambda\Psi$ with $\Psi \in \mathcal{H}$. These column vectors can't be simultaneously zero (the trace of $S(\lambda)$ is equal to 1, σ a.e.). Hence σ almost all λ of I is an eigenvalue and σ is pure point on I. For an eigenvalue λ letting $\Psi_o = \sigma_o(\lambda)$ and $\Psi_{-1} = \sigma_{o,-1}(\lambda)$ we obtain $|\Psi_n| \leq C\rho^{|n|}$.

REMARKS 5.2 (i) In the above proof, the hypothesis of 5.1 has only to hold for m = 0 and m = -1, but as it's easily seen these two statements are equivalent.

(ii) The same result holds when we replace ${}^{\Lambda}\tilde{\sigma}_{m,n}$ by ${}^{\Lambda}\sigma^x_{m,n}$ (for a fixed x or a bounded sequence x_{Λ}) in the hypothesis of theorem 5.1, but checking such a property seems much more difficult, the measures ${}^{\Lambda}\sigma^x_{m,n}$ being not as tractable as the ${}^{\Lambda}\tilde{\sigma}_{m,n}$ for which we have a nice formula (4.4).

(iii) If the sequence b is bounded (in which case $\mathcal{H}$ is isomorphic to $\ell^2(\mathbb{Z})$) the assumption $\Sigma\, b_n^{1/2} \rho^{|n|} < +\infty$ is vacuous.

Exercise 5.3 Let I be an open subset of $\mathbb{R}$, then if $\sum_n b_n^{1/2} |\tilde{\sigma}_{m,n}|(I) < +\infty$

the spectrum of H is pure point on I.

The criterion 5.3 can be obtained as in [37] using the following "Ruelle formula" proved in [53] :

$$||\Pi_c \Psi||^2 = \lim_{\Lambda \to \mathbb{Z}} \lim_{T \to +\infty} \frac{1}{T} \int_0^T ||1_{\Lambda^c} e^{itH} \Psi||^2 dt$$

where Ψ is in $\mathcal{H}$, Π_c is the projection from $\mathcal{H}$ onto the subspace $\mathcal{H}^a + \mathcal{H}^s$ and 1_{Λ^c} is the projection operator $[1_{\Lambda^c}(\phi)]_n = \phi_n$ if $n \notin \Lambda$

$= 0$ if $n \in \Lambda$

Exercise 5.4 Let I be an open subset of $\mathbb{R}$.

(i) Prove that if for each integer m (or only for m = -1 and m = 0) we have $||\Pi_c E_I e^m|| = 0$, then the spectrum of H is pure point on I.

(ii) Prove that if for each m (or m = 0 and m = -1) $\Sigma\, b_n(|\sigma_{m,n}|(I))^2 < +\infty$ then the spectrum of H is pure point on I.

Hint : $\frac{1}{T} \int_0^T ||1_{\Lambda^c} e^{itH} \Psi||^2 dt \leq \sum_{n \notin \Lambda} b_n(|\sigma_{m,n}|(I))^2$

(iii) Observe that the criterion obtained by restriction of (ii) to boxes is stronger than 5.3.

I.6 Singularity of the spectrum

Osseledec's theorem 1.1 together with proposition 3.2 about slowly increasing generalized eigenfunctions, are powerful tools to obtain singular properties of the spectrum. The two following lemmas and the added remark actually contain essential ideas developed in the sequel. We assume in this section that $\lim_{n \to \pm\infty} \frac{1}{|n|} \text{Log}(1 + |a_n| + b_n) = 0$.

We say that the sequence $Y_n(\lambda)$ has an hyperbolic behavior if there exist two strictly positive numbers $\gamma_+(\lambda)$ and $\gamma_-(\lambda)$ such that :

$$\lim_{n \to \pm\infty} \frac{1}{|n|} \text{Log}||S_n(\lambda)|| = \gamma_\pm(\lambda)$$

If the sequence $Y_n(\lambda)$ has an hyperbolic behavior, Osseledec's theorem 1.1 insures the existence of two contractive vectors $V^+(\lambda)$ and $V^-(\lambda)$.

<u>LEMMA 6.1</u> *Let* m *be a positive continuous measure and* A *a Borel subset of* $\mathbb{R}$, *such that the sequence* $Y_n(\lambda)$ *has an hyperbolic behavior* m a.e. *on* A. *Then the spectral measure* σ *is orthogonal to* m *on* A.

Proof : Osseledec's theorem 1.1 implies existence of contractive vectors $V^+(\lambda)$ and $V^-(\lambda)$ m a.e. on A. Such λ are eigenvalues of H iff $V^+(\lambda)$ and $V^-(\lambda)$ are proportional from the hypothesis on the sequence b_n. The measure m being continuous such countable set of λ is m negligible. This implies that each non zero solution of $H\Psi = \lambda\Psi$ is exponentially growing in at least one direction of $\mathbb{Z}$ for m almost all λ in A. Since for σ almost λ there exists a non zero solution of $H\Psi = \lambda\Psi$ which is slowly increasing, the conclusion follows.

<u>LEMMA 6.2</u> *Let* A *be a Borel subset of* $\mathbb{R}$ *such that the sequence* $Y_n(\lambda)$ *has an hyperbolic behavior* σ a.e. *on* A. *Then the spectral measure* σ *is pure point on* A.

Proof : Osseledec's theorem 1.1 implies existence of contractive vectors $V^+(\lambda)$ and $V^-(\lambda)$ σ a.e. on A. Moreover for σ almost all λ, there exists a non zero slowly increasing function Ψ solution of $H\Psi = \lambda\Psi$ and thus $V^-(\lambda)$ and $V^+(\lambda)$ have to be proportional σ a.e. on A. Hence, on A, σ is supported by the eigenvalues of H (if any).

<u>REMARK 6.3</u> Lemmas 6.1 and 6.2 can hardly be directly applied to a deterministic operator H. But we shall see in the sequel that in the "random case" and under broad conditions, for a given measure m, the hypothesis of 6.1 are satisfied for "almost all samples". Thus we could think that occurence of the hypothesis of 6.2 is highly improbable... Nevertheless we shall see that, in the independent case, this hypothesis holds with probability one ! This is certainly the most amazing fact in localization theory.

CHAPTER II

ERGODIC SCHRÖDINGER OPERATORS

II.1 Definition and examples

We now suppose that (a_n, b_n), $n \in \mathbb{Z}$, is a stationary random process. This means that the real random variables $(a_n(\omega), b_n(\omega))$ are defined on some complete probability space $(\Omega, \mathcal{A}, \mathbb{P})$ and that there exists an invertible measurable transformation θ of Ω leaving $\mathbb{P}$ invariant and such that $a_{n+1} = a_n \circ \theta$, $b_{n+1} \circ \theta$. In general we don't write the variable ω and when a property depends only upon the common law of the sequence (a_n, b_n) we omit the index n and speak of the variables (a) and (b). We say that the family $H(\omega)$ of associated operators on $\mathcal{H}$ is ergodic if a θ invariant measurable subset of Ω is of zero or one $\mathbb{P}$ measure. It's easily seen that $H \circ \theta = U^{-1} H U$ where U is the shift operator on $\mathbb{Z}$, $(U\Psi)_n = \Psi_{n-1}$.

The most important example in this book will be the "independent case", that is (a_n, b_n) is an independent identically distributed sequence of random variables with values in $\mathbb{R}^2$. Then $(\Omega, \mathcal{A}, \mathbb{P})$ is constructed as the $\mathbb{Z}$ power of a probability space $(\mathbb{R}^2, \mathcal{F}, \mu)$. For $\omega = \{(x_n, y_n)\}$, $n \in \mathbb{Z}$, we define $\theta(\omega) = \{(x_{n+1}, y_{n+1})\}$ and $a_n(\omega) = x_n$, $b_n(\omega) = y_n$. An other example is the "quasi periodic case" where $(\Omega, \mathcal{A}, \mathbb{P})$ is the one dimensional torus $[0, 2\Pi]$ endowed with the Lebesgue probability, θ is the "rotation" $\theta(\omega) = \omega + 2\Pi\alpha$ with α an irrational number. The sequences $a_n = a \circ \theta^n$, $b_n = b \circ \theta^n$ are constructed from two real continuous functions a and b on Ω. A particular case is the "Mathieu operator" associated to the functions $b(\omega) = 1$, $a(\omega) = \beta \cos\omega$ where β is a real number. A degenerated example is given by the "periodic case" where $(\Omega, \mathcal{A}, \mathbb{P})$ is the finite set $\{0,1,2, \ldots, r-1\}$ with the uniform distribution. The "rotation" modulo r acts by $\theta(\omega) = \omega+1$. More general examples of almost periodic or limit periodic cases are also of great interest [57]. Ergodicity does not require any disorder as shows the periodic case (we don't assume any mixing property) nevertheless that is a sufficient assumption to imply $\mathbb{P}$ a.e. properties of the spectrum of $H(\omega)$ as a set. If we want to obtain stronger properties, as absence of absolutely continuous part in the spectrum, some randomness is necessary (the spectrum is absolutely continuous in the periodic case). In some sense the Lyapunov exponent defined in the section 3 is a good indicator for the required disorder as it is shown in the section 5.

Throughout this chapter we assume that $H(\omega)$ is an ergodic family of random operators.

II.2 General spectral properties

We assume in this section that the random variable (b) is strictly positive and bounded away from zero, then by I.2.2 for $\mathbb{P}$ almost all ω the operator $H(\omega)$ is self adjoint on the Hilbert space $\mathcal{H}(\omega)$ (note that the scalar product is itself random). The shift operator U is a unitary map from $\mathcal{H} \circ \theta$ onto $\mathcal{H}$ thus H and $H \circ \theta$ are unitary equivalent. For ϕ, Ψ in $\mathcal{H} \circ \theta$ we have the equality $\langle (E_\Delta \circ \theta)\, \phi, \Psi\rangle = \langle E_\Delta\, U\phi\, , U\Psi\rangle$ and thus $\sigma_n \circ \theta = \sigma_{n+1}$.

LEMMA 2.1 *Let Δ be a Borel subset of $\mathbb{R}$, m and n two integers, then the function $\sigma_{m,n}(\Delta)$ defined $\mathbb{P}$ a.e. on Ω is mesurable.*

Proof : By a monotone class argument it's enough to prove this property for an open interval $\Delta =]-\infty\ \alpha[$. For such a Δ we have

$$\sigma_{m,n}(\Delta) = \lim_{\varepsilon \downarrow 0} \lim_{\delta \downarrow 0} \frac{1}{\Pi} \int_{-\infty}^{\alpha-\delta} \Im m \langle R_{t+i\varepsilon}\, c^m, c^n\rangle\, dt$$

thank's to : $\lim_{\varepsilon \downarrow 0} \frac{1}{\pi} \int_a^b \mathfrak{I}m \langle R_{t+i\varepsilon} e^m, e^n \rangle = \sigma_{m,n}(]a,b[) + \frac{1}{2}(\sigma_{m,n}(a) + \sigma_{m,n}(b))$

By virtue of I.4.4 we obtain :

$$\langle R_{t+i\varepsilon} e^m, e^n \rangle = \lim_{\Lambda \to \mathbb{Z}} \frac{1}{\pi} \int (\lambda - (t+i\varepsilon))^{-1} \frac{p_n \; p_m}{p_N^2 + p_{N+1}^2} d\lambda$$

This last expression is obviously jointly measurable in (ω,t) and this gives the result.

The following lemma due to L.A. Pastur [50] has a well known consequence in the "periodic case" since it implies that such operators have no eigenvalues.

LEMMA 2.2 *Let* λ *be a fixed real number. For* $\mathbb{P}$ *almost all* ω*,* λ *is not an eigenvalue of* $H(\omega)$.

Proof : Defining the subset Ω_λ of Ω by $\Omega_\lambda = \{\sigma(\lambda) > 0\}$ the lemma 2.1 implies that Ω_λ is measurable. The relation $\sigma_n \circ \theta = \sigma_{n+1}$ added to the fact that σ and $(\sigma_n + \sigma_{n+1})$ have the same null sets proves that Ω_λ is θ invariant thus $\mathbb{P}(\Omega_\lambda) = 0$ or 1. If we denote by $\operatorname{tr} E_\lambda$ the trace of the projection E_λ onto the eigenspace associated to λ, then $\operatorname{tr} E_\lambda = 1$ if $\omega \in \Omega_\lambda$, $\operatorname{tr} E_\lambda = 0$ if $\omega \notin \Omega_\lambda$ and $\operatorname{tr} E_\lambda = \Sigma\, b_n \sigma_n(\lambda)$.
Taking expectation we obtain :
$\mathbb{P}(\Omega_\lambda) = \Sigma\, \mathbb{E}(b_n \sigma_n(\lambda))$. Stationarity implies that $\mathbb{E}(b_n \sigma_n(\lambda))$ is a constant which has to be zero and thus $\mathbb{P}(\Omega_\lambda) = 0$.

The two following lemmas are proven by H. Kunz and B. Souillard in [37].

LEMMA 2.3 *There exists a closed subset* $\mathfrak{S}$ *of* $\mathbb{R}$ *such that for* $\mathbb{P}$ *almost all* ω*,* $\mathfrak{S}(\omega) = \mathfrak{S}$.

Proof : Following the lines of [37], let R_n be the countable set of open intervals of $\mathbb{R}$ with rational endpoints and length $\frac{1}{n}$. For $I \in R_n$ the subset $\Omega_I = \{\sigma(I) > 0\}$ is θ invariant (see the proof of 2.2) and hence of zero or one $\mathbb{P}$ measure. Let $S_n = \{I \in R_n \;/\; \mathbb{P}(\Omega_I) = 1\}$ and

$\mathfrak{S} = \bigcap_{n=1}^{\infty} \bigcup_{I \in S_n} I$; it's easy to check that for $\mathbb{P}$ almost all ω,

$\mathfrak{S}(\omega) = \mathfrak{S}$ (and hence $\mathfrak{S}$ is closed !).

In the quasi periodic case this result holds for every ω in Ω [3] and in the independent case the set $\mathfrak{S}$ has a more explicit description.

LEMMA 2.4 *In the independent case, assume that* $b_n = 1 \ \forall n \in \mathbb{Z}$. *Then for* $\mathbb{P}$ *almost all* ω, $\mathfrak{S}(\omega) = [-2, +2] + S\mu$ *where* $S\mu$ *is the topological support of the law* μ *of* (a).

Proof : We denote by $\overset{\circ}{H}$ and V the operators $(\overset{\circ}{H}\Psi)_n = -\Psi_{n+1} - \Psi_{n-1}$, $(V\Psi)_n = a_n \Psi_n$ such that $H = \overset{\circ}{H} + V$. It's easily seen that the spectrum of V is for $\mathbb{P}$ a.e. equal to $S\mu$, that $\|\overset{\circ}{H}\| = 2$ and that the spectrum of $\overset{\circ}{H}$ is equal to $[-2, +2]$. Applying the result of the exercise 2.5 below (the proof can be found in [31]) we obtain $\mathfrak{S}(\omega) = [-2, +2] + S\mu$.
For the converse we give the proof when $S\mu$ is a finite set, the general case handled in [37] being a straighforward extension of the main ideas given below. Let $\lambda \in [-2, +2]$, by Weyl's criterion there exists a sequence (Ψ^k), $k \in \mathbb{N}$, in $\mathcal{H}$ such that $\|\Psi^k\| = 1$, $\lim_{k \to +\infty} \|(\overset{\circ}{H} - \lambda I)\Psi^k\| = 0$.
$\overset{\circ}{H}$ is a bounded operator so we can choose Ψ^k to be zero outside of a box Λ_k.
If $\alpha \in S\mu$ and if N is a positive integer, the set $\Omega_\alpha = \{a_n = \alpha / n=0,1,..,N\}$ is of positive $\mathbb{P}$ measure and by "Poincare's recurrence lemma" we know that $\mathbb{P}\,(\overline{\lim_{n \to +\infty}}\ \theta^{-n}(\Omega_\alpha)) = 1$. Then for $\mathbb{P}$ almost all ω, there exists for each positive integer N a box of length N such that $a_n(\omega) = \alpha$ for n in this box. This implies that for $\mathbb{P}$ almost all ω there exists for each k a box $\tilde{\Lambda}_k$ of same length as Λ_k and such that $a_n(\omega) = \alpha$ for $n \in \tilde{\Lambda}_k$. Let U_k be the translation of $\mathbb{Z}$ such that $U_k(\tilde{\Lambda}_k) = \Lambda_k$, taking in account that $\overset{\circ}{H}$ is translation invariant, that is $U_k\overset{\circ}{H} = \overset{\circ}{H}U_k$, we obtain letting $\phi^k = U_k \Psi^k$: $(H - (\lambda+\alpha)I)\,\phi^k = U_k\,(\overset{\circ}{H} - \lambda I)\,\Psi^k + (V - \alpha I)\,\phi^k$.
The last term of this equality is zero and a new application of Weyl's criterion using the sequence ϕ^k gives the result.

An important consequence of the lemma 2.4 is that such operators have no discrete spectrum (but can have pure point spectrum !). Furthermore we see that their spectrum depend upon the law μ only by its support $S\mu$, so we can't expect any continuity property of the spectrum with respect to μ without strong assumptions on the convergence of supports.

Exercise 2.5 Let A be a bounded operator, B a self adjoint operator and $H = A+B$. Prove that if λ is in the spectrum of H then $d(\lambda,\sigma(B)) \leq \|A\|$ where $d(\lambda,\sigma(B)) = \inf\{|\lambda - y| / y \in \text{spectrum of } B\}$.

Hint : See [31] chapter V theorem 4.10.

Exercise 2.6 Compute the spectrum of the periodic operator

$(H\Psi)_n = -\Psi_{n-1} - \Psi_{n+1} + (-1)^n \Psi_n$

Hint : Perform a Fourier transform and for the general periodic case see [55] theorem 1.

II.3 The Lyapunov exponent in the general ergodic case

In this section we assume that $\mathbb{E}\,\mathrm{Log}(1+|a|+|b|) < +\infty$ the random variable (b) taking any real value.

We recall in the particular case of Schrödinger matrices the definition of the Lyapunov exponent given in A.I.2 Let λ be a complex number and $\gamma_n(\lambda) = \mathbb{E}\,\mathrm{Log}\,\|S_n\|$, $n \in \mathbb{Z}$, (see I.1). It is readily seen that for positive integers m,n :

(i) $\gamma_{m+n} \leq \gamma_m + \gamma_n$

(ii) $\gamma_0 \leq \frac{1}{2}\mathbb{E}\,\mathrm{Log}(2 + (a-\lambda b)^2) < +\infty$

(iii) $\gamma_{-n} = \gamma_{n-1} \quad (n \geq 1)$

Thus γ_n is a positive subadditive sequence and the Lyapunov exponent $\gamma(\lambda)$ is defined by :

$$\gamma(\lambda) = \lim_{|n| \to +\infty} \frac{1}{|n|}\gamma_n(\lambda) = \inf_{n \geq 1} \frac{1}{n}\gamma_n(\lambda)$$

A function f on $\mathbb{C}$ with values in $[-\infty, +\infty[$ is called subharmonic if f is upper semicontinuous and if $f(\lambda)$ is less than the mean value of f over the circles of center λ. Observing that the same submean property is true for discs and taking limits when the radius is going to zero, we see that two subharmonic functions almost everywhere equal (with respect to the Lebesgue measure on $\mathbb{C}$) are equal. The role of subharmonicity of the Lyapunov index has been first pointed out by M. Herman in [26] and used in the general ergodic case by W. Craig and B. Simon in [11].

PROPOSITION 3.1 *The function* $\gamma(\lambda)$ *is subharmonic.*

Proof : $S_n(\lambda)$ is an holomorphic matrix in the variable λ, thus $\mathrm{Log}\|S_n(\lambda)\|$ is subharmonic and the same is true for $\gamma_n(\lambda)$. The infinimum of a sequence of locally bounded subharmonic functions is itself subharmonic and this yields the result.

As it was shown in A.I for the independent case, γ can be recovered as a pointwise almost everywhere limit :

PROPOSITION 3.2 *Let* λ *be a fixed complex number. Then the sequence* $\frac{1}{|n|}\mathrm{Log}\|S_n\|$ *converges* $\mathbb{P}$ *a.e to* $\gamma(\lambda)$ *when* $|n|$ *goes to* $+\infty$.

Proof : This is just the subadditive ergodic theorem thank's to the relation, for positive integers m,n :

$\mathrm{Log}\|S_{n+m}\| \leq \mathrm{Log}\|S_n\| + (\mathrm{Log}\|S_m\|)\circ\theta^n$. See [42] for a proof.

It is readily seen that for each fixed λ the sequence $\frac{1}{|n|}\mathrm{Log}\|Y_n\|$ converges $\mathbb{P}$ a.e to zero, thus the proposition 3.2 together with Osseledec's theorem (I.1.1) implies that for each fixed λ such that $\gamma(\lambda) > 0$, there exists for $\mathbb{P}$ almost all ω, an hyperbolic structure in the two directions of $\mathbb{Z}$. If we think to the $\mathbb{P}$ a.e pure point spectrum, we see that we must obtain the a.e in the reverse order, namely : for $\mathbb{P}$ almost all ω there exist hyperbolic structures in the two directions of $\mathbb{Z}$ with the same contractive vector for $\sigma(\omega)$ almost all λ. This inversion of a.e properties is the most important problem discussed in the sequel.

As we are mainly concerned in this book with the independent case we give without proof a result of S. Kotani [35] asserting that γ is "often positive". We say that the process (a_n,b_n) is deterministic if (a_o,b_o) is measurable with respect to the subsigma algebras generated by $\{(a_k,b_k)/k \leq -1\}$.

PROPOSITION 3.3 *Assume that* $\{\lambda \in \mathbb{R}/\ \gamma(\lambda) = 0\}$ *is of positive Lebesgue measure, then the process* (a_n,b_n) *is deterministic.*

The proof given by B. Simon [58] assume that (a) is bounded, but the general case is handled in [33]. A typical example of deterministic

process is given by the Mathieu operator and the next proposition shows that the converse of 3.3 is not true.
The very easy proof given here is due to to M. Herman [27].

PROPOSITION 3.4 : *Let* $H(\omega)$ *be the Mathieu operator*

$$[H(\omega)\,\Psi]_n = -\Psi_{n-1} - \Psi_{n+1} + \beta\,\Psi_n \cos(2\Pi\, n\alpha + \omega)$$

where α,β *are real numbers with* α *irrational.*
Then for any complex number λ , $\gamma(\lambda) \geq \mathrm{Log}|\frac{\beta}{2}|$.

Proof : For a complex number z we set $z_n = z\, e^{2\Pi i n\alpha}$ and we define the matrix Z_n by

$$Z_n(z) = \begin{bmatrix} \frac{\beta}{2}(z_n^2+1) - \lambda z_n & -z_n \\ z_n & 0 \end{bmatrix}$$

Observing that $Z_n(e^{i\omega}) = z_n Y_n(\omega,\lambda)$ we have

$$\|Z_n(e^{i\omega}) \ldots Z_o(e^{i\omega})\| = \|S_n(\omega,\lambda)\|$$

and thus :

$$\gamma_n(\lambda) = \mathbb{E}\,\mathrm{Log}\|S_n(\omega,\lambda)\| = \frac{1}{2\Pi}\int_0^{2\Pi} \mathrm{Log}\|Z_n(e^{i\omega}) \ \ldots\ Z_o(e^{i\omega})\|\, d\omega$$

The function $z \to \mathrm{Log}\|Z_n(z) \ \ldots\ Z_o(z)\|$ is subharmonic and this implies:

$$\gamma_n(\lambda) \geq \mathrm{Log}\|Z_n(0) \ldots Z_o(0)\| = n\,\mathrm{Log}|\tfrac{\beta}{2}|.$$

Exercise 3.5 Compute the Lyapunov exponent for the operator $(H\Psi)_n = -\Psi_{n-1} - \Psi_{n+1} + (1)^n\,\Psi_n$ and compare $\{\lambda/\gamma(\lambda) = 0\}$ with the spectrum of H computed in 2.6 .

II.4 The Lyapunov exponent in the independent case

In this section it is assumed that we are in the independent case and that $\mathbb{E}\,\mathrm{Log}(1+|a|+|b|) < +\infty$ the random variable (b) taking any real value

In the independent case the Lyapunov exponent is given by the

Furstenberg's formula involving an "invariant measure" and this allows us to prove extra properties of positivity or continuity for this exponent.

We recall there some definitions given in the first part of the book. For a non zero vector x of $\mathbb{R}^2$, $\bar{x}$ is the corresponding point of the compact projective line denoted by X. A matrix g of $SL(2,\mathbb{R})$ acts on X by $g.\bar{x} = \overline{gx}$ and for two probability measures μ, ν on $SL(2,\mathbb{R})$ and X respectively, we define the "convolution" $\mu \overset{\bullet}{*} \nu$ by

$\mu \overset{\bullet}{*} \nu(f) = \int f(g.\bar{x})\, d\mu(g)\, d\nu(\bar{x})$ where f is a continuous function on X.

We say that ν is μ invariant if $\mu \overset{\bullet}{*} \nu = \nu$. For a real λ , μ_λ is the law of the matrix

$Y = \begin{bmatrix} a-\lambda b & -1 \\ 1 & 0 \end{bmatrix}$. Let ν be a probability measure on X such that

$\nu(\bar{x}_o) = 0$ for $x_o = \begin{pmatrix}1\\0\end{pmatrix}$ then we can define the image $\tilde{\nu}$ of ν on $\mathbb{R}$ by the application $\bar{x} \to \frac{u}{v}$ where $x = \begin{pmatrix}u\\v\end{pmatrix}$.

We shall say in the sequel that we are in the "Helmotz case" when there exists a value λ^* of λ and a constant c such that the law of (a,b) is supported by the line $a - \lambda^* b = c$. In this case the operator H can be written :

$$(H\Psi)_n = b_n^{-1}\left[-\Psi_{n-1} - \Psi_{n+1} + c\Psi_n\right] + \lambda^* \Psi_n$$

and the matrix Y has the form :

$$Y = \begin{bmatrix} c-(\lambda-\lambda^*)\, b & -1 \\ 1 & 0 \end{bmatrix}$$

<u>PROPOSITION 4.1</u> *If the law of* (a,b) *is not concentrated on a single point of* $\mathbb{R}^2$ *then for* $\lambda \in \mathbb{R}$ ($\lambda \neq \lambda^*$ *in the Helmotz case)* :

(i) There exists a unique μ_λ *invariant probability measure* ν_λ *, and* $\nu_\lambda(\bar{x}) = 0$, $\forall \bar{x} \in X$.

(ii) $\gamma(\lambda) > 0$ *and* $\gamma(\lambda) = \iint \text{Log} \frac{\|gx\|}{\|x\|}\, d\mu_\lambda(g)\, d\nu_\lambda(\bar{x})$

(iii) If there exists $\alpha > 0$ *such that* $\mathbb{E}(|a|^\alpha + |b|^\alpha) < +\infty$ *then*

$\gamma(\lambda) = \int \text{Log}\, |t|\, d\tilde{\nu}_\lambda(t)$

Proof : Looking at the theorem A.II.4.4 it's enough in order to prove (i) and (ii), to verify that the closed subgroup G_λ of $SL(2,\mathbb{R})$ generated by the support of μ_λ leaves no probability measure fixed on X Under the hypothesis of 4.1 G_λ contains at least two matrices $g_1 = \begin{bmatrix} \alpha & -1 \\ 1 & 0 \end{bmatrix}$ and $g_2 = \begin{bmatrix} \beta & -1 \\ 1 & 0 \end{bmatrix}$ with $\alpha \neq \beta$. Letting $\delta = \alpha-\beta$ we see that G_λ contains the matrices $u = g_1\, g_2^{-1} = \begin{bmatrix} 1 & \delta \\ 0 & 1 \end{bmatrix}$ and $v = g_1^{-1} g_2 = \begin{bmatrix} 1 & 0 \\ \delta & 1 \end{bmatrix}$. If we set $x_o = \begin{pmatrix} 1 \\ 0 \end{pmatrix}$ then $u\overline{x_o} = \overline{x_o}$ and for every $\bar{x}$ of X the sequence $u^n \bar{x}$ converges to $\overline{x_o}$. Thus any fixed probability measure under the action of G_λ must be concentrated on $\overline{x_o}$, but $v.\overline{x_o} \neq \overline{x_o}$ and such a probability measure does not exist.

The hypothesis of (iii) implies that $\|g\|^\alpha$ is μ_λ integrable and we know from the theorem A.VI.4.3 that there exists $\beta > 0$ such that $\sup_{\bar{y}\in X} \int \delta^{-\beta}(\bar{x},\bar{y})\, d\nu_\lambda(\bar{x}) < +\infty$ where δ is the usual distance on X defined by $\delta(\bar{x},\bar{y}) = \dfrac{\|x \wedge y\|}{\|x\|\,\|y\|}$. Choosing $y = \begin{pmatrix} 1 \\ 0 \end{pmatrix}$ we obtain for $x = \begin{pmatrix} u \\ v \end{pmatrix}$:

$$\int \left|\frac{u^2+v^2}{v^2}\right|^{\beta/2} d\nu_\lambda(\bar{x}) < +\infty \quad \text{and hence} \quad \int \operatorname{Log}\left(\frac{u^2+v^2}{v^2}\right) d\nu_\lambda(\bar{x}) < +\infty .$$

Let now m and ℓ be the positive measures on X such that $\tilde{m}$ and $\tilde{\ell}$ are the Cauchy and Lebesgue measures on $\mathbb{R}$.

We have the relation $\dfrac{\|x\|^2}{\|gx\|^2} = \dfrac{dg^{-1}m}{dm}(\bar{x}) = \dfrac{dg^{-1}m}{dg^{-1}\ell} \cdot \dfrac{dg^{-1}\ell}{d\ell} \cdot \dfrac{d\ell}{dm}(\bar{x})$

Letting $f(\bar{x}) = \dfrac{dm}{d\ell}(\bar{x}) = \dfrac{1}{\pi}\,\dfrac{v^2}{u^2+v^2}$, the above result implies that Log $f(\bar{x})$ is ν_λ integrable and remarking that $\dfrac{dg^{-1}m}{dg^{-1}\ell}(\bar{x}) = f(g.\bar{x})$ we obtain :

$$\iint \left(\operatorname{Log} \frac{dg^{-1}m}{dg^{-1}\ell} + \operatorname{Log} \frac{d\ell}{dm}\right) d\mu_\lambda\, d\nu_\lambda = 0$$

Applying now Furstenberg's formula :

$$\gamma(\lambda) = -\frac{1}{2}\iint \operatorname{Log} \frac{dg^{-1}m}{dm}\, d\mu_\lambda\, d\nu_\lambda = -\frac{1}{2}\iint \operatorname{Log} \frac{dg^{-1}\ell}{d\ell}\, d\mu_\lambda\, d\nu_\lambda$$

$$= -\frac{1}{2}\int \operatorname{Log}\left(\frac{v^2}{u^2}\right) d\nu_\lambda\,(\bar{x}) = \int \operatorname{Log} |t|\, d\tilde{\nu}_\lambda(t)$$

REMARK 4.2 The exponent γ is easily computed in the Helmotz case at the point λ^* as the spectral radius of the matrix

$\begin{bmatrix} c & -1 \\ 1 & 0 \end{bmatrix}$ and we find :

$$\gamma(\lambda^*) = 0 \quad \text{if } |c| \leq 2$$

$$\gamma(\lambda^*) = \text{Log}\,\frac{1}{2}(|c| + \sqrt{c^2-4}) \quad \text{if } |c| \geq 2$$

In the general ergodic case we have seen that γ is uppersemicontinuous (3.1). In the independent case γ is actually continuous and we divide the proof in two parts : the case $\lambda \neq \lambda^*$ (which is easy) and the case $\lambda = \lambda^*$ (which is more involved).

<u>LEMMA 4.3</u> *Let λ_n be a sequence of real numbers distincts from λ^* and with limit λ_o. Then for any subsequence extracted from ν_{λ_n} and weakly convergent to ν the function $\gamma(\lambda)$ has a limit along this subsequence equal to :*

$$I(\lambda_o,\nu) = \iint \text{Log}\,\frac{\|gx\|}{\|x\|}\, d\mu_{\lambda_o}(g)\, d\nu(\bar{x})$$

Moreover ν is μ_{λ_o} invariant.

Proof : Define for each $\lambda \in \mathbb{R}$ the function $f_\lambda(\bar{x}) = \int \text{Log}\,\frac{\|gx\|}{\|x\|}\, d\mu_\lambda(g)$. From $|\text{Log}\,\frac{\|gx\|}{\|x\|}| \leq \text{Log}\|g\|$, it is readily seen that f_λ is continuous on X. If we set $g_\lambda = \begin{bmatrix} a-\lambda b & -1 \\ 1 & 0 \end{bmatrix}$ then :

$$\sup_{\bar{x}\in X} |f_{\lambda_1}(\bar{x}) - f_{\lambda_2}(\bar{x})| \leq \mathbb{E}\left|\text{Log}\,\frac{\|g_{\lambda_1}x\|}{\|g_{\lambda_2}x\|}\right| \leq \mathbb{E}\ \text{Log}\|g_{\lambda_1} g_{\lambda_2}^{-1}\|$$

This yields the first result writing :

$$\gamma(\lambda_n) - I(\lambda_o,\nu) = \nu_{\lambda_n}(f_{\lambda_n} - f_{\lambda_o}) + (\nu_{\lambda_n} - \nu_{\lambda_o})\, f_{\lambda_o}$$

In order to see that ν is μ_{λ_o} invariant we note that $\mu_{\lambda_n} \dot{*}\, \nu_{\lambda_n} = \nu_{\lambda_n}$ and taking in account that μ_{λ_n} converges weakly to μ_{λ_o} on $SL(2,\mathbb{R})$ and that X is compact the proof is complete.

<u>COROLLARY 4.4</u> *The function $\gamma(\lambda)$ is continuous at each point λ_o distinct from λ^*.*

Proof : Let λ_n be a sequence with limit $\lambda_o \neq \lambda^*$. From the lemma 4.3 we know that each limit point of ν_{λ_n} is μ_{λ_o} invariant, hence equal to ν_{λ_o}.

This implies that ν_{λ_n} converges weakly to ν_{λ_o} and $\gamma(\lambda_n)$ converges to $I(\lambda_o, \nu_{\lambda_o})$ which is equal to $\gamma(\lambda_o)$ by 4.1.

COROLLARY 4.5 *In the Helmotz case γ is continuous at the point λ^* (Assuming that* (b) *is bounded when* $|c| > 2$).

Proof : We know from 3.1 that γ is an uppersemicontinuous non negative function, hence each point λ with $\gamma(\lambda) = 0$ is a continuity point of γ and this gives the result for $|c| \leqslant 2$ in view of 4.2. In the hyperbolic case $|c| > 2$, the matrix $\begin{bmatrix} c & -1 \\ 1 & 0 \end{bmatrix}$ has two real eigenvalues e_1 and e_2 with $|e_1| > 1 > |e_2|$ and we denote by $x_1 = \begin{pmatrix} e_1 \\ 1 \end{pmatrix}$, $x_2 = \begin{pmatrix} e_2 \\ 1 \end{pmatrix}$ the associated eigenvectors. Each μ_λ invariant probability measure ν is a convex combination of $\delta_{\bar{x}_1}$ and $\delta_{\bar{x}_2}$, and we see from 4.2 that $I(\lambda^*, \delta_{\bar{x}_1}) = \gamma(\lambda^*)$. Thus continuity of γ at the point λ^* follows from 4.3 if we prove that any limit point ν of a sequence ν_{λ_n} with $\lambda_n \to \lambda^*$ is such that $\nu(\bar{x}_2) = 0$. Let T be the closed subset of X defined by $T = \{\bar{x}/u \geqslant v \geqslant 0 \text{ or } u \leqslant v \leqslant 0\}$. It is readily seen that T is invariant under the action of any matrix $\begin{bmatrix} d & -1 \\ 1 & 0 \end{bmatrix}$ with $d > 2$. Assuming $c > 2$ and the variable (b) bounded, T is invariant under the action of $\begin{bmatrix} c-(\lambda-\lambda^*)b & -1 \\ 1 & 0 \end{bmatrix}$ for λ near λ^*. Thus for λ near λ^*, ν_λ is supported by T and any limit point ν is also supported by T. Since we have $e_1 > 1 > e_2 > 0$ we see that $\bar{x}_2 \notin T$. The same works for $c < -2$ replacing v by $-$v in the definition of T.

REMARK 4.6 (i) Corollaries 4.4 and 4.5 are direct consequences of general theorems about perturbations of random product of matrices which have been proven by H. Hennion [25], H. Furstenberg, Y. Kifer [21]. In the case $|c| > 2$ and (b) bounded, D. Ruelle [54] has proven that γ is actually analytic at $\lambda = \lambda^*$. But when $|c| = 2$, analyticity can break down as shows the exercise 4.12 (iv). Such singularities have been studied by B. Derrida [16].

(ii) In the case $c = 2$ ($c = -2$) the matrix $\begin{bmatrix} c & -1 \\ 1 & 0 \end{bmatrix}$ has an unique invariant probability measure $\delta_{\bar{x}_1}$ with $x_1 = \begin{pmatrix} 1 \\ 1 \end{pmatrix}$ ($\delta_{\bar{x}_{-1}}$ with $x_{-1} = \begin{pmatrix} 1 \\ -1 \end{pmatrix}$).

(iii) In the case $|c| < 2$, the matrix $\begin{bmatrix} c & -1 \\ 1 & 0 \end{bmatrix}$ can have a lot of invariant probability measures ν but the integral $I(\lambda^*,\nu)$ does not depend on the choice of ν and is equal to zero. In order to see this, we remark that such an integral does not depend on the choice of the norm on $\mathbb{R}^2$ since if we set $f(\bar{x}) = \mathrm{Log}\,\frac{\|x\|}{|||x|||}$ where $|||x|||$ is any norm on $\mathbb{R}^2$ we have

$$\iint (f(g.\bar{x}) - f(\bar{x}))\, d\nu(\bar{x})\, d\mu_{\lambda^*}(g) = 0$$

Now there exists a real inversible matrix Q such that $Q\begin{bmatrix} c & -1 \\ 1 & 0 \end{bmatrix}Q^{-1}$ is a rotation and choosing the norm $|||x||| = \|Qx\|$ we obtain $I(\lambda^*,\nu) = 0$.

We now turn our attention to a result of E. Le Page [43] improving the corollary 4.4. We say that a function f defined on an open subset U of $\mathbb{R}$ is locally Hölder continuous if for any compact $T\subset U$ there exists a strictly positive $\alpha = \alpha(T)$ and a constant $C(T)$ such that

$$|f(t) - f(s)| \leq C(T)\, |t-s|^{\alpha} \quad , \quad s,t \in T$$

PROPOSITION 4.7 *Assume that there exists a strictly positive constant* β *such that* $\mathbb{E}\,\mathrm{Log}(|a|^{\beta} + |b|^{\beta}) < +\infty$. *Then the function* $\gamma(\lambda)$ *is locally Hölder continuous on* $\mathbb{R}\setminus\{\lambda^*\}$.

Proof : We assume that the law of (a,b) is not concentrated on a point of $\mathbb{R}^2$, in which case the result is obvious.

The most important part of the work is to prove that the function $\lambda \to \nu_\lambda$ is, in some sens, locally Hölder continuous. Using this property Furstenberg's formula will give the result for $\gamma(\lambda)$. The proof uses heavily the Hölder spaces $\mathcal{L}_\alpha$ introduced in A.V and the spectral properties of the operator $P_\lambda f(\bar{x}) = \int f(g.\bar{x})\, d\mu_\lambda(g)$ on $\mathcal{L}_\alpha$, for α small enough. Our essential goal is obtain some uniformity (with respect to λ in a compact set) of the spectral decomposition of the operator P_λ. Unfortunately this property is not easy to check since the function $\lambda \to P_\lambda$ is in general not continuous from $\mathbb{R}$ to the space of bounded operators on $\mathcal{L}_\alpha$. We give below the essential steps of the proof of 4.7. In all the lemmas, T is a fixed compact subset of $\mathbb{R}\setminus\{\lambda^*\}$.

LEMMA 4.8 *There exists* $\alpha(T) > 0$ *such that for* $\alpha \in \,]0, \alpha(T)]$, $\lambda \in T$, *the spectrum of* P_λ *acting on* $\mathcal{L}_\alpha$ *is contained in a disc of radius strictly less than one, except the eigenvalue 1 which is simple. Moreover for* $f \in \mathcal{L}_\alpha$

$P_\lambda f = \nu_\lambda(f)\ \mathbb{1} + Q_\lambda(f)$ *where the spectral radius of* Q_λ *is strictly less than one and* $\mathbb{1}(\bar{x}) = 1 \quad \forall \bar{x} \in X$, $\nu_\lambda(Q_\lambda f) = 0$, $Q_\lambda(\mathbb{1}) = 0$.

Proof : For each fixed λ this is just the theorem A.V.4.3 . The uniformity with respect to $\lambda \in T$ is a direct consequence of the uniform convergence of $\frac{1}{n}\mathbb{E}\ \mathrm{Log}\|S_n(\lambda,\omega)\,\bar{x}\|$ to $\gamma(\lambda)$ on $\{\lambda,\bar{x}\} \in T \times X$. (We remark that this result gives again the continuity of $\gamma(\lambda)$ on T). In order to obtain this uniformity we remark that for a sequence $(\bar{x}_n, \lambda_n)$ converging to $(\bar{x}_o, \lambda_o)$ the sequence $\nu_n = \frac{1}{n}\sum_{k=0}^{n-1} \mu_{\lambda_n}^{k} \,\dot{*}\, \delta_{\bar{x}_n}$ converges veakly to ν_{λ_o} (each limit point is μ_{λ_o} invariant). Moreover looking at the proof of 4.3 we obtain that $\frac{1}{n}\mathbb{E}\ \mathrm{Log}\|S_n(\lambda_n,\omega)\bar{x}_n\| = \nu_n(f_{\lambda_n})$ converges to $\nu_{\lambda_o}(f_{\lambda_o}) = \gamma(\lambda_o)$ and this yields the result.

<u>LEMMA 4.9</u> *There exists* $\beta(T) > 0$ *and for* $\alpha \in \,]0,\beta(T)]$ *a constant* $C(\alpha)$ *such that for* $(s,t) \in T$, $f \in \mathcal{L}_\alpha$ *:*

$$\|P_s f - P_t f\|_{\alpha/2} \leq C(\alpha)\|f\|_\alpha\ |t-s|^{\alpha/2}$$

Proof Using the notations of A.V.2.4 and 4.3 we have for $f \in \mathcal{L}_\alpha$ $(\alpha > 0)$ $s,t \in T$:

(i) $|P_s f(\bar{x}) - P_t f(\bar{x})| \leq m_\alpha(f)\ \mathbb{E}[\delta^\alpha(g_s\cdot\bar{x},\ g_t\cdot\bar{x})]$

Since $\delta(g_s\cdot\bar{x},\ g_t\cdot\bar{x}) \leq |b|\ \|g_s\|\ \|g_t\|\ |s-t|$ we obtain from the integrability condition in 4.7 the existence (for α small enough) of $A(\alpha)$ such that :

$$|P_s f(\bar{x}) - P_t f(\bar{x})| \leq m_\alpha(f)\ A(\alpha)\ |s-t|^\alpha$$

(ii) $|P_s f(\bar{x}) - P_s f(\bar{y})| \leq m_\alpha(f)\ \mathbb{E}\left[\dfrac{\delta^\alpha(g_s\cdot\bar{x},\ g_s\cdot\bar{y})}{\delta^\alpha(\bar{x},\ \bar{y})}\right]$

Since $\dfrac{\delta(g_s\cdot\bar{x},\ g_s\cdot\bar{y})}{\delta(\bar{x},\ \bar{y})} \leq \|g_s\|^2$ we obtain the existence (for α small enough) of $B(\alpha)$ such that :

$$|P_s f(\bar{x}) - P_s f(\bar{y})| \leq m_\alpha(f)\ B(\alpha)\ \delta^\alpha(\bar{x},\bar{y})$$

(iii) From (i) and (ii) we obtain that

$$|(P_s-P_t)f(\bar{x}) - (P_s-P_t)f(\bar{y})| \leq 2m_\alpha(f)\ \max[B(\alpha),\ A(\alpha)]\ \min(\delta^\alpha(\bar{x},\bar{y}),|t-s|^\alpha)$$

and the result of 4.9 follows from

$$\min(\delta^{\alpha/2}(\bar{x},\bar{y}),\ |s-t|^{\alpha}\ \delta^{-\alpha/2}(\bar{x},\bar{y})) \leq |s-t|^{\alpha/2}$$

Remark that the lemma 4 implies that $\lambda \to P_\lambda$ is continuous if we think of P_λ as a bounded operator from $\mathcal{L}_\alpha$ to $\mathcal{L}_{\alpha/2}$.

The following lemma is technical and the proof can be found in [43]. (Remark that the assertions of 4.10 would be obvious if $\lambda \to P_\lambda$ was continuous from $\mathbb{R}$ to the bounded operators on $\mathcal{L}_\alpha$).

LEMMA 4.10 *For* $\alpha \in]0, \alpha(T)]$ *(lemma 4.8) we have*

(i) $\sup_{\lambda \in T} \|(Q_\lambda - I)^{-1}\|_\alpha < +\infty$

(ii) $\sup_{\lambda \in T,\ z \in K} \|(P_\lambda - zI)^{-1}\|_{\alpha/2} < +\infty$ *where K is a compact subset contained in all the resolvent sets of* P_λ, $\lambda \in T$

LEMMA 4.11 *There exists* $\delta(T) > 0$ *and for* $\alpha \in]0, \delta(T)]$ *a constant* $D(\alpha)$ *such that for* $f \in \mathcal{L}_\alpha$, $(s,t) \in T$:

$$|\nu_t f - \nu_s f| \leq D(\alpha)|t-s|^{\alpha/2}\ \|f\|_\alpha$$

Proof : From 4.10 (i) we know that there exists a circle Γ in $\mathbb{C}$ with center 1 such that intersection of the spectrum of P_λ, $\lambda \in T$, with the associated disc is equal to the eigenvalue 1 (when $\alpha < \alpha(T)$). Moreover using the relation

$R_t(z) - R_s(z) = R_s(z)\ (P_s - P_t)\ R_t(z)$ where $R_t(z) = (P_t - z\,I)^{-1}$ and the lemmas 4.10 (ii), 4.9, we see that for $\alpha \leq \alpha(T) \wedge \beta(T) = \delta(T)$ and $f \in \mathcal{L}_\alpha$, $(s,t) \in T$, there exists $K(\alpha)$ with

$$\|R_t(z)\,f - R_s(z)f\|_{\alpha/2} \leq \|f\|_\alpha\ |t-s|^{\alpha/2} K(\alpha)$$

The "Cauchy" formula $\nu_t(f)\ \mathbb{1} = \frac{1}{2\pi i}\int_\Gamma R_t(z)(f)\,dz$ gives the result.

The proof of the proposition 4.7 is now easy since with the notations of 4.3

$$\gamma(t) - \gamma(s) = \nu_t(f_t - f_s) + (\nu_s - \nu_t)(f_s)$$

and we have only to check that the function $(\lambda,\bar{x}) \to f_\lambda(\bar{x})$ has a continuous derivative on $\mathbb{R}\times X$.

Exercise 4.12 (Example A.II.5.2 revisited)

Suppose that (a) and (b) are independent random variables distributed according to the Cauchy laws $C_{\alpha_1+i\beta_1}$ and $C_{\alpha_2+i\beta_2}$ with $\beta_1 > 0$, $\beta_2 > 0$, $\beta_1+\beta_2 > 0$ (We recall that a Cauchy law $C_{\alpha+i\beta}$ with $\beta > 0$ has the density $\frac{\beta dt}{(t-\alpha)^2+\beta^2}$ and when $\beta = 0$, C_α is the Dirac distribution δ_α)

(i) Prove that $\vartheta_\lambda = C_z$ where z is the unique solution with $\mathfrak{Im}\, z > 0$ of the equation :

$$z + \frac{1}{z} = (\alpha_1 - \lambda\,\alpha_2) + i\,(\beta_1 + |\lambda|\;\beta_2)$$

(In the Helmotz case (that is when $\beta_1 = 0$) we assume $\lambda \neq 0$)

(ii) Prove that $\gamma(\lambda) = \mathrm{Log}|z|$, z being defined in (i).

(iii) In the case $\alpha_2 = 1$, $\beta_2 = \alpha_1 = 0$, $\beta_1 = \delta > 0$ prove that

$$\mathrm{Ch}\,\gamma(\lambda) = \frac{1}{4}\left(\sqrt{(2+\lambda)^2 + \delta^2} + \sqrt{(2-\lambda)^2 + \delta^2}\right)$$

and therefore

$$\gamma(\lambda) = \gamma(-\lambda) \geqslant \gamma(0) = \mathrm{Arg\,sh}\,(\tfrac{\delta}{2})$$

(iv) In the Helmotz case $\alpha_1 = 2$, $\beta_1 = \alpha_2 = 0$, $\beta_2 = \delta > 0$ prove that

$$\mathrm{Ch}\,\gamma(\lambda) = \delta\,\frac{|\lambda|}{4} + \sqrt{1 + \frac{\delta^2\lambda^2}{16}}$$

and hence for λ near zero $\gamma(\lambda)$ has the singular behaviour $\gamma(\lambda) \sim |\lambda|^{1/2}\sqrt{\delta/2}$

In various problems of random walks in random environment the corollary 4.5 needs to be stated in a more precise way. From 4.3 and 4.6 (ii) we know that in the Helmotz case with $c = 2$, if $\lambda_n \to \lambda^*$ then the sequence ν_{λ_n} converges weakly to $\delta_{\bar{x}_I}$. We may hope that with a new scale on the parameter λ and the space X , it's possible to obtain a less degenerated limit. We give below the essential features of a theorem of W.R. Schneider [56] in the simplest case. In this purpose it's more convenient to use the change of variable $\begin{pmatrix} u \\ v \end{pmatrix} \to \begin{pmatrix} u-v \\ v \end{pmatrix}$ on X and to change the sign of λ in order to work with the matrix

$$Z = \begin{bmatrix} \lambda b+1 & \lambda b \\ 1 & 1 \end{bmatrix} = \begin{bmatrix} 1 & -1 \\ 0 & 1 \end{bmatrix} \begin{bmatrix} 2+\lambda b & -1 \\ 1 & 0 \end{bmatrix} \begin{bmatrix} 1 & 1 \\ 0 & 1 \end{bmatrix}$$

If we assume that (b) and λ are non negative, then the matrix Z leaves invariant the subset $X^+ = \{\bar{x} \in X / x = \binom{u}{v}, \ uv \geqslant 0\}$. Denoting always by μ_λ the law of Z and ν_λ the invariant measure (for $\lambda > 0$) we see that $\tilde{\nu}_\lambda$ is supported by $[0, +\infty[$ and $\tilde{\nu}_\lambda$ converges weakly to δ_0 when $\lambda \downarrow 0$. We also assume that $\beta = \mathbb{E}(b) < +\infty$.

Exercise 4.13 With the above hypothesis and notations let θ_λ be the probability measure on X^+ defined by $\theta_\lambda(f) = \int f(\begin{bmatrix} 1 & 0 \\ 0 & \lambda \end{bmatrix} \cdot \vec{x}) \, d\nu_{\frac{\lambda^2}{\beta}}(\bar{x})$.

i) Prove that θ_λ is the invariant measure for the matrix

$$\begin{bmatrix} \frac{\lambda^2 b}{\beta} + 1 & \frac{\lambda b}{\beta} \\ \lambda & 1 \end{bmatrix}$$

This shows that the problem of finding the weak limit of θ_λ when $\lambda \downarrow 0$ is not trivial since the limit matrix is the identity matrix and we are in the most degenerated case.

ii) For $t > 0$ we denote by $\hat{\theta}_\lambda(t)$ and $\hat{\rho}(t)$ the Laplace transforms $\hat{\theta}_\lambda(t) = \int e^{-ts} \, d\tilde{\theta}_\lambda(s)$, $\hat{\rho}(t) = \mathbb{E}[e^{-tb}]$.

Prove that the invariance equation for θ_λ yields :

$$\hat{\theta}_\lambda(t) = \hat{\rho}(\tfrac{t\lambda}{\beta}) \int \exp(-\frac{ts}{\lambda s+1}) \, d\tilde{\theta}_\lambda(s)$$

and

$$\hat{\theta}_\lambda(t) \left[1 - \hat{\rho}(\tfrac{t\lambda}{\beta})\right] \frac{1}{\lambda} = \hat{\rho}(\tfrac{t\lambda}{\beta}) \int \phi(s,t,\lambda) \, d\tilde{\theta}_\lambda(s) \text{ where}$$

$$\phi(s,t,\lambda) = \frac{1}{\lambda}\left[\exp(-\frac{ts}{\lambda s+1}) - \exp(-ts)\right]$$

iii) Prove the following relations :

$$\lim_{\lambda \downarrow 0} \frac{1}{\lambda}(1 - \hat{\rho}(\tfrac{t\lambda}{\beta})) = t$$

$$|\phi(s,t,\lambda) - ts^2 \exp(-ts)| \leqslant C \frac{\lambda}{t^2}.$$

(where C is a numerical constant).

(iv) Prove that if $\tilde{\nu}$ is a vague limit point of $\tilde{\theta}_\lambda$ when $\lambda \downarrow 0$ then the Laplace transform $\hat{\nu}$ of $\tilde{\nu}$ satisfies $\hat{\nu}(t) = \frac{d^2}{dt^2} \hat{\nu}(t)$

Conclude that $\tilde{\nu}$ is proportional to ∂_1.

(v) Let now ν be a weak limit point of θ_λ when $\lambda \downarrow 0$. From (iv) we know that $\nu = \alpha \, \delta_{\bar{x}_1} + (1-\alpha)\delta_{\overline{x_\infty}}$ where $\alpha \in [0,1]$, $x_1 = \binom{1}{1}$ and $x_\infty = \binom{1}{0}$.

For $\bar{x} \in X^+$ we define $f(\bar{x}) = \frac{u}{u+v}$. Applying the invariance equation to f and substracting the function $\frac{u}{v+(1+\lambda)u}$ prove that $\theta_\lambda(\Psi_\lambda) = 0$ where

$$\Psi_\lambda = \frac{1}{(v+(1+\lambda)u)}\left[\frac{u^2}{u+v} - (v+\lambda u)^2\, \mathbb{E}\, \frac{b}{\beta(v+(1+\lambda)u) + \lambda b(u+\lambda v)}\right]$$

Furthermore prove that

$\operatorname{Sup}_{\bar{x}\in X^+} |\Psi_\lambda(\bar{x}) - \frac{u-v}{u+v}|$ converges to zero when $\lambda \downarrow 0$.

This implies that $\alpha = 1$, thus θ_λ converges weakly to $\delta_{\bar{x}_1}$ when $\lambda \downarrow 0$.

II.5 Absence of absolutely continuous spectrum

We assume in this section that (b) is strictly positive, bounded away from zero, and that $\mathbb{E}\operatorname{Log}(1+|a|+b) < +\infty$. The following results are mainly due to L.A. Pastur [50] and K. Ishii [29].

THEOREM 5.1 *Let* A *be a Borel subset of* $\mathbb{R}$ *such that the Lyapunov exponent* $\gamma(\lambda)$ *is strictly positive almost everywhere on* A *with respect to the Lebesgue measure.*
Then $\sigma^a(A) = 0$ $\mathbb{P}$ a.e.

Proof : Define the measurable subset W of $\Omega\times\mathbb{R}$ by :

$$W = \left\{ (\omega,\lambda) \;\middle/\; \begin{array}{l} \omega\in\Omega,\ \lambda\in A,\ \lim_{|n|\to\infty} \frac{1}{|n|}\operatorname{Log}\|Y_n(\omega,\lambda)\| = 0 \\ \lambda \text{ is not an eigenvalue of } H(\omega) \\ \lim_{|n|\to\infty} \frac{1}{|n|}\cdot \operatorname{Log}\|S_n(\omega,\lambda)\| = \gamma(\lambda) \end{array} \right\}$$

The lemma 2.2 and proposition 3.2 implie that for each $\lambda \in A$, $\mathbb{P}(\omega/(\omega,\lambda) \in W) = 1$ and thus using Fubini's theorem, for $\mathbb{P}$ almost all ω the set $A_\omega = \left\{ \lambda \in A \;\middle/\; \begin{array}{l} (\omega,\lambda) \in W \\ \gamma(\lambda) > 0 \end{array} \right\}$
is of full Lebesgue measure in A.

For $\lambda \in A_\omega$, let V^+ and V^- be the contractive vectors given by I.1.1 applied in the two directions of $\mathbb{Z}$. These vectors are not proportional (otherwise λ would be an eigenvalue of $H(\omega)$) therefore each non zero solution of $H\Psi = \lambda\Psi$ is growing with an exponential rate at least in one direction of $\mathbb{Z}$. Looking at the proposition I.3.2 we obtain

$\sigma(\omega)(A(\omega)) = 0$ and thus $\sigma^a(\omega)\ (A) = 0$.

REMARK 5.2 The above theorem can be stated in the more general form : "Let A be a Borel subset of $\mathbb{R}$, m a positive measure on $\mathbb{R}$, such that γ is strictly positive m almost everywhere on A. Then for $\mathbb{P}$ almost all ω, $\sigma(\omega)$ is orthogonal to m on A". As a consequence, if we know that γ is strictly positive everywhere, taking a fixed spectral measure $\sigma(\omega_0)$ as a measure m, we see that $\sigma(\omega)$ is orthogonal to $\sigma(\omega_0)$ for $\mathbb{P}$ almost all ω. This means that the dependence of σ upon ω is in general extremely "chaotic", although the support of σ is $\mathbb{P}$ a.e. constant (2.3).

REMARK 5.3 We have proven in 5.1 that for $\mathbb{P}$ almost all ω, for almost all λ of A, there exists an hyperbolic structure in both directions of $\mathbb{Z}$ with distinct contractive vectors. This suggests that the existence of pure point spectrum is some what unexpected since in this case we ought to obtain that for $\mathbb{P}$ almost all ω, for $\sigma(\omega)$ almost all λ of A there exists an hyperbolic structure in both directions of $\mathbb{Z}$ with the same contractive vector ! (But there is no contradiction since σ is $\mathbb{P}$ a.e. orthogonal to the Lebesgue measure).

COROLLARY 5.4 *The absolutely continuous part of the spectrum vanishes* $\mathbb{P}$ a.e. *in (at least) the three following cases :*

(i) The independent case, assuming that the law of (a,b) *is not supported by a single point of* $\mathbb{R}^2$.

(ii) The Mathieu equation when $|\beta| > 2$.

(iii) In the general ergodic case assuming that the process (a_n, b_n) *is not deterministic.*

REMARK 5.5 In the general ergodic case it has been proved in [58] that for each Borel subset A of positive Lebesgue measure and contained in $\{\lambda | \gamma(\lambda) = 0\}$ then $\sigma^a(A) > 0$ $\mathbb{P}$ a.e. This result added to 5.1 shows that $\mathfrak{S}^a = \overline{\{\gamma = 0\}}$ $\mathbb{P}$ a.e. where the closure means essential closure with respect to the Lebesgue measure.

On the half positive axis, the first proof of absence of absolutely continuous spectrum was given by K. Ishii in [29] and we outline his method in the following exercise.

Exercise 5.6 Let ^{+}H be the operator H restricted to the positive half axis as explained in the exercise I.2.10.

(i) Prove that for $\Im m\, \lambda \neq 0$

$$<{}^{+}R_{\lambda}\, e^{o}, e^{o}> = \int (t-\lambda)^{-1}\, d\, {}^{+}\sigma(t) = \lim_{n\to\infty} -\frac{q_n}{p_n} = \sum_{n=0}^{+\infty} \frac{1}{p_n\, p_{n+1}}$$

Hint : use the lemma I.4.2

(ii) Let A be a Borel subset of $\mathbb{R}$ such that γ is strictly positive almost everywhere on A and :

$$W = \left\{ (\omega,\lambda) \Bigg/ \begin{array}{l} \omega \in \Omega,\ \lambda \in A \\ \lim\limits_{n\to\infty} \frac{1}{n} \operatorname{Log} \| S_n(\omega,\lambda) \| = \gamma(\lambda) \\ \lim\limits_{n\to+\infty} \frac{1}{n} (|a_n| + |b_n|) = 0 \end{array} \right\}$$

prove that for $\mathbb{P}$ almost all ω, the set

$$A_{\omega} = \{\lambda \in A \,/\, (\omega,\lambda) \in W \text{ and } \gamma(\lambda) > 0\}$$

is of full Lebesgue measure in A.

(iii) Prove that for $\lambda \in A_{\omega}$:

$$\lim_{\varepsilon \downarrow 0} \Im m\left(\sum_{n=0}^{+\infty} \frac{1}{p_{n+1}(\lambda + i\varepsilon)\, p_n(\lambda + i\varepsilon)}\right) = 0$$

Hint : observe that $|p_n(\lambda)| \leqslant |p_n(\lambda + i\varepsilon)|$ and that if $p_n(\lambda) = 0$ we can use the identity ($\Im m\, z \neq 0$)

$$\frac{1}{p_n(z)\, p_{n+1}(z)} + \frac{1}{p_n(z)\, p_{n-1}(z)} = (a_n - z b_n) \frac{1}{p_{n+1}(z)\, p_{n-1}(z)}$$

Up to now we have only shown that the positivity of γ implies absence of absolutely continuous spectrum and we may ask wether this positivity implies the pure point spectrum property. It is known that it is false in general, in example the Mathieu operator has no point spectrum when the rotation is given by a Liouville number α [57]. Hence even when γ has a strictly positive lower bound on $\mathbb{C}$ (when $|\beta| > 2$ see 3.4) we can't assert the absence of singular continuous spectrum. However S. Kotani [36] has obtained a weaker result on the positive half axis and we follow his proof in the next exercise.

Exercise 5.7 Let $^{+}H^{y}$ be the operator on the positive half axis with boundary condition y at the point zero (see I.4.7) and A be a Borel subset of $\mathbb{R}$ such that γ is strictly positive almost everywhere on A. We want to prove that for $\mathbb{P}$ almost all ω, for Lebesgue almost all y, the spectrum of ${}^{+}H^{y}$ is pure point on A (if any) and that the eigenfunctions are exponentially decaying.

(i) Defining W in the same way than in the exercise 5.6 and looking at the result (ii) of I.4.7 prove that for $\mathbb{P}$ almost all ω, for Lebesgue almost all y, for ${}^{+}\sigma^{y}$ almost all λ of A we have :

$$\lim_{n\to+\infty} \frac{1}{n} \operatorname{Log}\|S_n(\omega,\lambda)\| = \gamma(\lambda)$$

(ii) Observe that under the same almost sure assumptions than in (i) the sequence $\|S_n(\omega,\lambda)\binom{1}{y}\|$ is either exponentially growing or exponentially decaying (with rate $\gamma(\lambda)$).

Hint : use Ossedelec's theorem.

(iii) Looking at the proposition I.3.2 adapted to the operator ${}^{+}H^{y}$ and ${}^{+}\sigma^{y}$ (see exercise I.3.3) prove under the same almost sure assumptions that only the second alternative can take place and conclude.

II.6 Distribution of states. Thouless formula.

In this section we assume that (b) is strictly positive and bounded away from zero.

The distribution of states is the probability measure on $\mathbb{R}$ defined by $k(\Delta) = \mathbb{E}[b_n \ \sigma_n(\Delta)]$. This expectation does not depend on the variable n thank's to the relations $(b_n\sigma_n)\circ\theta = b_{n+1}\ \sigma_{n+1}$ and $k(\mathbb{R}) = 1$ (since $\sigma_n(\mathbb{R}) = b_n^{-1}$). This probability measure k is sometimes called "integrated density of states" but we don't use this expression since in many cases k is not absolutely continuous !.

We say that a sequence of boxes Λ goes to infinity if the boxes are increasing and if the length of Λ denoted by $|\Lambda|$ goes to infinity (compare with the definition of a sequence going to $\mathbb{Z}$ in I.4). Generally we choose $\Lambda = [0,N]$ with N going to infinity as sequence of boxes going to infinity.

For a box Λ we define the probability measure k_Λ by $k_\Lambda = \frac{1}{|\Lambda|} \sum_{n\in\Lambda} b_n\sigma_n$.

PROPOSITION 6.1 *Let f be a k integrable function. Then for any sequence of boxes going to infinity the random sequence* $k_\Lambda(f)$ *converges* $\mathbb{P}$ a.e *to* $k(f)$ *and* $\mathbb{E}[k_\Lambda(f)]$ *converges to* $k(f)$.

Proof : This is just the Birkhoff ergodic theorem.

COROLLARY 5.2 *For any sequence of boxes going to infinity the random sequence* k_Λ *converges weakly* $\mathbb{P}$ a.e *to* k, *and* $\mathbb{E}[k_\Lambda]$ *converges weakly to* k.

Proof : We use 6.1 looking at a countable dense subset of the space of continuous functions with compact support on $\mathbb{R}$.

PROPOSITION 6.3 *The topological support of* k *is* $\mathbb{P}$ a.e *equal to the spectrum of* $H(\omega)$. *Moreover* k *is a continuous measure.*

Proof : $k = \frac{1}{2}\,\mathbb{E}(b_o\sigma_o + b_{-1}\sigma_{-1})$ and the support of $(b_o\sigma_o + b_{-1}\sigma_{-1})$ is equal to the support of σ. Then by the lemma 2.3 we obtain the first part of 6.3. For the last statement see the lemma 2.2.

Usually the distribution of states is defined as the limit of the distribution of eigenvalues of the operator H restricted to boxes and it's observed that this limit does not depend upon the boundary conditions given at the endpoints of the boxes. In order to prove this result let ${}^\Lambda H^{y,x}$ be the operator H restricted to a box Λ with boundary conditions y and x as explained in I.4.6. If we denote by $({}^\Lambda\lambda_n^{y,x})_{n\varepsilon\Lambda}$ the eigenvalues of ${}^\Lambda H^{y,x}$ we define the eigenvalues distribution :

$${}^\Lambda k^{y,x} = \frac{1}{|\Lambda|}\sum_{n\varepsilon\Lambda} b_n\,{}^\Lambda\sigma_n^{y,x} = \frac{1}{|\Lambda|}\sum_{n\varepsilon\Lambda} \partial_{{}^\Lambda\lambda_n^{y,x}}$$

PROPOSITION 6.4 *Assuming that the random variable* (a) *is bounded, for fixed* $x,y \in \mathbb{R}$ *the random sequence* ${}^\Lambda k^{y,x}$ *converges weakly to* k $\mathbb{P}$ a.e. *and* $\mathbb{E}[{}^\Lambda k^{y,x}]$ *converges weakly to* k, *when* Λ *goes to infinity.*

Proof : Under the hypothesis on the random variable (a) the sequences ${}^\Lambda k^{y,x}$ and k_Λ are supported by the same compact subset of $\mathbb{R}$. Then by corollary 6.2 it's enough to prove that for each positive integer m the sequence $\int t^m({}^\Lambda k^{y,x} - k_\Lambda)(dt)$ converges to zero when Λ goes to infinity. This last integral is equal to

$$\frac{1}{|\Lambda|}\sum_{n\varepsilon\Lambda} b_n < (({}^\Lambda H^{y,x})^m - H^m)\, e^n,\, e^n>

In this sum the only non zero terms are those of index n with $M \leqslant n \leqslant M+m$ and $N-m \leqslant n \leqslant N$ (where $\Lambda = [M,N]$). Since the operators ${}^{\wedge}H^{y,x}$, H are uniformly bounded the proof is complete.

The definition given here of the distribution of states and the construction by the proposition 6.1 can be found in [3]. Some other authors [34] define the distribution of states as the weak limit of the distribution of eigenvalues of the operator H restricted to boxes by means of the subadditive and superadditive ergodic theorem. As usual it's possible to obtain the convergence in 6.1 with an exponential rate assuming ϕ mixing properties of the dynamical system $(\Omega, \mathfrak{a}, \mathbb{P}, \theta)$. We have the same speed of convergence under this hypothesis for the distribution of eigenvalues, but the proof is more involved [34]. We may ask what are the functions f for which the proposition 6.1 apply. Indeed bounded measurable function are available since k is a probability measure, and when the random variable a is bounded any measurable locally bounded function works since k has a compact support (by example a continuous function). In the general case following [12] we prove :

<u>LEMMA 6.5</u> *Assume that* $\mathbb{E}\,\mathrm{Log}(1+|a|) < +\infty$ *then for* $\mathfrak{Im}\,\lambda \neq 0$ *the function* $f(x) = \mathrm{Log}|\lambda - x|$ *is* k *integrable.*

Proof : It's enough to prove that $\mathrm{Log}(1+|x|)$ is k integrable. Let $V^{\pm}$ be the two diagonal (and hence self adjoint) operators $(V^{\pm}\Psi)_n = b_n^{-1}(a_n \pm 2)\Psi_n$. $V^{\pm}$ and H are self adjoint with the same domain D_1 (see I.2) and it's readily seen that $V^- \leqslant H \leqslant V^+$.
The distributions of states $k^{\pm}$ of $V^{\pm}$ (defined in the same way than for H) are the laws of the random variables $b^{-1}(a \pm 2)$. Thus for any increasing (decreasing) positive function g^+ (g^-) we have

$$k(g^+) \leqslant k^+(g^+) \ , \ k(g^-) \leqslant k^-(g^-)$$

Taking $g^+ = \mathrm{Log}(1+|x|) \,.\, \mathbb{1}_{\{x \geqslant 0\}}$

$$g^- = \mathrm{Log}(1+|x|) \,.\, \mathbb{1}_{\{x \leqslant 0\}}$$

we obtain the expected result since b^{-1} is bounded.

In the independent case the distribution of states is related to the invariant measure ν_λ on the projective line X given by the proposition 4.1. Let X^- be the subset of X defined by

$X^- = \{\bar{x} \;/\; x = \begin{pmatrix} u \\ v \end{pmatrix},\ uv \leq 0\}$

PROPOSITION 6.6 *In the independent case, assume that the law of* (a,b) *is not concentrated on a single point of* $\mathbb{R}^2$*, and that the random variable* (a) *is bounded. Then for each real* λ ($\lambda \neq \lambda^*$ *in the Helmotz case)*

$$k(]-\infty, \lambda]) = \nu_\lambda(X^-)$$

Proof : From 4.1 we know that for $\lambda \neq \lambda^*$ there exists a unique μ_λ invariant probability ν_λ on X. Let x_o be the vector $\begin{pmatrix} 1 \\ 0 \end{pmatrix}$ and for $N \geq 0$

$\nu_{\lambda,N} = \frac{1}{N+1} \sum_{k=0}^{N} \mu_\lambda^k \;\dot{*}\; \delta_{\bar{x}_o}$ where μ_λ^k is the kth power of convolution of μ_λ on $SL(2,\mathbb{R})$. Each weak limit point of $\nu_{\lambda,N}$ when $N \to +\infty$ is μ_λ invariant and hence equal to ν_λ. This implies that the sequence $\nu_{\lambda,N}$ converges weakly to ν_λ when $N \to +\infty$.

If we denote by $\mathbb{1}_{X^-}$ the indicator function of the set X^-, it is readily seen that :

$$\nu_{\lambda,N}(X^-) = \frac{1}{N+1} \mathbb{E}\Big[\sum_{k=0}^{N} \mathbb{1}_{X^-}(\bar{x}_k)\Big] \quad \text{where } x_k = \begin{pmatrix} p_k \\ p_{k-1} \end{pmatrix}$$

But $\sum_{k=0}^{N} 1_{X^-}(\bar{x}_k)$ is the number of change of sign of the sequence p_k for $0 \leq k \leq N$. Let $\lambda_o, \lambda_1, \ldots, \lambda_N$ be the roots of the polynom p_{N+1} then by the Sturm theorem [2] (see also [15]) this number of change of sign is equal to 0 if $\lambda < \lambda_o$, r+1 if $\lambda_r < \lambda < \lambda_{r+1}$, N if $\lambda > \lambda_N$. Thus if we note by ${}^N H$ the operator H restricted to the box $[0,N]$ with zero boundary conditions at each endpoint and ${}^N k$ its distribution of states, we have :

$$|\nu_{\lambda,N}(X^-) - \mathbb{E}[{}^N k(]-\infty,\lambda])]| \leq \frac{1}{N+1}$$

We know that $\nu_{\lambda,N}$ converges weakly to ν_λ, that $\mathbb{E}[{}^N k]$ converges weakly to k (by 6.4) and that ν_λ and k are continuous measures thus

$$\lim_{N\to+\infty} \nu_{\lambda,N}(X^-) = \nu_\lambda(X^-) = \lim_{N\to+\infty} \mathbb{E}[\,k(]-\infty,\lambda])] = k(]-\infty,\lambda])$$

We now turn our attention to a remarkable relation between the Lyapunov exponent and the distribution of states first pointed out by D. Thouless [62] and hence called the "Thouless formula". We follow here the proof given by J. Avron, B. Simon [3] and W. Craig, B. Simon [12].

PROPOSITION 6.7 *Assume that the random variable* a *is bounded and* $\mathbb{E}(\text{Log } b) < +\infty$ *, then for each complex number* λ

$$\gamma(\lambda) = \mathbb{E}(\text{Log } b) + \int \text{Log}|t-\lambda| \, dk(t)$$

Proof : Let λ be a complex number with $\Im m \lambda \neq 0$ and $N \geq 1$. The roots of the polynom p_{N+1} , (q_{N+1}), are the eigenvalues of the operator H restricted to the box $[0,N]$ $([1,N])$ with zero boundary condition at each endpoint. Denoting by ${}^{N}k$ $({}_{N}k)$ the distribution of these eigenvalues it is readily seen that

$$\int \text{Log}|t-\lambda| \, d^{N}k(t) + \frac{1}{N+1} \sum_{k=0}^{N} \text{Log } b_k = \frac{1}{N+1} \text{Log}|p_{N+1}(\lambda)|$$

$$\int \text{Log}|t-\lambda| \, d_{N}k(t) + \frac{1}{N} \sum_{k=1}^{N} \text{Log } b_k = \frac{1}{N} \text{Log}|q_{N+1}(\lambda)|$$

The random variable (a) is bounded, thus the probability measures ${}^{N}k$, ${}_{N}k$ are supported by the same compact set. The Birkhoff ergodic theorem and 6.4 implie that :

$$\lim_{N\to+\infty} \frac{1}{N+1} \text{Log } |p_{N+1}(\lambda)| = \lim_{N\to+\infty} \frac{1}{N} \text{Log}|q_{N+1}(\lambda)|$$

$$= \int \text{Log}|t- \lambda| \, dk(t) + \mathbb{E}(\text{Log } b) , \ \mathbb{P} \text{ a.e.}$$

We have thus proved that each entry of the matrix $S_N = \begin{bmatrix} p_{N+1} & q_{N+1} \\ p_N & q_N \end{bmatrix}$ has the same asymptotic behaviour. Since we know from 3.2 that

$$\lim_{N\to+\infty} \frac{1}{N} \text{Log}||S_n|| = \gamma(\lambda) \quad \mathbb{P} \text{ a.e.}$$

we can conclude that for $\Im m \lambda \neq 0$

$$\gamma(\lambda) = \mathbb{E}(\text{Log } b) + \int \text{Log}|t-\lambda| \, dk(t)$$

Each function in the above formula is a subharmonic function on $\mathbb{C}$: for γ this is the proposition 3.1 and the subharmonicity of $\int \text{Log}|t-\lambda| \, dk(t)$ is straightforward. Since two subharmonic functions almost everywhere equal, with respect to the Lebesgue measure on $\mathbb{C}$, are everywhere equal, we obtain the result.

REMARK 6.8 : We have assumed that the random variable (a) is bounded in the proof of the Thouless formula. But if we only assume that $\mathbb{E} \text{ Log}(1 + |a| + b) < +\infty$ then $\gamma(\lambda)$, $\mathbb{E}$ Log b and $\int \text{Log}|t-\lambda| \, dk(t)$ are well defined (the latter for $\Im m \lambda \neq 0$ by 6.5). This suggests that such a formula should be true under this weaker hypothesis...

We know from 6.3 that the distribution function $K(\lambda) = k(]-\infty,\lambda])$ is continuous. We list below som extra regularity properties of K. K is said locally Log Hölder continuous if for all compact subset T of $\mathbb{R}$ there exists $C_T > 0$ such that for $s,t \in T$, $|s-t| \leq \frac{1}{2}$ we have $|K(s) - K(t)| \leq C_T (\mathrm{Log}|s-t|^{-1})^{-1}$.

PROPOSITION 6.9

(i) In the general ergodic case assume that (a) *and* (b) *are bounded, then* K *is locally* Log *Hölder continuous.*

(ii) In the independent case assume that (a) *is bounded and that there exists* $\alpha > 0$ *with* $\mathbb{E}(b^\alpha) < +\infty$ *then* K *is locally Hölder continuous on* $\mathbb{R} \setminus \lambda^*$.

(iii) In the independent case assume that $b_n = 1$, $\forall n \in \mathbb{Z}$, $\mathbb{E}(a^2) < +\infty$ *and that the law of* (a) *has a density on* $\mathbb{R}$, *then* K *has a continuous derivative.*

(iv) In the independent case, assume that the conditional law of (a) *given* (b) *has a bounded density* ϕ_b *on* $\mathbb{R}$, *such that* $\mathbb{E}(b\|\phi_b\|) < +\infty$. *Then* k *has a bounded density with respect to the Lebesgue measure.*

Proof

(i) Since (b) is bounded there exists β such that $\beta b \leq 1$ and then by the Thouless formula :

$$\gamma(\beta\lambda) - \mathbb{E}\,\mathrm{Log}\,\beta b = \int \mathrm{Log}|t-\lambda|\, d\tilde{k}(t)$$

where $\tilde{K}(t) = K(\beta t)$. Thus this latter integral is positive for all $\lambda \in \mathbb{C}$ and this implies that $\tilde{K}$ and hence K must be locally Log Hölder continuous by standart analysis arguments [12].

(ii) The result follows from 4.7 and the properties of the Hilbert transform [47]. For $\Psi \in L^2(\mathbb{R})$ we denote by $\tilde{\Psi}$ its Hilbert transform and if $]A,B[$ contains the support of k, we choose $\Psi(\lambda) = K(\lambda)$ if $\lambda \in]A,B[$, $\Psi(\lambda) = 0$ if $\lambda \notin]A,B[$. By integration by parts and the Thouless formula:

$\tilde{\Psi}(\lambda) = \frac{1}{\Pi}(\gamma(\lambda) - \mathrm{Log}|B-\lambda|)$ for almost all λ with respect to the Lebesgue measure.

Letting $\Phi(\lambda) = \frac{1}{\Pi}(\gamma(\lambda) - \mathrm{Log}|B-\lambda|)$, Φ is locally Hölder continuous on $\mathbb{R}\setminus\{B,\lambda^*\}$, hence $\tilde{\Phi}$ is also locally Hölder continuous and $\tilde{\Phi} = \Psi$ a.e hence $\tilde{\Phi} = \Psi$ on $\mathbb{R}\setminus\{B,\lambda^*\}$. Since B can be chosen outside of any compact we obtain the result.

(iii) This result will be proven in III.4.1.

(iv) Let λ be a number with $\Im m\,\lambda > 0$, N an integer. We denote by ${}^{N}_{\pm}R_{\lambda}$ the resolvent of the operator H restricted to the half axis $[N, +\infty[$ or $]-\infty, N]$.

Following the lines of I.2.12 and I.2.14 we define

$$ {}^{N}_{\pm}m(\lambda) = \langle {}^{N}_{\pm}R_{\lambda}\, e^{N}, e^{N} \rangle \;, \quad {}^{N}_{\pm}\Psi_{n} = \langle {}^{N}_{\pm}R\, e^{N}, e^{n} \rangle $$

where $n \geq N$ or $n \leq N$. The functions ${}^{N}_{\pm}\Psi$ can be extended to the whole of $\mathbb{Z}$ in such a way that these extensions satisfy $H\Psi = \lambda\Psi$, and we find : ${}^{N}_{\pm}\Psi = ({}^{N}_{\pm}m)({}^{N}_{\pm}p) + ({}^{N}_{\pm}q)$, where ${}^{N}_{\pm}p$ and ${}^{N}_{\pm}q$ are solutions of $H\Psi = \lambda\Psi$ with the boundary conditions

$$ {}^{N}_{+}p_{N} = {}^{N}_{+}q_{N-1} = {}^{N}_{-}p_{N} = {}^{N}_{-}q_{N+1} = 1 $$

$$ {}^{N}_{+}p_{N-1} = {}^{N}_{+}q_{N} = {}^{N}_{-}p_{N+1} = {}^{N}_{-}q_{N} = 0 $$

Then it's easy to verify that :

$$ \langle R_{\lambda}\, e^{N}, e^{N} \rangle = \frac{-1}{({}^{N+1}_{+}m) + ({}^{N-1}_{-}m) + \lambda b_{N} - a_{N}} $$

The functions $({}^{N+1}_{+}m)$ and $({}^{N-1}_{-}m)$ are functions of the process (a_n, b_n) for $n \geq N+1$, $n \leq N-1$ respectively and have positive imaginary parts. Taking expectation in the above formula, first integrating with respect to (a_N, b_N), we obtain :

$$ \Im m \int \frac{dk(t)}{t-\lambda} \leq \mathbb{E}[b\, \|\phi_{b}\|] < +\infty $$

This yields the expected result, using Lebesgue's convergence theorem in the formula :

$$ k(h) = \lim_{\varepsilon \downarrow 0} \frac{1}{\Pi} \int h(x) \left(\Im m \int \frac{dk(t)}{t-x-i\varepsilon}\right) dx $$

where h is a continuous function with compact support.

Let $L^{1,\alpha}$ be the space of functions $f \in L^{1}(\mathbb{R})$ such that there exists $g \in L^{1}(\mathbb{R})$ with $\hat{f}(t)(1+t^{2})^{\alpha/2} = \hat{g}(t)$ (where $\hat{f}$ is the Fourier transform of f). B. Simon and M. Taylor [61] have recently proven that in the independent case with $b_n = 1$, $n \in \mathbb{Z}$, if we assume that (a) is bounded with a density in $L^{1,\alpha}$ for a strictly positive α, then K is a

C^∞ function. A typical example is given by the uniform distribution on a compact interval.
When the density of (a) has stronger regularity, namely when $|\mathbb{E}(e^{ita})| \leq \alpha\, e^{-\beta|t|}$ for some positive constants α and β such that $\alpha < \beta$, then K has an analytic continuation to a strip $\{z \in \mathbb{C} \,/\, |\Im m\, z| < \delta\}$ for some $\delta > 0$. This result is proven by R. Carmona in [9] using the "Molcanov formula". Further results about Lifschitz exponents can also be found in this course.

Exercise 6.10 In the general ergodic case assume that (a) is bounded. Prove that the sequence of probability measures $\overset{N\sim}{k}$ $(N \geq 1)$ converges weakly almost surely to k where

$$\frac{d\overset{N\sim}{k}}{d\lambda} = \frac{1}{\Pi(N+1)} \quad \frac{1}{P_N^2 + P_{N+1}^2} \sum_{k=0}^{N} b_k\, P_k^2$$

The two following exercises show that the distribution of states can be the same in two models which have in general very different spectral properties ([59] and [18]).

Exercise 6.11 "The Maryland model", see [18]. Let $H = \overset{\circ}{H} + V$. $\overset{\circ}{H}$ is the Laplace operator $\overset{\circ}{H}\Psi = -\Psi_{n-1} - \Psi_{n+1}$ and for a positive number β and irrational α we define $(V\Psi)_n = \beta\, \Psi_n \;\mathrm{tg}(\Pi n\alpha + \omega)$ where $\omega \in [0,\Pi] \setminus \{\Pi/2 - \Pi n\alpha\}_{n\in\mathbb{Z}} = \Omega$. This is not exactly a quasi-periodic model since a is not bounded but a lot of explicit computations can be performed in this "solvable model".
Let U be the operator $(U\Psi)_n = e^{2i\Pi n\alpha}\Psi_n$ and z be the complex number $e^{2i\omega}$.

(i) Prove that $H = \overset{\circ}{H} + i\beta(zU - I)(zU + I)^{-1}$

(ii) Prove that for $\Im m\,\lambda > 0$

$$R_\lambda = (zU + I)(I + z\,C\,U)^{-1}\, \overset{\circ}{R}_{\lambda+i\beta}$$

where $C = \overset{\circ}{R}_{\lambda+i\beta}(\overset{\circ}{H} - (\lambda - i\beta)\, I)$

(iii) Observing that $\|C\| < 1$ expand $(I+z\,C\,U)^{-1}$ in power series and prove that

$$\mathbb{E}[R_\lambda] = \overset{\circ}{R}_{\lambda+i\beta}$$

Conclude that k has a density and $k = \overset{\circ}{k} * C_{i\beta}$ ($C_{i\beta}$ is the Cauchy law and see I.4.5 for $\overset{\circ}{k}$).

Hint : for λ real.

$$\frac{dk}{d\lambda} = \lim_{\varepsilon \downarrow 0} \mathfrak{Im} \int \frac{dk(t)}{t-(\lambda+i\varepsilon)} = \mathfrak{Im} \int \frac{d\overset{\circ}{k}(t)}{t-\lambda-i\beta}$$

Exercise 6.12 The "Lloyd model", see [60]. Let $H = \overset{\circ}{H} + V$ as in the exercise 6.11 but now $(V\Psi)_n = a_n \Psi_n$ where a_n is a sequence of real independent random variables with a common law $C_{i\beta}$ (see 4.12)

(i) Let I be a finite set of integers (perhaps not all distinct) and s_k, $k \in I$, a set of positive numbers with $\sum_{k \in I} s_k = 1$.

Prove that $\mathbb{E}\left[\exp it \sum_{k\in I} s_k a_k\right] = \exp(-\beta|t|)$

(ii) "Dyson's formula". Let A and B be two squares matrices with complex entries and t a real number.

Prove that $e^{t(A+B)} = e^{tB} + \int_0^t e^{s(A+B)} A\, e^{(t-s)B}\, ds$

Iterating this relation, expand $e^{t(A+B)}$ in a convergent series.

(iii) Applying (i) and (ii) to the operators $\hat{H}$ and $\hat{\overset{\circ}{H}}$

prove that $\mathbb{E}\left[e^{it\hat{H}}\right] = e^{-\beta|t|}\, e^{it\hat{\overset{\circ}{H}}}$

and $\mathbb{E}\left[e^{it H}\right] = e^{-\beta|t|}\, e^{it \overset{\circ}{H}}$

(iv) Conclude from (iii) that $k = \overset{\circ}{k} * C_{i\beta}$

II.7 The pure point spectrum. Kotani's criterion

We assume in this section that (b) is strictly positive, bounded away from zero and that $\mathbb{E}\,\mathrm{Log}(1 + |a| + b) < +\infty$.

Let $\mathfrak{a}_+(\mathfrak{a}_-)$ be the subsigma algebra of $\mathfrak{a}$ generated by the process (a_n, b_n) for $n \geq N$ $(n \leq M)$ and integers M,N. For a measurable positive function Φ defined on $(\Omega, \mathfrak{a})$, we denote by $\mathbb{E}[\phi \mid \mathfrak{a}_\pm]$ the conditional expectation with respect to $\mathfrak{a}_\pm$ and for each Borel subset Δ of the real line we set $\sigma_\pm(\Delta) = \mathbb{E}\left[\sigma(\Delta) \mid \mathfrak{a}_\pm\right]$. $\sigma_\pm(\Delta)$ are $\mathfrak{a}_\pm$ measurable functions defined $\mathbb{P}$ almost everywhere. It's possible to construct a topology on Ω such that Ω becomes a Polish space and $\mathcal{A}$ its Borel sigma algebra. Thus there exists a choice of $\sigma_{\pm}(\Delta)$ such that $\sigma_{\pm}$ are positive measures on $\mathbb{R}$ with $\sigma_{\pm}(\mathbb{R}) = \sigma(\mathbb{R})$. By a monotone class argument we see that if $F_{\pm}$ are measurable functions on $\Omega \times \mathbb{R}$ with respect to the product of $\mathcal{A}_{\pm}$ with the Borel sigma algebra of $\mathbb{R}$ then

$$\mathbb{E}\left[\int F_{\pm}\, d\sigma \mid \mathcal{A}_{\pm}\right] = \int F_{\pm}\, d\sigma_{\pm} \qquad \text{and}$$

$$\mathbb{E}\left[\int F_{\pm}\, d\sigma\right] = \mathbb{E}\left[\int F_{\pm}\, d\sigma_{\pm}\right]$$

(take care that $\sigma_{\pm}$ have nothing to do with $^{\pm}\sigma$ of I.2.10) We are now able to prove Kotani's criterion [36].

<u>PROPOSITION 7.1.</u> *Let* A *be a Borel subset of* $\mathbb{R}$ *and* m *a positive measure on* A. *We assume that* $\gamma(\lambda) > 0$ m.a.e *on* A *and that* $\sigma_{\pm}$ *are absolutely continuous on* A *with respect to* m, $\mathbb{P}$ *almost everywhere. Then for* $\mathbb{P}$ *almost all* ω *the exponent* $\gamma(\lambda)$ *is strictly positive* $\sigma(\omega)$ *almost everywhere on* A, *the spectrum of* $H(\omega)$ *is pure point on* A *(if any) and the eigenfunctions decay with the exponential rate* $-\gamma(\lambda)$.

Proof : Define the subsets $W_{\pm}$ of $\Omega \times \mathbb{R}$ by :

$$W_{\pm} = \left\{ (\omega,\lambda) \;\middle/\; \begin{array}{l} \omega \in \Omega,\ \lambda \in A,\ \lim\limits_{n\to\pm\infty} \frac{1}{|n|} \operatorname{Log}\|Y_n\| = 0 \\ \lim\limits_{n\to\pm\infty} \frac{1}{|n|} \operatorname{Log}\|S_n\| = \gamma(\lambda) > 0 \\ \sigma_{\pm} \text{ are absolutely continuous with respect to } m \end{array} \right\}$$

The sets $W_{\pm}$ are measurable with respect to the products of $\mathcal{A}_{\pm}$ by the Borel sigma algebra of $\mathbb{R}$ and we know from the ergodic theorem 3.2 and the hypothesis that $W_{\pm}$ are of full $\mathbb{P} \otimes m$ measure.

Let $F_{\pm}$ be the functions defined by

$$F_{\pm}(\omega,\lambda) = 1 \quad \text{if} \quad (\omega,\lambda) \notin W_{\pm}$$
$$\qquad\quad = 0 \quad \text{if} \quad (\omega,\lambda) \in W_{\pm}$$

Using the preceding discussion and the absolute continuity of $\sigma_{\pm}$ with respect to m :

$$\mathbb{E}\left[\int F_{\pm}\, d\sigma\right] = \mathbb{E}\left[\int F_{\pm}\, d\sigma_{\pm}\right] = 0$$

Then by Fubini's theorem, for $\mathbb{P}$ almost all ω, for $\sigma(\omega)$ almost all λ in A, each solution of $H\Psi = \lambda\Psi$ is, in each direction of $\mathbb{Z}$, growing or decaying exponentially fast (thank's to Osseledec's theorem I.1.1). Looking at the proposition I.3.2, for $\sigma(\omega)$ all λ there exists a slowly increasing solution of $H\Psi = \lambda\Psi$ and thus this solution must be exponentially decaying in each direction of $\mathbb{Z}$ with rate $-\gamma(\lambda)$. From the integrability condition $\mathrm{Log}(1+b) < +\infty$ we know that for $\mathbb{P}$ almost all ω, $|b_n(\omega)| \leq n$ for n large enough. This implies that $\sigma(\omega)$ is supported by the eigenvalues of $H(\omega)$, hence $\sigma(\omega)$ is pure point and the eigenfunctions decay with the rate $-\gamma(\lambda)$.

REMARK 7.2 The above matching method of two solutions of the difference equation was employed by S.A. Molchanov [45] and R. Carmona [8].

If the process (a_n, b_n) is deterministic then the hypothesis of 7.1 does not hold because in this case $\sigma_{\pm} = \sigma$ and we know from 5.2 that if $\gamma > 0$ m a.e. on A then for $\mathbb{P}$ almost all ω, $\sigma(\omega)$ is orthogonal to m. Thus in order to check the hypothesis of 7.1 some randomness is inevitable. We shall see in III sufficient conditions in order to deal with criterion 7.1.

II.8 Asymptotic properties of the conductance in the disordered wire

In physical litterature there exist several approaches to the notion of conductance. We only give there a formal definition and we refer for instance to J.L. Pichard [51] for physical background. We call "plane waves" two sequences u and v of the form $u_n = e^{in\theta}$, $v_n = e^{-in\theta}$ where θ is a real number different from a multiple of Π. Then any sequence Ψ of complex numbers can be "projected" on the plane waves u and v according to :

$$\begin{bmatrix}\Psi_{n+1}\\ \Psi_n\end{bmatrix} = \alpha_n \begin{bmatrix}u_{n+1}\\ u_n\end{bmatrix} + \beta_n \begin{bmatrix}v_{n+1}\\ v_n\end{bmatrix} = T_n \begin{bmatrix}\alpha_n\\ \beta_n\end{bmatrix}$$

where $T_n = \begin{bmatrix}u_{n+1} & v_{n+1}\\ u_n & v_n\end{bmatrix}$ is an invertible matrix. In order to compute $\begin{bmatrix}\alpha_n\\ \beta_n\end{bmatrix}$, $n \geq 0$, from $\begin{bmatrix}\alpha_{-1}\\ \beta_{-1}\end{bmatrix}$, we set $Z_n = T_n^{-1} Y_n T_{n-1}$, $\Pi_n = Z_n Z_{n-1} \cdots Z_o$.

It is readily seen that $\Pi_n = T_n^{-1} S_n T_{-1}$ and thus if Ψ satisfies the difference equation $(H\Psi)_n = \lambda\Psi_n$, $n \in [0, \ldots, N]$, we obtain the relation :

$$\begin{pmatrix} \alpha_N \\ \beta_N \end{pmatrix} = \Pi_N \begin{pmatrix} \alpha_{-1} \\ \beta_{-1} \end{pmatrix}$$

This means that if we "input" the wave of "coordinates" $\begin{pmatrix} \alpha_{-1} \\ \beta_{-1} \end{pmatrix}$ into the disordered box $\Lambda = [0, \ldots, N]$, then the "output" wave have coordinates $\begin{pmatrix} \alpha_N \\ \beta_N \end{pmatrix}$.

The reflexion and transmission coefficients t_N and r_N of the disordered box $\Lambda = [0, \ldots, N]$ are defined by the equation :

$$\begin{pmatrix} t_N \\ 0 \end{pmatrix} = \Pi_N \begin{pmatrix} 1 \\ r_N \end{pmatrix}$$

and the conductance of Λ is the ratio $C_N = |\frac{t_N}{r_N}|^2$. Our problem is to find the assymptotic behavior of C_N when $N \to +\infty$.

Let $SU(1,1)$ be the subgroup of matrices in $SL(2,\mathbb{C})$ of the form $\begin{pmatrix} \alpha & \beta \\ \bar{\beta} & \bar{\alpha} \end{pmatrix}$. It's an easy exercise to verify that if $Y \in SL(2,\mathbb{R})$ then $T_n^{-1}\, Y\, T_{n-1}$ is in $SU(1,1)$. Thus Z_N and therefore Π_N are in $SU(1,1)$. As a consequence we may write :

$C_N = \frac{1}{2}\,(\|\Pi_N x\|^2 - 1)$ where x is the vector $\begin{pmatrix} 1 \\ 0 \end{pmatrix}$ of $\mathbb{R}^2$. Since $\|\Pi_N x\| = \|T_n^{-1} S_n T_{-1} x\| = \|T_n^{-1} S_n y\|$ where $y = T_{-1}x$, the asymptotic behavior of C_N is obtained from the asymptotic behavior of the columns of S_N which is well known in the independent case.

<u>THEOREM 8.1</u> *Assume that the sequence* (a_n,b_n), $n \geq 0$, *is an* i.i.d *sequence with values in* $\mathbb{R}^2$ *and that its law is not supported by a single point of* $\mathbb{R}^2$. *Then for* $\lambda \in \mathbb{R}$ *(*$\lambda \neq \lambda^*$ *in the Helmotz case) :*

(i) The sequence $\frac{1}{n}\operatorname{Log} C_n$ *converges* $\mathbb{P}$ a.e. *to* $-2\gamma(\lambda)$ *when* $n \to +\infty$. *Moreover* $\gamma(\lambda)$ *is strictly positive.*

(ii) Assume that there exists $\alpha > 0$ *with* $\mathbb{E}(|a|^\alpha + |b|^\alpha) < +\infty$, *then there exists a* <u>*strictly positive*</u> *number* $\sigma^2(\lambda)$ *such that the law of the sequence* $\frac{1}{\sqrt{n}}(\operatorname{Log} C_n + 2n\,\gamma(\lambda))$ *converges weakly to a centered Gaussian law with variance* $\sigma^2(\lambda)$.

Proof : From the inequalities :

$$\|T_n\|^{-1} \|S_n y\| \leqslant \|T_n^{-1} S_n y\| \leqslant \|T_n^{-1}\| \|S_n y\|$$

we can write $\text{Log} \|T_n^{-1} S_n y\| = \text{Log}\|S_n y\| + U_n$ where the random sequence U_n is bounded since $\|T_n\| \leqslant 2$, $\|T_n^{-1}\| \leqslant |\sin\theta|^{-1}$.
Thus we obtain :

$$\frac{1}{n} \text{Log } C_n = \frac{V_n}{n} - \frac{2}{n} \text{Log } \|S_n y\| \quad \text{where}$$

$$V_n = \left[\text{Log } 2 - 2\text{Log } U_n - \text{Log } (1 - (\|S_n y\| + U_n)^{-1})\right]$$

We know from 4.1 that $\gamma(\lambda) > 0$ and from A.II.3.6 that $\frac{1}{n}$ Log $\|S_n y\|$ converges $\mathbb{P}$ a.e. to $\gamma(\lambda)$. This implies that $\frac{V_n}{n}$ converges $\mathbb{P}$ a.e. to zero and the result (i) follows.

In order to prove (ii) we write :

$$\frac{1}{\sqrt{n}} (\text{Log } C_n + 2n\,\gamma(\lambda)) = -\frac{2}{\sqrt{n}} (\text{Log } \|S_n y\| - n\,\gamma(\lambda)) + \frac{V_n}{\sqrt{n}} .$$

The result (ii) now is a direct application of A.V.4.3 and of the $\mathbb{P}$ a.e. convergence of $\frac{Vn}{n}$ to zero.

REMARKS

(i) The same result was already obtained by A.J. O'Connor [48] and T. Verheggen [63] under stronger assumptions on the law of (a,b). Actually they suppose that this law has a density with respect to the Lebesgue measure on $\mathbb{R}$.

(ii) Extensions to the strip, using results of chapter III and A.IV. can be obtained in a similar way and this is what is done by R. Jhonston and H. Kunz in [30].

CHAPTER III

THE PURE POINT SPECTRUM

We now prove the strongest singularity property for the spectrum of Schrödinger operators, that is the pure point spectrum property. Until the end of the year 1984 the only known proofs were very close from the original one given by the Russian school (see the introduction). In most of these works it was not clear what was really necessary in order to obtain the pure point spectrum property, since they dealed essentially with the independent or Markov cases under strong assumptions on these processes. A very clarifying idea of Kotani introduced in II.7 and applied to the independent case in the first section gives very easily the expected result, at least in the classical Schrödinger case, that is when $b_n = 1$, $\forall n \in \mathbb{Z}$. It's not clear that such a procedure can be easily applied to the Helmotz case. Furthermore Kotani's theory does not use any approximation of the spectral properties of H by means of "thermodynamic limits" along sequences of boxes going to $\mathbb{Z}$. Thus if we want to obtain some information about the speed of convergence of these spectral measures computed in boxes, we have to use an other way. In this purpose the second section is devoted to the study of the Laplace transform on $SL(2,\mathbb{R})$. The corresponding operators were already introduced in A.V. In section 4, using the spectral properties of these operators,we prove again the pure point spectrum property in the independent case, thank's to the "good" approximations of the spectral measures of H obtained in I. These results are also used in section 5, in order to obtain a representation formula for the density of states.

III.1 The pure point spectrum, first proof.

In this section we suppose that (a_n,b_n) is a random process defined on some complete probability space $(\Omega,\mathcal{A},\mathbb{P})$ but we don't require ergodicity or stationarity. We only assume that for $\mathbb{P}$ almost ω :
(i) The sequence $b_n(\omega)$ is strictly positive and bounded away from zero.
(ii) $\lim\limits_{n\to\pm\infty} \frac{1}{|n|} \operatorname{Log}(1+|a_n(\omega)| + b_n(\omega)) = 0$
This last assumption is used to check the hypothesis (ii) in Osseledec's theorem I.1.1 and to establish that for $0 < r < 1$ the series $\Sigma\, b_n r^n$ is $\mathbb{P}$ a.e convergent.

Carefull readers notice that the proof in II.7.1 has nothing to do with ergodicity if we assume the existence of γ properly. Thus we can prove exactly in the same way as in II.7.1 :

PROPOSITION 1.1 *Let* A *be a Borel subset of* $\mathbb{R}$ *and* m *a positive measure on* A. *We assume that there exist* $\mathbb{P}\times m$ a.e *strictly positive functions* $\gamma_\pm(\omega,\lambda)$ *such that* $\lim\limits_{n\to\pm\infty} \frac{1}{|n|} \operatorname{Log} |S_n| = \gamma_\pm$ $\mathbb{P}\otimes m$ a.e. *Furthermore we suppose that* $\sigma_\pm$ *(defined in* II.7) *are absolutely continuous with respect to* m *on* A.
Then for $\mathbb{P}$ *almost all* ω, $\gamma_\pm(\omega,\lambda)$ *are strictly positive for* $\sigma(\omega)$ *almost all* λ, *the spectrum of* $H(\omega)$ *is pure point on* A *(if any) and the eigenfunctions decay with the exponential rate* $-\gamma_\pm(\omega,\lambda)$.

As remarked by S. Kotani [36] and F. Delyon, Y. Levy, B. Souillard [14], if we suppose that the process (a_n,b_n) is an independent sequence of random variables with values in $\mathbb{R}^2$ (not necessarily identically distributed), then the assumption on $\sigma_\pm$ is easily checked :

COROLLARY 1.2 *Let* (a_n,b_n) *be an independent process such that there exist two functions* $\gamma_{\pm}(\lambda)$, m a.e *strictly positive and that for each* λ, $\lim_{n\to\pm\infty} \frac{1}{|n|} \operatorname{Log}|S_n| = \gamma_{\pm}$ $\mathbb{P}$ a.e. *(where* m *is the Lebesgue measure on* $\mathbb{R}$*)*.

Furthermore we suppose that there exist two consecutive indexes $n = n_o$, n_o+1, *such that the conditional law of* a_n *given* b_n *has a bounded density* ϕ_{b_n} *with* $\mathbb{E}(\|\phi_{b_n}\|) < +\infty$.

Then the conclusions of 1.1 *hold, replacing* σ *by* $\sigma_{n_o} + \sigma_{n_o+1}$.

Proof : In view of proposition 1.1 and exercise I.2.9 we have only to prove that for $n = n_o$ and $n = n_o+1$ $\mathbb{E}[\sigma_n | \mathfrak{a}_{\pm}]$ is absolutely continuous with respect to the Lebesgue measure m. Choosing $N = n_o+2$ and $M = n_o-1$ in the definition of $\mathfrak{a}_{\pm}$ in II.7 these conditional expectations are nothing else than expectations with respect to the random variables (a_n, b_n) for $n = n_o$ and $n = n_o+1$. Looking at the proof of II.6.9 (iv) we know that for $\mathfrak{I}m\,\lambda > 0$ and any n :

$$<R_\lambda e^n,\ e^n> = -\left[({}^{n+1}_{+}m) + ({}^{n-1}_{-}m) + \lambda b_n - a_n\right]^{-1}$$

where ${}^{n+1}_{+}m(\lambda)$ is a function of (a_k, b_k) for $k \geq n+1$,

and ${}^{n-1}_{-}m(\lambda)$ is a function of (a_k, b_k) for $k \leq n-1$,

with strictly positive imaginary parts.

Denoting by $\mathbb{E}_n$ the expectation with respect to the single random variable (a_n, b_n) we obtain :

$$\mathfrak{I}m \int \frac{[\mathbb{E}_n(\sigma_n)](dt)}{t-\lambda} \leq \mathbb{E}[\|\phi_{b_n}\|]$$

Thus $\mathbb{E}_n(\sigma_n)$ has a bounded density with respect to the Lebesgue measure on $\mathbb{R}$ and the result follows.

Note that for each λ the limit of $\frac{1}{|n|} \operatorname{Log}\|S_n\|$ is $\mathbb{P}$ a.e. constant and thus we can choose $\gamma_{\pm}$ as functions of λ only.

<u>COROLLARY 1.3</u> *Let* (a_n, b_n) *be an independent identically distributed random process such that* $\mathbb{E}\operatorname{Log}(1 + |a| + b) < +\infty$ *and that the law of* (a) *given* (b) *has a bounded density* ϕ_b *with* $\mathbb{E}[\|\phi_b\|] < +\infty$. *Then the conclusions of* 1.1 *hold.*

Proof : From II.4.1 we know that $\gamma(\lambda) > 0$ everywhere (the assumption on the law of (a) excludes the Helmotz case).

Thus the ergodic theorem II.3.2 and corollary 1.2 yield the result

(Remark that the hypothesis (ii) of the begining of this section is satisfied since $\mathbb{E}\,\mathrm{Log}\,(1 + |a| + b) < +\infty$).

REMARK 1.4

(i) In the Helmotz case, in example for the operator $(H\Psi)_n = b_n^{-1}(-\Psi_{n-1} - \Psi_{n+1})$, we obtain in the proof of 1.2

$$\langle R_\lambda e^n, e^n\rangle = -\left[({}^{n+1}_{\ +}m) + ({}^{n-1}_{\ -}m-) + \lambda b_n\right]^{-1}$$

and if we integrate with respect to the variable b_n we don't obtain the absolute continuity of $\sigma_\pm$ in general. (See however the remark of S. KOTANI in the appendix).

(ii) In the independent case, it would be very tempting to choose $\mathcal{a}_\pm$ as the tail sigma algebras generated by the process (a_n, b_n). Thus $\sigma_\pm$ would be obtained as expectations rather than as conditional expectations . Unfortunately it's far from evidence that the mesurability property for the sets $W_\pm$ introduced in the proof of II.7.1 can be easily checked....

III.2 The Laplace transform on SL(2,$\mathbb{R}$)

Let G be the unimodular group SL(2,$\mathbb{R}$) , X the one dimensional real projective space, C(X) the Banach space of complex continuous functions on X endowed with the supremum norm. As explained in II.4, G acts continuously on X if we use the usual norm topology for the two by two matrices and the distance δ on X defined by

$$\delta(\bar{x},\bar{y}) = \frac{\|x \wedge y\|}{\|x\|\,\|y\|},$$ since it is readily seen that :

$$\delta(g_1.\bar{x}\ , g_2.\bar{y}) \leq \delta(\bar{x},\bar{y})\ \|g_2\|\ \|g_1\| + \|g_1 - g_2\|\ \|g_2\|$$

Moreover a sequence $\overline{x_n}$ converges to $\bar{x}_o$ in X iff there exists a sequence g_n in G such that g_n converges to the identity in G and $\overline{x_n} = g_n.\bar{x}_o$. In order to check this last property we observe that for $x = \begin{pmatrix} r\cos\theta \\ r\sin\theta \end{pmatrix}$, $r \in \mathbb{R} \setminus \{0\}$, $\theta \in [0,\Pi[$ we can write $\bar{x} = g_\theta.\bar{x}_\infty$ where g_θ is the rotation of angle θ and $x_\infty = \begin{pmatrix} 1 \\ 0 \end{pmatrix}$.

Let now μ be a probability measure on G and t a real number. We define the operator T_t by :

$$T_t f(\bar{x}) = \int \left(\frac{\|x\|}{\|gx\|}\right)^t f(g.\bar{x})\, d\mu(g) \quad , \; f \in C(X).$$

It is readily seen that if $\int \|g\|^{\eta}\, d\mu(g) < +\infty$ for $\eta > 0$, then T_t is a positive bounded operator on C(X) for $t \in [-\eta, +\eta]$ and the Markovian operator T_o is just the transition operator for the left random walk of law μ on X. We denote by T_t^* the dual operator, acting on the space M(X) of complex measures on X by $(T_t^* \nu)(f) = \nu(T_t f)$, $\nu \in M(X)$, $f \in C(X)$. A probability measure ν such that $T_o^* \nu = \nu$ is called a μ invariant probability measure, such measures always exist from the Markov-Kakutani theorem (see A.I.3.5) but are not unique in general.
Most of the sequel is devoted to the study of spectral properties of the operators T_t. These operators have already been introduced in A.V but we have to notice that we use them on C(X) rather than on Hölder spaces. (see remark 2.10.(ii)).

PROPOSITION 2.1 *Let* η *be a strictly positive number such that* $\int \|g\|^{\eta}\, d\mu(g) < +\infty$. *Then the map* $t \to T_t$ *is analytic from* $[-\eta, +\eta]$ *to the space of bounded operator on* C(X). *More precisely if*

$$D_t f(\bar{x}) = \int \left(\text{Log} \frac{\|x\|}{\|gx\|}\right)\left(\frac{\|x\|}{\|gx\|}\right)^t f(g.\bar{x})\, d\mu(g)$$

then $$\left|\left(\frac{T_t^n - T_s^n}{t-s} - D_s^n\right) f(\bar{x})\right| \leq \|f\| \frac{|t-s|}{2\varepsilon}\left(\|T_{t+\varepsilon}^n\| + \|T_{t-\varepsilon}^n\| + \|T_{s+\varepsilon}^n\| + \|T_{s-\varepsilon}^n\|\right)$$

for $f \in C(X)$, $n \geq 1$, $\varepsilon > 0$, $(s,t) \in [-\eta, +\eta]$.

Proof : We first remark that the n th power of the above operators are the operators associated with the n th power of convolution μ^n of μ on $SL(2,\mathbb{R})$ and that

$$\int \|g\|^{\eta}\, d\mu^n(g) \leq \left[\int \|g\|^{\eta}\, d\mu(g)\right]^n$$

Thus it's enough to prove the above formula for n = 1.
As in the classical proof of analyticity of the Laplace transform, the result follows from the Taylor formula (where σ is strictly positive, $r \in]s,t[$) :

$$\sigma^t - \sigma^s = (t-s)(\text{Log}\,\sigma)\sigma^s + \frac{(t-s)^2}{2} (\text{Log}\,\sigma)^2 \sigma^r$$

and the bounds :

$$\sigma^r \leqslant \sigma^s + \sigma^t$$

$$|\mathrm{Log}\,\sigma| \leqslant \frac{1}{\varepsilon}(\sigma^{\varepsilon} + \sigma^{-\varepsilon}) \quad \text{for any } \varepsilon > 0$$

PROPOSITION 2.2 *Assume that :*

(i) there exists a strictly positive number η *such that* $\int ||g||^{\eta}\, d\mu(g) < +\infty$

(ii) μ *has a unique invariant probability measure* ν

(iii) $\gamma = \iint \mathrm{Log} \frac{||gx||}{||x||}\, d\mu(g)\, d\nu(\bar{x}) > 0$

Then there exists a strictly positive number t_o *such that the spectral radius of* T_t *is strictly less than* 1 *for* $t \in]0,t_o]$.

Proof : Define $f_n(\bar{x}) = \frac{1}{n} \int \mathrm{Log} \frac{||x||}{||gx||}\, d\mu^n(g)$.

As in the proof of II.4.8, the unicity of ν implies the uniform convergence, of the sequence f_n to $-\gamma$ on the compact space X. Applying 2.1 to the function $\mathbb{1}(\bar{x}) = 1$, $\forall \bar{x} \in X$, $s = 0$, $t \in]0,\eta/2]$, $\varepsilon = \eta/2$ yields :

$$T_t^n \mathbb{1}(\bar{x}) \leqslant 1 + t\left[n f_n(\bar{x}) + \frac{4t}{\eta} K^n\right]$$

where $K = \int ||g||^{\eta}\, d\mu(g)$

Now we can choose an integer n_o such that $\sup_{\bar{x}\in X} f_{n_o}(\bar{x}) \leqslant -\frac{\gamma}{2} < 0$

and then there exists $t_o > 0$ such that for $t \in]0,t_o]$

$$\sup_{\bar{x}\in X} T_t^{n_o} \mathbb{1}(\bar{x}) = ||T_t^{n_o}|| < 1.$$

REMARK 2.3 The result 2.2 can be obtained from the general theory of analytical perturbations of bounded operators, if we know that the spectrum of T_o except the eigenvalue 1, is contained in a disc of radius strictly less than one. In this case if $r(t)$ denote the spectral radius of T_t it can be proven that $\left.\frac{d}{dt} r(t)\right|_{t=0} = -\gamma$ and the result follows from $r(0) = 1$ and $\gamma > 0$.

But such an assumption on the spectrum on T_o requires strong regularity properties on μ as we shall see in the sequel (Remark 2.7).

In order to investigate the spectral properties of T_o we prove the following , where $T(\mu)$ is the closed subsemigroup of G generated by the support of μ :

PROPOSITION 2.4 : *The eigenvalue* 1 *is the only eigenvalue of modulus* 1 *for* T_o *and the associated eigenspace is one dimensional, in (at least) the two following cases :*

(i) The orbits of $T(\mu)$ *are dense in* X.

(ii) There exists a unique μ *invariant probability measure* ν.

Proof : Let $f \in C(X)$ with $T_o f = \lambda f$, $|\lambda| = 1$, $\|f\| = 1$ and $x_o \in X$ with $|f(\bar{x}_o)| = 1$. The invariance equation $T_o^n f = \lambda^n f$ implies that $f(g.\bar{x}_o) = f(\bar{x}_o)$ for $g \in T(\mu)$. The proof is complete in the case (i). Assume now that there exists a unique μ invariant probability ν of topological support S. As it has been often remarked, for any $\bar{x}$ in X, the sequence $\nu_{n,\bar{x}} = \frac{1}{n}\sum_{k=1}^{n} \mu^{k} * \delta_{\bar{x}}$ converges weakly to ν. Then if $\bar{y}$ is a point of S and U an open set containing $\bar{y}$ we have $\underline{\lim}\, \nu_{n,x_o}(U) \geqslant \nu(U) > 0$. This implies the existence of g in $T(\mu)$ such that $g.\bar{x}_o \in U$ and hence $f(\bar{y}) = f(\bar{x}_o)$ $\forall \bar{y} \in S$.
Moreover for $\bar{x} \in X$ we have :

$$\nu_{n,\bar{x}}(f) = \frac{1}{n}\sum_{k=1}^{n} T_o^{k} f(\bar{x}) = \frac{1}{n}\left(\sum_{k=1}^{n}\lambda^{k}\right) f(\bar{x})$$

and this sequence converges to $\nu(f) = f(\bar{x}_o) \neq 0$. This implies that $\lambda = 1$ and $f(\bar{x}) = f(\bar{x}_o)$. We also remark that the assumption (i) implies (ii).

In the sequel m will be the "Cauchy law" on X, that is the invariant probability measure on X under the action of S O (2) and we have already used in II.4.1 the relation :

$$\frac{dg^{-1}m}{dm}(\bar{x}) = \frac{\|x\|^2}{\|gx\|^2}$$

We denote by $\check{\mu}$ the image measure of μ by $g \longrightarrow g^{-1}$ and $\check{T}_t$ stands for the operator T_t associated to the measure $\check{\mu}$. If V is a subspace of C(X), we denote by Vm the subspace of M(X) defined by $Vm = \{fm \,/\, f \in V\}$ (Remark that $f \longrightarrow fm$ is one to one from C(X) to M(X)).

PROPOSITION 2.5 *Assume that* $\int \|g\|^2 \, d\mu(g) < +\infty$, *then :*

(i) $(\check{T}_t f)m = T^*_{2-t}(fm)$, $t \in [0,2]$, $f \in C(X)$

(ii) $\forall k \in \mathbb{N}$, $\forall \lambda \in \mathbb{C}$, $\forall t \in [0,2]$:

$$[\mathrm{Ker}(\check{T}_t - \lambda I)^k]m \subset \mathrm{Ker}(T^*_{2-t} - \lambda I)^k$$

Proof : Let f,h be in C(X) :

$$\left[T^*_{2-t}(fm)\right](h) = \iint f(\bar{x})\left(\frac{\|x\|}{\|gx\|}\right)^{2-t} h(g.\bar{x})\, d\mu(g)\, dm(\bar{x})$$

$$= \iint f(\bar{x})\, h(g^{-1}.\bar{x})\left(\frac{\|x\|}{\|g^{-1}x\|}\right)^{-t} dg^{-1}m(\bar{x})\, d\check{\mu}(g)$$

$$= \iint f(g.\bar{x})\, h(\bar{x})\left(\frac{\|x\|}{\|gx\|}\right)^{t} dm(\bar{x})\, d\check{\mu}(g)$$

$$= \left[(\check{T}_t f)\, m\right](h)$$

The assertion (ii) is a simple consequence of (i) and we may remark that (i) implies that $T^*_2 m = m$.

PROPOSITION 2.6 *Assume that* μ *has a density with respect to the Haar measure on* G *and that* $\int \|g\|^2\, d\mu(g) < +\infty$. *Then* T_t *is a compact operator on* C(X) *for* $t \in [0,2]$.

Proof : Let dg be an Haar measure on G and $d\mu = \phi\, dg$, then for u in G :

$$T_t f(u.\bar{x}) = \int f(g.\bar{x})\left(\frac{\|ux\|}{\|gx\|}\right)^{t} \phi(g u^{-1})\, dg$$

and :

$$|T_t f(u.\bar{x}) - T_t f(\bar{x})| \leq \|f\| \int \left(\frac{\|x\|}{\|gx\|}\right)^{t} \left|\phi(g) - \phi(gu^{-1})\left(\frac{\|ux\|}{\|x\|}\right)^{t}\right| dg$$

Thus if ϕ is continuous with a compact support the result follows from Ascoli's lemma.

In the general case, there exists a sequence Ψ_n of continuous function with compact support, which converges to $\phi(g)\|g\|^t$ in $L_1(\mu)$. Taking $\phi_n(g) = \|g\|^{-t}\, \Psi_n(g)$ and denoting by ${}^nT_t$ the operator associated to $\mu_n = \phi_n\, dg$, we obtain

$$\|{}^nT_t - T_t\| \leq \int |\phi(g)\, \|g\|^t - \Psi_n(g)|\, dg$$

and T_t is compact as limit in norm of compact operators.

REMARK 2.7 μ is said "spread out" on G, if there exists n such that μ^n is not orthogonal to the Haar measure on G. In such a situation

T_o is a quasi compact operator (See A. Brunel and D. Revuz [6] for more details). Hence T_o has a finite number of eigenvalues of modulus one and the remainder of its spectrum is contained in a disc of radius strictly less than one.

Looking at 2.4 and 2.3 this is enough to use the perturbation theory for T_o (If μ is pread out, then ν is unique from A.II.4.3).

Unfortunately it's not true in general that T_t $(t > 0)$ has to be quasi-compact, when μ is spread out and since we need the spectral properties of T_2 we have to assume the existence of a density for μ.

COROLLARY 2.8 *Assume that* $\int \|g\|^2 \, d\mu(g) < +\infty$ *and that* μ *has a density with respect to the Haar measure on* G. *Then for* $k \in \mathbb{N}$, $\lambda \in \mathbb{C}\setminus\{0\}$ *we have*

(i) $[\operatorname{Ker}(\check{T}_t - \lambda I)^k]\, m = \operatorname{Ker}(T^*_{2-t} - \lambda I)^k$

(ii) $\dim(\operatorname{Ker} \check{T}_t - \lambda I)^k = \dim \operatorname{Ker}(T_{2-t} - \lambda I)^k$

(iii) The unique μ *invariant probability measure* ν *has a continuous density* $\check{\rho}$ *on* X *with respect to* m *and* $\check{T}_2 \check{\rho} = \check{\rho}$

Proof : We apply 2.5 (ii) to the values t and 2-t replacing μ by $\check{\mu}$ for the latter. Looking at 2.6 we have only to use the fact that for a compact operator T, the characteristic spaces have the same (finite) dimension for T and T^* and that the map $f \rightarrow fm$ is one to one. If μ has a density, then $G_\mu = G$ thus ν is unique and (i) applied to $t = 2$ implies that $\nu = \check{\rho}m$ with $\check{T}_2\check{\rho} = \check{\rho}$.

THEOREM 2.9 *Assume that* $\int \|g\|^2 \, d\mu(g) < +\infty$ *and that* μ *has a density with respect to the Haar measure on* G. *Then there exist unique* μ, $(\check{\mu})$, *invariant probability measures* ν, $(\check{\nu})$, *with continuous densities* $\check{\rho}$, (ρ) *with respect to* m, *and operators* Q_o *and* Q_2 *of spectral radius strictly less than one, such that :*

$$T_o^n(f) = \nu(f)\, \mathbb{1} + Q_o^n(f)$$

$$T_2^n(f) = m(f)\, \rho + Q_2^n(f)$$

Moreover the spectral radius $r(t)$ *of* T_t *is such that* $r(o) = r(2) = 1$ *and* $r(t) < 1$ *for* $t \in \,]0,2[$.

Proof : T_o is compact and μ has a unique invariant probability measure

ν thus by 2.4, except the eigenvalue 1, the spectrum of T_o is contained in a disc of radius strictly less than one, and we obtain the first decomposition. This result is also true for $\check{T}_o$ and by 2.8 the spectrum of T_2 has the same structure. We already know that $r(0) = r(2) = 1$ and Hölder's inequality implies that the function $t \longrightarrow \operatorname{Log} r(t)$ is convex on $[0,2]$. From 2.2 this function is strictly negative at a point in $]0,2[$ and hence is strictly negative on $]0,2[$.

REMARK 2.10 (i) Most of the results of this section are true in a much more general setting. Actually it's possible to traduce all of them to the case of the action of a locally compact group G acting continuously on a compact metric space X. If we replace $\operatorname{Log} \frac{\|x\|}{\|gx\|}$ by an additive cocycle σ introduced in A.I.3.2, the Laplace transform is given by

$$T_t f(\bar{x}) = \int \sigma^t(g,\bar{x})\, f(g.\bar{x})\, d\mu(g)$$

See for instance [41] and chapter IV.

(ii) In view of A.V.2.5 the operator T_o acting on $\mathcal{L}_\alpha$ possesses the same decomposition as in theorem 2.9, without any density assumption on μ. By perturbation theory such a decomposition holds for T_t if t is small enough (A.V.4.3). But we need a decomposition of T_2 and this is the reason why we don't use the Hölder space $\mathcal{L}_\alpha$ and work with C(X).

Exercise 2.11 It's not always true that ν has X as support, even when μ has a density. (Think of μ supported by the matrices of $SL(2,\mathbb{R})$ with non negative entries, then ν is supported by the invariant set $X^+ = \{\bar{x} \,/\, x = \binom{u}{v} \,/\, uv \geq 0\})$. Suppose that μ is supported by the matrices of the form $\begin{bmatrix} t & -1 \\ 1 & 0 \end{bmatrix}$ where t is a real random variable such that the topological support of the law of t contains an open set of $]-2, +2[$. We know from the proof of II.4.1 that ν is unique. Prove that the support of ν is equal to X.

Exercise 2.12 Without any assumption on μ prove that T_1 is a bounded operator of $L_2(m)$ and is self adjoint if $\mu = \check{\mu}$. Denoting by γ the limit of the subadditive sequence $\frac{1}{n}\int \operatorname{Log} \|g\|\, d\mu^n(g)$ (assuming that $\int \operatorname{Log} \|g\|\, d\mu(g) < +\infty$) prove that the logarithm of the spectral radius of T_1 acting on $L_2(m)$ is greater than $-\gamma$.

Hint : $\frac{1}{n} \operatorname{Log} \|T_1^n\| \geq \frac{1}{n} \operatorname{Log} \int \frac{\|x\|}{\|gx\|} \, d\mu^n(g) \, dm(\bar{x}) \geq -\frac{1}{n} \int \operatorname{Log} \|g\| \, d\mu^n(g)$

As a consequence if we know that this spectral radius is strictly less than one (this is the case when G_μ is not amenable, see Y. Guivarch and Y. Derrienic [17]) then γ is strictly positive.

Exercise 2.13 Suppose that μ is supported by the matrices $\begin{bmatrix} t & -1 \\ 1 & 0 \end{bmatrix}$ where t is a real random variable distributed according to the Cauchy law $C_{\alpha+i\beta}$ with $\beta > 0$. Compute the eigenfunctions ρ and $\check{\rho}$ of T_2 and $\check{T}_2$ with $m(\rho) = m(\check{\rho}) = 1$.

III.3 The pure point spectrum, second proof.

In order to obtain the pure point spectrum property in the general independent equidistributed case, we apply the results of the preceding section together with the criterion given in I.5.1.
We denote by μ_λ the law of the matrix $Y = \begin{bmatrix} a-b\lambda & -1 \\ 1 & 0 \end{bmatrix}$, $T_{t,\lambda}$ the operator T_t associated to the law μ_λ and by ν_λ a μ_λ invariant probability measure.

PROPOSITION 3.1 *Assume that* $\mathbb{E}(a^2+b^2) < +\infty$. *Let* J *be a compact subset of* $\mathbb{R}$ *and suppose that for* $\lambda \in J$, *the law of* $a - b\lambda$ *has a density* Ψ_λ *with respect to the Lebesgue measure of* $\mathbb{R}$ *such that* $\lambda \longrightarrow \Psi_\lambda$ *is continuous from* J *to* $L_1(\mathbb{R})$. *Then there exists positive constants* A_J, B_J, C_J *with* $A_J < 1$ *and such that :*

(i) $\|T_{1,\lambda}^n\| \leq B_J \, A_J^n$, $n \in \mathbb{N}$, $\lambda \in J$

(ii) $\|T_{2,\lambda}^n\| \leq C_J$, $n \in \mathbb{N}$, $\lambda \in J$

Proof : An easy Jacobian computation shows that μ_λ^3 has a density ϕ_λ with respect to the Haar measure of G and that $\lambda \longrightarrow \phi_\lambda$ is continuous from J to $L_1(dg)$. Then following the lines of 2.6 it's readily seen that $\lambda \longrightarrow T_{t,\lambda}^3$ is continuous from J to the space of bounded operators on C(X) for $t \in [0,2]$. We now observe that it's enough to prove (i) and (ii) for $T_{t,\lambda}^3$ rather for $T_{t,\lambda}$ since

$\|T_{t,\lambda}\| \leq \int \|g\|^t \, d\mu_\lambda = \mathbb{E}([2 + (a-\lambda b)^2]^{t/2})$

Looking at 2.9 we know that the spectral radius of $T^3_{1,\lambda}$ is strictly less than one and that $T^{3n}_{2,\lambda}(f) = m(f)\,\rho_\lambda + Q^n(f)$ where ρ_λ is the continuous density with respect to m of $\check{\nu}_\lambda$, and the spectral radius of Q_λ is strictly less than one.

The continuous perturbation theory of compact operators (see Kato [31]) asserts that the maping $\lambda \longrightarrow Q_\lambda$ is continuous from J to the space of bounded operators on C(X) and $\lambda \longrightarrow \rho_\lambda$ is continuous from J to C(X). The conclusion of 3.1 now follows from the next lemma :

LEMMA 3.2 *Let* $\lambda \to T_\lambda$ *be a continuous mapping from a compact set* J *to the space of bounded operators of a Banach space, such that for each* $\lambda \in J$ *the spectral radius of* T_λ *is strictly less than* 1. *Then there exists* $A_J < 1$ *and* n_o, *such that for* $n \geq n_o$, $\lambda \in J$, $\|T^n_\lambda\| \leq A^n_J$.

Proof : The Logarithm of the spectral radius $r(\lambda)$ of T_λ is an uppersemicontinuous function which attains its maximum value ℓ on the compact set J, thus $\ell < 0$. Furthermore $\ell = \mathrm{Sup}(\ell, \mathrm{Log}\, r(\lambda))$ is the pointwise limit of the decreasing sequence $\mathrm{Sup}(\ell, \inf\limits_{1\leq k\leq n} \frac{1}{k} \mathrm{Log}\|T^k_\lambda\|)$ of continuous functions on J, and the result is a consequence of Dini's lemma.

THEOREM 3.3 *Let* H *be the Schrödinger operator associated to an independent identically distributed process with values in* $\mathbb{R}^2$ *such that the variable* (b) *is strictly positive, bounded away from zero and* $\mathbb{E}(a^2 + b^2) < +\infty$. *Let* I *be an open subset of* $\mathbb{R}$ *with a compact closure* J *such that for* $\lambda \in J$ *the law of* $a - \lambda b$ *has a density* Ψ_λ *with respect to the Lebesgue measure on* $\mathbb{R}$ *and* $\lambda \longrightarrow \Psi_\lambda$ *is continuous from* J *to* $L_1(\mathbb{R})$.

Then for almost all ω *the spectrum of* $H(\omega)$ *is pure point on* I *(if any) and the eigenfunctions fall off exponentially fast.*

Proof : Let m, n be two integers and $\Lambda = [M,N]$ a box with $M < m < n < N$. We denote by Z_3, Z_2, Z_1 the products

$Z_3 = Y_{m-1}\, Y_{m-2} \cdots Y_M$

$Z_2 = Y_{n-1}\, Y_{n-2} \cdots Y_m$

$Z_1 = Y_N\, Y_{N-1} \cdots Y_n$

From the formula I.4.4 we can write :

$$\mathbb{E}(|\hat{\tilde{\sigma}}_{m,n}|(I)) \leqslant \frac{1}{\Pi}\int_I (\mathbb{E}\ \frac{\|Z_3\ x_\infty\|\ \|Z_2\ Z_3\ x_\infty\|}{\|Z_1\ Z_2\ Z_3\ x_\infty\|^2})\ d\lambda$$

where $x_\infty = \begin{pmatrix}1\\0\end{pmatrix}$.

Integrating with respect to Z_1 we obtain that the above expectation is less than :

$$\mathbb{E}\left|\frac{\|Z_3\ x_\infty\|}{\|Z_2\ Z_3\ x_\infty\|}\right| \|T^{N-n+1}_{2,\lambda}\|$$

Integrating with respect to Z_2 this latter expression is less than :

$$\|T^{n-m}_{1,\lambda}\|\ \ \|T^{N-n+1}_{2,\lambda}\|$$

Looking at the proposition 3.1 we obtain

$$\mathbb{E}(|\hat{\tilde{\sigma}}_{m,n}|(I)) \leqslant \frac{\lambda(I)}{\Pi}\ B_J\ C_J\ A_J^{|n-m|}$$

Then for $A_J < D_J < 1$ we have for $\mathbb{P}$ almost all ω and a finite $F_J(\omega)$:

$$|\hat{\tilde{\sigma}}_{m,n}|(I) \leqslant F_J(\omega)\ D_J^{|n-m|}$$

The theorem I.5.1 gives the result since the integrability of b^2 implies that $\Sigma\ b_n^{1/2}\ D_j^{|n|}$ is $\mathbb{P}$ almost everywhere finite.

<u>LEMMA 3.4</u> *With the notations of* 3.3 *the law of* $a - \lambda b$ *has a density* Ψ_λ *with* $\lambda \longrightarrow \Psi_\lambda$ *continuous from* J *to* $L_1(\mathbb{R})$ *in (at least) the two following cases :*

(i) In the classical Schrödinger case (that is $b_n = 1\ \ \forall n \in \mathbb{Z}$*), when the law of the random variable* (a) *has a density on* $\mathbb{R}$ *and* J *is any closed interval.*

(ii) In the Helmotz case when the law of the random variable (b) *has a density on* $\mathbb{R}$ *and* J *is any closed interval not containing* λ^*.

<u>Proof</u> Let Ψ be the density of (a) or (b) then the density Ψ_λ of $a - \lambda b$ is given by $\Psi_\lambda(t) = \Psi(t+\lambda)$ in the case (i) and $\Psi_\lambda(t) = \frac{1}{\lambda-\lambda^*}\ \Psi(\frac{t+c}{\lambda-\lambda^*})$ in the case (ii).

Approximating Ψ in $L_1(\mathbb{R})$ by continuous functions with compact support, the result is straightforward.

<u>REMARKS 3.5</u> (i) In the classical Schrödinger case we have obtained the pure point spectrum property assuming that the law of (a) has a density

and $\mathbb{E}(a^2) < +\infty$. These hypothesis are slightly different from those of corollary 1.3 in which it is assumed that the law of (a) has a bounded density.

(ii) It's easy to see that lemma 3.4 applies to more general situations, in example if we suppose that (a) and (b) are independent random variables each law being absolutely continuous.

(iii) The proof given in 3.2 applies to random Schrödinger operators in a half axis (As well as the results of the section 1).

III.4 The density of states

We assume in this last section that (a_n), $n \in \mathbb{Z}$, is an independent identically distributed sequence of real random variables, that the law of (a) has a density and $\mathbb{E}(a^2) < +\infty$. In this situation we shall prove that the distribution of states introduced in II.6 has a continuous density constructed from the invariant function of the operator $T_{2,\lambda}$ defined in the sections 2 and 3.

PROPOSITION 4.1 *Under the above hypothesis the distribution of states of* $(H\Psi)_n = -\Psi_{n+1} - \Psi_{n-1} + a_n\Psi_n$ *has a density with respect to the Lebesgue measure given by*

$$\frac{dk}{d\lambda} = \frac{1}{2\Pi}\int \rho_\lambda(\bar{x})\,\check{\rho}_\lambda(\bar{x})\,dm(\bar{x})$$

where ρ_λ *and* $\check{\rho}_\lambda$ *are the unique invariant functions of* $T_{2,\lambda}$ *and* $\check{T}_{2,\lambda}$ *with* $m(\rho_\lambda) = m(\check{\rho}_\lambda) = 1$, *and* m *is the Cauchy measure on* X.

Proof : Let ϕ be a continuous function with compact support on $\mathbb{R}$, Λ the box $[-N,N]$. We know from I.4.4 that

$$k(\phi) = \frac{1}{2}\,\mathbb{E}\left[\sigma_o + \sigma_1\right] = \frac{1}{2\Pi}\int \phi(\lambda)\lim_{N\to+\infty}\mathbb{E}\left(\frac{\|Y_oY_{-1}\cdots Y_{-N}\,x_\infty\|^2}{\|Y_NY_{N-1}\cdots Y_{-N}\,x_\infty\|^2}\right)d\lambda$$

where $x_\infty = \begin{pmatrix}1\\0\end{pmatrix}$. The latter expectation is equal to

$\mathbb{E}\left[T^N_{2,\lambda}\,\mathbb{1}\,(Y_oY_{-1}\cdots Y_{-N}\cdot\bar{x}_\infty)\right]$ where $\mathbb{1}$ is the function $\mathbb{1}(\bar{x}) = 1$.
Looking at the decomposition of T_2 given in 2.9 this latter expectation can be written :

$$T_{o,\lambda}^{N+1}\ \rho_\lambda(\bar{x}_\infty)\ +\ \mathbb{E}\left[Q_{2,\lambda}^{N}\ \mathbb{1}\ (Y_o\ Y_{-1}\ \cdots\ Y_{-N}\cdot\bar{x}_\infty)\right]$$

From the decomposition of T_o given in 2.9 we have :

$$T_{o,\lambda}^{N+1}\ \rho_\lambda(\bar{x}_\infty) = \nu_\lambda\ (\rho_\lambda) + Q_{o,\lambda}^{N+1}\ \rho_\lambda(\bar{x}_\infty).$$

Taking in account that the spectral radius of $Q_{o,\lambda}$ and $Q_{2,\lambda}$ are strictly less than one we obtain

$$k(\phi) = \frac{1}{2\Pi}\int \phi(\lambda)\ \nu_\lambda(\rho_\lambda)\ d\lambda$$

thus $\frac{dk}{d\lambda} = \frac{1}{2\Pi}\ \nu_\lambda(\rho_\lambda) = \frac{1}{2\Pi}\int \rho_\lambda(\bar{x})\ \check{\rho}_\lambda(\bar{x})\ dm(\bar{x})$

From lemma 3.4 and the proof of 3.1 we know that $\lambda \longrightarrow \rho_\lambda$ and $\lambda \longrightarrow \check{\rho}_\lambda$ are continuous in the norm topology of $C(X)$ thus $\frac{dk}{d\lambda}$ is continuous.

Exercise 4.2 In [43] we can find an other kind of formula for $\frac{dk}{d\lambda}$ namely if we denote by $\cos^2(\bar{x})$ the continuous function on X defined by $\cos^2(\bar{x}) = \frac{u^2}{u^2+v^2}$, $x = \begin{pmatrix} u \\ v \end{pmatrix}$,

then $\frac{dk}{d\lambda} = \frac{1}{\Pi}\int \cos^2(\bar{x})\ \rho_\lambda(\bar{x})\ \check{\rho}_\lambda(\bar{x})\ dm(\bar{x})$

Hint : Write $k(\phi) = \mathbb{E}\left[\sigma_o(\phi)\right]$ and follow the lines of the proof of 4.1. Moreover using the invariance equation for ν_λ it is readily seen that these two expressions of the density of states are identical.

Exercise 4.3 Using the result of exercise 2.13 compute $\frac{dk}{d\lambda}$ when the law of (a) is the Cauchy law $C_{i\beta}$ and compare to exercise II.6.12.

REMARK 4.4 The same computation of $\mathbb{E}\left[\sigma_o(\phi)\right]$ can be performed in the general case of an independent identically distributed process (a_n, b_n) but in this case we don't obtain the density of states since $k(\phi) = \mathbb{E}\left[b_o\ \sigma_o(\phi)\right]$.

CHAPTER IV

SCHRÖDINGER OPERATORS IN A STRIP

Most of the results obtained for the one dimensional Schrödinger operator in the preceding chapters can be adapted to the case of the strip. Deterministic properties are easily generalized, replacing at time, the Poincare half plane by the generalized Siegel half plane. For random operators, the theory of Lyapunov exponents is more involved and we refer heavily to part A chapter IV for related results. In contrast with the one dimensional case, positivity of some exponents is not easily checked and needs more assumptions on the "disorder". As in chapter III we give two proofs of the pure point spectrum property. The first proof is a straightforward adaptation of Kotani's criterion. In order to treat the Helmotz case and to obtain some information about "thermodynamic limits" we need some Laplace analysis on symplectic groups. Such results, easily obtained on $SL(2,\mathbb{R})$, require here much more work since we have to deal with the Poisson kernels associated to the boundaries of symplectic groups.

IV.1 The deterministic Schrödinger operator in a strip

Let ℓ be a positive integer (width of the strip) and $\mathcal{L}$ be the linear space of sequences $\Psi = (\Psi_n)$, $n \in \mathbb{Z}$ with $\Psi_n \in \mathbb{C}^{\ell}$. The Schrödinger operator H is associated to two given sequences of real symmetric square matrices of order ℓ, A_n and B_n, with B_n invertible, and acts on $\mathscr{L}$ by

$$(H\Psi)_n = B_n^{-1}(-\Psi_{n+1} - \Psi_{n-1} + A_n\Psi_n)$$

We say that H is a classical Schrödinger operator when $B_n = I$ and A_n is a tridiagonal Jacobi matrix whose entries (i,j) with $|i-j| = 1$ are equal to -1. For a complex number λ every solution of the difference equation $H\Psi = \lambda\Psi$ lies in a 2ℓ dimensional subspace of $\mathscr{L}$ and $\Psi_n = P_n\Psi_o + Q_n\Psi_{-1}$ where P_n and Q_n are solutions of the "matrix difference equation" $-M_{n-1} - M_{n+1} + A_n M_n = \lambda B_n M_n$ with the initial values $P_o = Q_{-1} = I$, $P_{-1} = Q_o = 0$. A solution Ψ of the difference equation is constructed from initial values Ψ_o and Ψ_{-1} by a product S_n of "transfer 'matrices" Y_n with :

$$Y_n = \begin{bmatrix} A_n - \lambda B_n & -I \\ I & 0 \end{bmatrix}$$

in the same way as in chapter I. We remark that the transfer matrices Y_n and therefore the products S_n belong to the symplectic group $SP(\ell,\mathbb{C})$ of 2ℓ by 2ℓ matrices g with complex entries, such that $g'Jg = J$ where $J = \begin{bmatrix} 0 & I \\ -I & 0 \end{bmatrix}$ and g' is the transpose of g (Remark that $SP(1,\mathbb{C}) = SL(2,\mathbb{C})$). As explained in A.IV we can consider the exterior powers $\Lambda^p g$ acting on $\Lambda^p(\mathbb{C}^{2\ell})$ and we have the following generalization of Osseledec's theorem [42] :

<u>PROPOSITION 1.1</u> *Let* (g_n), $n \in \mathbb{N}$ *be a sequence of matrices of* $SP(\ell,\mathbb{C})$ *such that*

(i) $$\lim_{n\to+\infty} \frac{1}{n} \operatorname{Log}\|\Lambda^p(g_n g_{n-1} \cdots g_1)\| = \gamma_1 + \cdots + \gamma_p$$

$$1 \leq p \leq \ell$$

(ii) $$\lim_{n\to+\infty} \frac{1}{n} \operatorname{Log}\|g_n\| = 0$$

Let $\lambda_1 > \lambda_2 > \ldots > \lambda_r$ *the strictly decreasing sequence of distinct elements of* $(\gamma_1, \gamma_2, \ldots, \gamma_\ell, -\gamma_\ell, -\gamma_{\ell-1}, \ldots, -\gamma_1)$. *Then there exists a strictly increasing sequence* $\{0\} = V_{r+1} \subset V_r \subset \ldots \subset V_1 = \mathbb{C}^{2\ell}$ *such that :*

(i) If $x \in V_i \setminus V_{i+1}$, $\lim_{n\to+\infty} \frac{1}{n} \operatorname{Log}\|g_n g_{n-1} \cdots g_1 x\| = \lambda_i$, $1 \leq i \leq r$

(ii) $\dim V_{i+1} - \dim V_i$ = *number of elements of the sequence* $(\gamma_1, \gamma_2, \ldots, -\gamma_1)$

which are equal to λ_i).

Actually this theorem tell us that if γ_ℓ is strictly positive, then for each non zero V in $\mathbb{C}^{2\ell}$ the sequence $\|g_n g_{n-1} \cdots g_1 V\|$ has an exponential behavior and is exponentially decreasing iff V belongs to a ℓ dimensional subspace of $\mathbb{C}^{2\ell}$ called the contractive subspace of the sequence g_n.

We now turn our attention to the Hilbertian operator induced by H. It's easily seen, following the lines of I, that H is essentially self adjoint on the Hilbert space $\mathcal{H} = \{\Psi \in \mathcal{L} / \sum \Psi_n' B_n \overline{\Psi_n} < +\infty\}$ assuming that B_n is a sequence of positive definite matrices such that the set of eigenvalues of the sequence B_n is bounded away from zero. This hypothesis will be in force throughout this chapter. We denote by $e^{k,n}$ $k \in [1, \ldots, \ell]$, $n \in \mathbb{Z}$, the orthogonal basis of $\mathcal{H}$ defined by $(e^{k,n})_m = 0$ if $m \neq n$, $(e^{k,n})_n = k$ th column of the matrix B_n^{-1}. We denote by $\sigma_{(i,n),(j,m)}$ the spectral measure associated to the vectors $e^{i,n}$, $e^{j,m}$, and by $\sigma_{(n,m)}$ the $\ell \times \ell$ matrix of measures $\sigma_{(i,n),(j,m)}$, moreover σ is the sum of the traces of the matrices $\sigma_{o,o}$ and $\sigma_{-1,-1}$. We remark that $\langle\Psi, e^{i,n}\rangle$ is the i th entry of Ψ_n and with the help of some algebraic computation we obtain the following formulas where $P_{n,(i,j)}$ is the entry of index (i,j) of the matrix P_n thus a polynomial function of λ :

PROPOSITION 1.2

(i) $\sum_{k=1}^{\ell} (P_{n,(i,k)}(H)\, e^{k,o} + Q_{n,(i,k)}(H)\, e^{k,-1}) = e^{i,n}$

(ii) $\sigma_{m,n} = (P_m, Q_n) \begin{bmatrix} \sigma_{o,o} & \sigma_{-1,o} \\ \sigma_{o,-1} & \sigma_{-1,-1} \end{bmatrix} \begin{pmatrix} P_m' \\ Q_n' \end{pmatrix}$

(iii) For each Ψ *in* $\mathcal{H}$ *the measure* $\sigma_{\Psi,\Psi}$ *is absolutely continuous with respect to* σ.

(iv) The spectrum of H *is the topological support of* σ *and* λ *is an eigenvalue of* H *iff* $\sigma(\lambda) > 0$.

Following the lines of I.3.2 we can prove :

PROPOSITION 1.3 *For σ almost all real λ there exists a generalized eigenfunction Ψ such that for each $\varepsilon > 0$:* $\lim_{|n|\to+\infty} \frac{||\Psi_n||}{|n|^{1/2+\varepsilon}} = 0.$

The spectral matrices $\sigma_{m,n}$ can be constructed from approximations computed in boxes and as in I.4 we introduce varying boundary conditions in order to obtain smooth approximations of $\sigma_{m,n}$.

For a box $\Lambda = \{(k,n)/k \in [1, \ldots, \ell]\ , n \in [M,N]\}$ and a real symmetric matrix X we define the operator $\hat{H}^X$ on the Hilbert space $\hat{\mathcal{H}}$ of sequences $\Psi = (\Psi_n)$, $n \in [M,N]$ with the scalar product inherited from $\mathcal{H}$ by :

$$(\hat{H}^X \Psi)_M = B_M^{-1}(-\Psi_{M+1} + A_M \Psi_M)$$

$$(\hat{H}^X \Psi)_N = (H\Psi)_n \quad \text{for} \quad M+1 < n < N-1$$

$$(\hat{H}^X \Psi)_N = B_N^{-1}(-\Psi_{N-1} + (A_N - X)\Psi_N)$$

Thus we can prove (with the notations introduced in I.4) :

PROPOSITION 1.4

(i) Let X, m,n *be fixed, then the sequence* $\hat{\sigma}^X_{m,n}$ *converges weakly to* $\sigma_{m,n}$ *when the box* Λ *goes to* $\mathbb{Z} \times \{1, \ldots, \ell\}$.

(ii) For $\mathfrak{I}m\lambda \neq 0$

$$<\hat{R}^X_\lambda e^M, e^M> = -\left[\hat{P}_{N+1} - X\hat{P}_N\right]^{-1}\left[\hat{Q}_{N+1} - X\hat{Q}_N\right]$$

where $<\hat{R}^X_\lambda e^M, e^M>$ *is the matrix* $<\hat{R}^X_\lambda e^{i,M} e^{j,M}>$

(iii) If X *and* Y *are real symmetric matrices of order* ℓ *with* Y *non negative then* $(\hat{P}_{N+1} - (X + iY)\hat{P}_N)$ *is invertible for* $\mathfrak{I}m\lambda > 0$ *(Take care that actually* $\hat{P}_N$ *is a function of* λ *!).*

We now want to compute the mean value of $\hat{\sigma}^X_{m,n}$ with respect to a probability measure on the symmetric matrices X analogous to the Cauchy measure in the one dimensional case. For this we recall some facts about the Poisson formula of the upper Siegel half plane (see [28]).

For $k = 0,1, \ldots, \ell$, we set $S_k = \{X + iY / X$ and Y are real symmetric matrices of order ℓ with Y non negative and of rank $k\}$. Then there exists a family $P(Z,dX)$ of probability measures on S_o, indexed by $Z \in \bigcup_{k=1}^{\ell} S_k$ called the "Poisson kernel". If f is an harmonic function (see [28] for the definition) in each S_k, $k = 1, \ldots, \ell$, which can be continuously extended to $S_o \cup \{\infty\}$ we have the representation theorem :

$$f(Z) = \int f(X)\, P(Z,dX) \quad , \; Z \in \bigcup_{k=1}^{\ell} S_k .$$

PROPOSITION 1.5 *The measure valued matrix* $\hat{\tilde{\sigma}}_{m,n} = \int \hat{\sigma}^X_{m,n}\, P(iI,dX)$ *has a density with respect to the Lebesgue measure on* $\mathbb{R}$ *given by*

$$\frac{d\hat{\tilde{\sigma}}_{m,n}}{d\lambda} = \frac{1}{\Pi}\, \hat{P}_m (\hat{\Delta})^{-1}\, \hat{P}'_n$$

where $\hat{\Delta} = \hat{P}'_{N+1}\, \hat{P}_{N+1} + \hat{P}'_N\, \hat{P}_N$

Proof : It's easily seen that $\hat{\sigma}^X_{m,n} = \hat{P}_m\, \hat{\sigma}^X_{M,M}\, \hat{P}'_n$, then it's enough to prove the above result for $m = n = M$. From proposition 1.4 we know that the function $[\hat{P}_{N+1} - Z\hat{P}_N]^{-1}[\hat{Q}_{N+1} - Z\hat{Q}_N]$ is analytic and hence harmonic, on $\bigcup_{k=1}^{\ell} S_k$, and extends continuously to $S_o \cup \{\infty\}$. Thus from the above discussion we obtain ($\Im m\, \lambda > 0$) :

$$\int \frac{d\hat{\tilde{\sigma}}_{MM}(t)}{t-\lambda} = - [\hat{P}_{N+1} - i\hat{P}_N][\hat{Q}_{N+1} - i\hat{Q}_N]$$

Taking in account that the matrix $\begin{bmatrix} \hat{P}_{N+1} & \hat{Q}_{N+1} \\ \hat{P}_N & \hat{Q}_N \end{bmatrix}$ is symplectic, we see that for each real λ :

$\lim_{\varepsilon \to o} \Im m \int \frac{d\hat{\sigma}_{MM}(t)}{t-(\lambda+i\varepsilon)} = (\hat{\Delta})^{-1}$. We remark that the matrix $\hat{\Delta}$ is invertible since :

$$\begin{bmatrix} \hat{P}_{N+1} & \hat{Q}_{N+1} \\ \hat{P}_N & \hat{Q}_N \end{bmatrix} = Y_N\, Y_{N-1} \cdots Y_M \begin{bmatrix} I & 0 \\ 0 & I \end{bmatrix}$$

Following I.5.1 we now establish a sufficient criterion implying the pure point spectrum property for the operator H. Let $S(\lambda)$ be matrix of density of $\begin{bmatrix} \sigma_{o,o} & \sigma_{-1,o} \\ \sigma_{o,-1} & \sigma_{-1,-1} \end{bmatrix}$ with respect to σ. If λ is an eigenvalue of H the terms of $S(\lambda)$ are well defined and let $\Psi^1, \ldots, \Psi^r$ an orthonormal basis of the associated eigenspace. We denote by T_o, (T_{-1}) the $\ell \times r$ matrix with entries $<e^{i,o}, \Psi^j>$, $(<e^{i,-1}, \Psi^j>)$. It is readily seen that

$$S(\lambda) = \frac{1}{\sigma(\lambda)} \begin{pmatrix} T_o \\ T_{-1} \end{pmatrix} (T'_o, T'_{-1})$$

The matrix $\begin{pmatrix} T_o \\ T_{-1} \end{pmatrix}$ is of rank r thus S is also of rank r. Furthermore from the relation $S'_n \; J \; S_n = J$ and if we denote by L_n the $2\ell \times r$ matrix $\begin{bmatrix} \Psi^1_{n+1}, & \Psi^2_{n+1}, & \ldots, & \Psi^r_{n+1} \\ \Psi^1_n, & \Psi^2_n, & \ldots\ldots, & \Psi^r_n \end{bmatrix}$ we obtain $L'_n \; J \; L_n = L'_o \; J \; L_o$. Since the eigenfunctions are going to zero at infinity, $L'_o \; J \; L_o = 0$ and this implies that the rank r of L_o is no greater than ℓ. Finally we obtain that the rank of $S(\lambda)$ is equal to the rank $r \leq \ell$ of the eigenspace associated to λ and that each column of $S(\lambda)$ can be taken as initial values of an eigenvector.

THEOREM 1.6 *Let* I *be an open subset of* $\mathbb{R}$ *for which there exists two positive constants* C *and* ρ *with* $\rho < 1$, *such that for each box* Λ, $(m,n) \in \Lambda$, $i,j \in [1, \ldots, \ell]$ *:*

$$|\hat{\tilde{\sigma}}_{(i,m),(j,n)}| \; (I) \leq C \; \rho^{|m-n|}$$

Then if $\Sigma ||B_n||^{1/2} \rho^{|n|} < +\infty$, *the spectrum of* H *is pure point on* I *and the associated eigenfunctions fall off exponentially at infinity.*

Proof : Following the lines of I.5.1 we see that for σ almost all λ each column of the matrix $S(\lambda)$ can be taken as initial values of an eigenvector. These column vectors can't be simultaneously zero (the trace of $S(\lambda)$ is equal to 1 σ.a.e) Hence σ is pure point and from the preceding discussion we can assert that all the eigenfunctions are exponentially decaying at infinity.

Exercise 1.7 We can define the Wronskian of two sequences of $\ell \times \ell$

matrices U_n and V_n by $W_n(U,V) = U_{n+1}V'_n - U_n V'_{n+1}$.
If (U_n) and (V_n) are two solutions of the matrix difference equation, prove that $W_n(U,V)$ is constant.

(A symplectic matrix g verifies $g' J g = g J g' = J$ and thus the "second" wronskian $W_n = U'_{n+1}V_n - U'_n V_{n+1}$ is also constant).

Exercise 1.8 It's possible to define the resolvent of H by means of m functions, using the same method than in the one dimensional case. We define the matrices ${}^N_\pm M(\lambda) = \langle {}^N_\pm R_\lambda e^N, e^N \rangle$, ${}^N_\pm U_n = \langle {}^N_\pm R_\lambda e^N, e^n \rangle$ where $n \geqslant N$ or $n \leqslant N$. The sequences ${}^N_\pm U_n$ can be extended to the whole of $\mathbb{Z}$ such that these extensions are solutions of the difference equation and we verify that ${}^N_\pm U_n = ({}^N_\pm P_n)({}^N_\pm M) + ({}^N_\pm Q_n)$, where ${}^N P$ and ${}^N Q$ are solutions of the difference equation with boundary conditions

$${}^N_+ P_N = {}^N_+ Q_{N-1} = {}^N_- P_N = {}^N_- Q_{N+1} = I$$

$${}^N_+ P_{N-1} = {}^N_+ Q_N = {}^N_- P_{N+1} = {}^N_- Q_N = 0$$

Let N and n be two integers and define the sequence V_m by

$$V_m = ({}^N_+ U_m)({}^N_- U'_n) \cdot W^{-1}({}^N_- U, {}^N_+ U) \quad , \; m \geqslant n$$

$$V_m = ({}^N_- U_m)({}^N_+ U'_n) \cdot W^{-1}({}^N_- U, {}^N_+ U) \quad , \; m \leqslant n$$

Prove that

$$\langle R_\lambda e^n, e^m \rangle = V_m$$

IV.2 Ergodic Schrödinger operators in a strip

We now suppose that the sequence (A_n, B_n) is defined on some ergodic dynamical system $(\Omega, \mathfrak{a}, \theta, \mathbb{P})$ such that $(A_{n+1}, B_{n+1}) = (A_n \circ \theta, B_n \circ \theta)$. Furthermore we assume that the law of the variable (B) is supported by a set of non negative symmetric matrices whose eigenvalues are uniformly bounded away from zero and that $\mathbb{E}\,\mathrm{Log}(1 + \|A\| + \|B\|) < +\infty$.

The Lyapunov exponents $\gamma_1(\lambda), \ldots, \gamma_\ell(\lambda)$ are defined for a complex number λ as limits of the subadditive sequences :

$$\gamma_1(\lambda) + \ldots + \gamma_p(\lambda) = \lim_{n\to\pm\infty} \frac{1}{|n|} \mathbb{E} \operatorname{Log} \|\Lambda^p S_n\|$$

for $1 \leq p \leq \ell$.

(The limits for $n = \pm\infty$ are equal since $\|S_n\| = \|S_n^{-1}\|$)

The subadditive ergodic theorem implies $\mathbb{P}$ almost everywhere convergence of $\frac{1}{|n|} \operatorname{Log}\|\Lambda^p S_n\|$ to $\gamma_1(\lambda) + \ldots \gamma_p(\lambda)$ (for a fixed λ). Moreover the integrability condition $\mathbb{E}\operatorname{Log}(1 + \|A\| + \|B\|)$ implies the $\mathbb{P}$ almost everywhere convergence of $\frac{1}{|n|} \operatorname{Log}\|Y_n\|$ to zero. Thus for each fixed λ, Osseledec's theorem 1.1 applies $\mathbb{P}$ a.e..Looking at the proof of II.3.1 we see that the sums $\gamma_1(\lambda) + \ldots + \gamma_p(\lambda)$ are subharmonic, but except for γ_1 we can't claim that the Lyapunov exponents are subharmonic.

The measurability property II.2.1 still holds and we obtain the Pastur's lemma about eigenvalues of H looking at the proof of II.2.2. The only change is that now, the trace of the spectral projector E_λ on the eigenspace associated to the eigenvalue λ is no greater than ℓ. Then the Ishii's, Pastur's theorem becomes :

THEOREM 2.1 *Let* A *be a Borel subset of* $\mathbb{R}$ *such that the Lyapunov exponent* $\gamma_\ell(\lambda)$ *is strictly positive almost everywhere on* A *with respect to the Lebesgue measure. Then* $\sigma^a(A) = 0$, $\mathbb{P}$. a.e.

Proof : We define the measurable subset W of $\Omega \times \mathbb{R}$ by

$$W = \left\{ (\omega,\lambda) \;\middle/\; \begin{array}{l} \omega \in \Omega\ ,\ \lambda \in A,\ \lim_{|n|\to\infty} \frac{1}{|n|} \operatorname{Log}\|Y_n(\omega,\lambda)\| = 0 \\ \lambda \text{ is not an eigenvalue of } H(\omega) \\ \lim_{|n|\to\infty} \frac{1}{|n|} \operatorname{Log}\|\Lambda^p S_n(\omega,\lambda)\| = \gamma_1(\lambda) + \ldots + \gamma_p(\lambda) \\ \text{for } p = 1, \ldots, \ell \end{array} \right\}$$

The proof is now the same than in II.5.1 remarking that $\gamma_\ell > 0$ implies that the two contractive subspaces V^+ and V^- are ℓ dimensional. Since λ is not an eigenvalue $V^+ \cap V^- = \{0\}$. Thus $\mathbb{R}^{2\ell} = V^+ \oplus V^-$ and each non zero solution of $H\Psi = \lambda\Psi$ is growing with an exponential rate at least in one direction of $\mathbb{Z}$.

REMARK 2.2 No converse of 2.1 is known for the strip. In the one dimensional case Kotani's proof uses heavily the Thouless formula. Such a formula has been proved in the strip by W. Craig, B. Simon [12] in the following form :

$$\gamma_1(\lambda) + \ldots + \gamma_\ell(\lambda) = \frac{1}{\ell} \int \mathrm{Log}|x-\lambda| \; dk(x)$$

They assume $B_n = I$, $\forall n \in \mathbb{Z}$, and that A is a bounded matrix. But it seems that no formula of this kind is available for the individual exponent $\gamma_\ell(\lambda)$ since it is believed that this exponent is not subharmonic in general. A weaker form of Kotani's converse would be : "Let A be a Borel subset of positive Lebesgue measure contained in $\{\lambda/\gamma_1(\lambda) = 0\}$ then $\sigma^a(A) > 0$, $\mathbb{P}$ a.e.". Such a result certainly can be proven using the above Thouless formula, the m functions introduced in exercise 1.8 and following the lines of [58].

Exercise 2.3 Ishii's proof of 2.1 (see exercise II.5.6) extends to the positive strip :

(i) Prove that for $\Im m\,\lambda > 0$

$$< {}^+R_\lambda\, e^o, e^o> = \lim_{n\to+\infty} - P_{n+1}^{-1} Q_{n+1} = \sum_{n=0}^{+\infty} (P_n' P_{n+1})^{-1}$$

(ii) Prove the expected result, following the lines of II.5.6

Exercise 2.4 Let Y be a real symmetric matrix of order ℓ and ${}^+H^Y$ the Schrödinger operator with boundary condition Y at the point zero. Prove that for $\Im m\,\lambda \neq 0$ and a box $\Lambda = [0,N]$:

(i) $< {}^\wedge R_\lambda^Y\, e^o, e^o > = - [P_{N+1} + Q_{N+1} Y]^{-1} Q_{N+1}$

(ii) $< {}^+R_\lambda^Y\, e^o, e^o> = M\,[I - YM]^{-1}$

(where $M = {}^o_+M$ introduced in exercise 1.8)

(iii) Denoting by ${}^+\sigma^Y$ the trace of ${}^+\sigma^Y_{o,o}$ prove that ${}^+\tilde{\sigma} = \int {}^+\sigma^Y\, P(iI,dY)$ has a bounded density with respect to the Lebesgue measure on $\mathbb{R}$.

(Hint : $M = U + iV$ with U and V symmetric, V positive definite).

(iv) Following the lines of exercise II.5.7 prove the generalization of Kotani's result :

"Let A be a Borel subset of $\mathbb{R}$ such that $\gamma_\ell(\lambda)$ is strictly positive almost everywhere on A. Then for $\mathbb{P}$ almost all ω, for almost all Y (with respect to the natural Lebesgue measure $\prod_{i\geq j} d\,y_{i,j}$ on the symmetric matrices) the spectrum of ${}^+H^Y$ is pure point on A (if any) and the eigenfunctions are exponentially decaying".

Hint : $P(iI, dY)$ is absolutely continuous with respect to $\prod_{i\geq j} dy_{i,j}$ (see [28]).

We now adapt Kotani's criterion II.7.1 to the strip :

PROPOSITION 2.5 *Let* A *be a Borel subset of* $\mathbb{R}$ *and* m *a positive measure on* A. *We assume that* $\gamma_\ell(\lambda) > 0$ m.a.e *on* A *and that* $\sigma_\pm$ *are absolutely continuous on* A *with respect to* m, $\mathbb{P}$ a.e. *Then for* $\mathbb{P}$ *almost all* ω, $\gamma_\ell(\lambda)$ *is strictly positive for* $\sigma(\omega)$ *almost all* λ, *the spectrum of* $H(\omega)$ *is pure point on* A *(if any) and the eigenfunctions fall off exponentially with a rate given by a Lyapunov exponent.*

Following the lines of II.7.1 we obtain that for $\mathbb{P}$ almost all ω, for $\sigma(\omega)$ almost all λ the contractive ℓ dimensional subspaces V^+ and V^- have a common non zero vector. But we don't know in which Osseledec's subspace of V^+ and V^- lies this common vector, neither if this common vector is unique. Furthermore given such a common vector the rate of decay of the associated eigenfunction has no reason to be the same in the two directions of $\mathbb{Z}$! From a physical point of view it seem's natural to think that this common vector has to be unique and that it does not belong to proper Osseledec's subspaces of V^+ and V^-. Thus the rate of decay would be given by $-\gamma_\ell(\lambda)$ in the two directions of $\mathbb{Z}$, but we don't know any proof of this ....

IV.3 Lyapunov exponents in the independent case. The pure point spectrum (first proof).

As seen in section 2, positivity of γ_ℓ is a crucial hypothesis to obtain absence of absolutely continuous spectrum or pure point spectrum. In contrast with the one dimensional case the only known proofs of this positivity require strong assumptions on the law of (A,B). In this section we suppose that (A_n, B_n) is an i.i.d. sequence of random variables. In order to avoid too complicated proofs, we also assume that (A) is a Jacobi matrix, (B) a diagonal matrix such that the ℓ random entries (a^i, b^i) of (A,B) are independent random variables with values in $\mathbb{R}^2$. We say that the "Helmotz case" occurs when, for some index i, there exists a constant c^i and a a value λ_i^* such that $a^i - \lambda_i^* b^i = c^i$.

Moreover we suppose that each (b^i) is positive, bounded away from zero and that $\mathbb{E} \operatorname{Log}(1 + \|A\| + \|B\|) < +\infty$. We denote by μ_λ the law of the matrix $\begin{bmatrix} A-\lambda B & -I \\ I & 0 \end{bmatrix}$.

PROPOSITION 3.1 *Assume that no* (a^i, b^i) *is supported by a single point of* $\mathbb{R}^2$. *Then* $\gamma_1(\lambda) > 0$ *(Assuming* $\lambda \neq \lambda_i^*$ *if the Helmotz case occurs).*

Proof : Following II.4.1 we see that the subgroup G_λ of $SP(\ell, \mathbb{R})$ generated by the support of μ_λ contains matrices of the form $u = \begin{bmatrix} I & \Delta \\ 0 & I \end{bmatrix}$ $v = \begin{bmatrix} I & 0 \\ \Delta & I \end{bmatrix}$ where Δ can be chosen diagonal with no nul diagonal entries.

We denote by E_1, (E_2) the projective subspace spanned by the ℓ first (last) basis vectors of $\mathbb{R}^{2\ell}$. Then for a vector $x = \begin{pmatrix} x_1 \\ x_2 \end{pmatrix}$ of $\mathbb{R}^{2\ell}$ we obtain $u^n x = \begin{pmatrix} x_1 + n\Delta x_2 \\ x_2 \end{pmatrix}$, $v^n x = \begin{pmatrix} x_1 \\ x_2 + n\Delta x_1 \end{pmatrix}$.

If $\bar{x} \in E_1$ then $u^n \cdot \bar{x} = \bar{x}$,

If $\bar{x} \notin E_1$ then $u^n \cdot \bar{x}$ converges to $\overline{\begin{pmatrix} \Delta x_2 \\ 0 \end{pmatrix}}$ which belongs to E_1. This implies that each fixed probability measure on $P(\mathbb{R}^{2\ell})$ under the action of G_λ is supported by E_1. The same is true for E_2 and as a consequence G_λ leaves no fixed probability measure on $P(\mathbb{R}^{2\ell})$ thus the result follows from A.III.7.5.

We have thus obtained under broad conditions positivity of γ_1. As seen in section 2 we actually need positivity of γ_ℓ and this is certainly the case if all the Lyapunov exponents are distinct. Unfortunately we have to work with contractive properties on the exterior powers of $\mathbb{R}^{2\ell}$ of the semi group generated by the support of μ_λ, as explained in A.IV. Such a property can hardly be directly checked for Schrödinger matrices and we are led to prescribe stronger assumptions the law of (A,B) in order to make use of a more tractable criterion given in A.IV.

Let F be the application of $\mathbb{R}^\ell$ in $G = SP(\ell, \mathbb{R})$ defined by

$$F(x) = \begin{bmatrix} A(x) & -I \\ I & 0 \end{bmatrix}$$ where A(x) is the Jacobi matrix whose diagonal elements are given by the coordinates of the vector x. In the one

dimensional case ($\ell = 1$) a straightforward computation proves that the application $\Phi(x_1,x_2,x_3) = F(x_3)\,F(x_2)\,F(x_1)$ of $\mathbb{R}^3$ in G is regular at each point of $\mathbb{R}^3$ with $x_2 \neq 0$ ($d\Phi$ is then of rank 3, which is the dimension of the Lie algebra $\mathcal{G}$ of G).

When $\ell \geq 2$, such a computation becomes tremendous thus we have to work in a more intrinsic way. This is what is done in [40] and we only give here a sketch of the proofs, for readers familiar with Lie algebras.

LEMMA 3.2 *Let* U *be an open set in* $\mathbb{R}^\ell$ *, then the closed subgroup* G_U *generated by* F(U) *is equal to* G.

Proof : As in the proof of 3.1 we see that G_U contains matrices of the form $u = \begin{bmatrix} I & \Delta \\ 0 & I \end{bmatrix}$ and $v = \begin{bmatrix} I & 0 \\ \Delta & I \end{bmatrix}$ where Δ is a diagonal matrix whose diagonal runs through an open set of $\mathbb{R}^\ell$. We can write $u = \exp \begin{bmatrix} 0 & \Delta \\ 0 & 0 \end{bmatrix}$, $v = \exp \begin{bmatrix} 0 & 0 \\ \Delta & 0 \end{bmatrix}$ and thus by varying Δ we can construct a family of elements of the Lie algebra $\mathcal{G}_U$ of G_U. For $\ell = 1$ this family generates $\mathcal{G}$ as a Lie algebra and the proof is complete. For $\ell > 1$ we have to verify that conjugates of these elements with respect to some matrices of F(U) generate $\mathcal{G}$.

LEMMA 3.3 *Let* U *be an open set in* $\mathbb{R}^\ell$, $(X_1, \ldots, X_r)$ *some non zero elements of* $\mathcal{G}$ *Then there exists* $(a_1, \ldots, a_{n_o})$ *in* U *such that the rank of the family (where* $g_k = F(a_k)$*)* :

$$\{\mathrm{Ad}(g_1^{-1})\,X_k\ ,\ \mathrm{Ad}(g_1^{-1}g_2^{-1})\,X_k,\ \ldots,\ \mathrm{Ad}(g_1^{-1}g_2^{-1},\ \ldots,\ g_{n_o}^{-1})\,X_k \ /\ k = 1, \ldots, r\}$$

is equal to the dimension of $\mathcal{G}$.

Proof : Consider the set of subspaces of $\mathcal{G}$ spanned by the above families for all the finite subsets $(a_1, \ldots, a_n)$ of U. It is easily seen that a subspace of maximum dimension in this set is invariant under $\mathrm{Ad}\,g^{-1}$ for g in F(U) and hence is an ideal of $\mathcal{G}$ from 3.2. $\mathcal{G}$ is a simple Lie algebra thus this maximal subspace must be equal to $\mathcal{G}$.

LEMMA 3.4 *For each open set* U *in* $\mathbb{R}^\ell$ *there exists an integer* n_o *and a set* $(a_1, \ldots, a_{n_o})$ *in* U *such that the map* $\Phi(x_1,x_2, \ldots, x_{n_o}) = F(x_{n_o}) \ldots F(x_2)\,F(x_1)$ *is a submersion near* $(a_1, \ldots, a_{n_o})$.

Proof : Choose as $(X_1, \ldots, X_\ell)$ matrices of the form $\begin{bmatrix} 0 & \Delta_i \\ 0 & 0 \end{bmatrix}$ where

Δ_i has zero entries except the term (i,i) which is equal to one. Let $(a_1, \ldots, a_{n_o})$ be the subset of U given by lemma 3.3. For $t \in \mathbb{R}^\ell$ and $T = \sum_{i=1}^{\ell} t_i X_i$ we obtain $F(a+t) = (\exp T)\, F(a)$ and thus :

$$\Phi(a_1 + t_1, \ldots, a_{n_o} + t_{n_o}) = (\exp T_{n_o})\; F(a_{n_o}) \ldots \exp(T_1)\, F(a_1)$$

The result follows from 3.3 and the computation of the differential of Φ at the point $(a_1, \ldots, a_{n_o})$.

REMARK 3.5 It's certainly possible to choose explicitely $(a_1, \ldots, a_{n_o})$ and to see that $n_o = 2\ell + 1$, but the computation is tremendous ! (The lower bound $n_o \geqslant 2\ell+1$ is obvious since $\dim \mathcal{G} = \ell(2\ell+1)$.

We now formulate the expected results of this algebraic manipulation :

THEOREM 3.6 *(i) Suppose that for* i = 1, ..., ℓ *the supports of the laws of* $(a_i - \lambda b_i)$ *contain an open set of* $\mathbb{R}$, *then there exists an integer* n_o *such that the support of* $\mu_\lambda^{n_o}$ *contains an open set of* $SP(\ell, \mathbb{R})$.

(ii) Suppose that for i = 1, ..., ℓ *the laws of* $(a_i - \lambda b_i)$ *have a density on* $\mathbb{R}$, *then there exists an integer* n_o *such that* $\mu_\lambda^{n_o}$ *has a density with respect to be Haar measure of* $SP(\ell, \mathbb{R})$.

Proof : Let U_i be an open set contained in the support of the law of $(a_i - \lambda b_i)$ and $U = \prod_{i=1}^{\ell} U_i$ then by 3.3, $\Phi(U^{n_o})$ contains an open set. Furthermore Φ is an analytical mapping and thus its Jacobian function is an analytical function which is not always zero. This implies that the zeroes of this function are contained in a countable union of lower dimensional subspaces of $\mathbb{R}^{\ell n_o}$, hence negligible with respect to the Lebesgue measure.

COROLLARY 3.7 *Suppose that for* i = 1, ..., ℓ *the supports of the law of* $(a_i - \lambda b_i)$ *contain an open set of* $\mathbb{R}$ *or that the law of* $(a_i - \lambda b_i)$ *have a density with respect to the Lebesgue of* $\mathbb{R}$, *then* $\gamma_\ell(\lambda) > 0$.

Proof : direct application of 3.6 and A.IV.3.5.

Before proving the main theorem of this section we give a useful deterministic lemma. If X is a symmetric real matrix of order n then each spectral measure is a function of the $\frac{n(n+1)}{2}$ entries of X. We denote by σ_k the spectral measure associated to the vector e_k of the orthonormal basis $(e_1, \ldots, e_n)$ and x_k the diagonal term of order (k,k) of X.

LEMMA 3.8 *With the above notations*

$$\int \sigma_k(dt)\, dx_k = \Pi\, dt$$

Proof : It's possible to suppose k = 1, then X can be written $X = \begin{bmatrix} x_1 & u' \\ u & \tilde{X} \end{bmatrix}$, where u is a (n-1) dimensional vector. For $\Im m\, \lambda > 0$ we set $\Psi = (X - \lambda I)^{-1} e_1$ and from the decomposition $\Psi = \begin{pmatrix} \Psi_1 \\ \phi \end{pmatrix}$ we obtain :

$$x_1 \Psi_1 + u'\phi = \lambda \Psi_1 + 1$$

$$\Psi_1 u + \tilde{X}\phi = \lambda\phi$$

Let $h = (\tilde{X} - \lambda I)^{-1} u$, then

$\phi = -\Psi_1 h$ and $\Psi_1 = (x_1 - u'h - \lambda)^{-1}$

Taking in account that $u'h = \int \frac{\tilde{\sigma}_u(dt)}{t-\lambda}$ we see that $\Im m(u'h) > 0$, hence:

$$\Im m \iint \frac{\sigma_1(dt)dx_1}{t-\lambda} = \Im m \int (x_1 - u'h - \lambda)^{-1}\, dx_1 = \Pi$$

THEOREM 3.9 *Suppose that each random variable* (a_i), $i = 1, \ldots, \ell$, *possesses a bounded density with respect to the Lebesgue measure. Then if* H *is the classical independent Schrödinger operator, the Lyapunov exponent* $\gamma_\ell(\lambda)$ *is strictly positive* $\forall\lambda \in \mathbb{R}$. *Moreover for* $\mathbb{P}$ *almost all* ω *the spectrum of* $H(\omega)$ *is pure point and the eigenfunctions fall off exponentially with a rate given by a Lyapunov exponent.*

Proof : Looking at 3.7 and 2.5 we have only to prove that $\sigma_\pm$ are absolutely continuous with respect to a fixed measure.
We choose $\mathfrak{a}_+$ generated by $(A_n)_{n\geq 1}$ and $\mathfrak{a}_-$ generated by $(A_n)_{n\leq -2}$. In a box $\Lambda = [-N,N]$, denoting by $\mathbb{E}_{(i,n)}$ the expectation with respect to the random variable $a_{i,n}$ we obtain from 3.8 :

$$\frac{d}{d\lambda} \mathbb{E}_{(i,n)} (\sigma_{(i,n)(i,n)}) \leq \Pi \, \|\phi_i\| \quad , n \in (0,-1),\ i \in (1, \ldots, \ell).$$

where ϕ_i is the density of the law of (a_i). Then letting Λ going to $\mathbb{Z}$ we obtain that $\mathbb{E}_{i,n}(\sigma_{(i,n)(i,n)})$ has a bounded density and thus the same is true for $\sigma_{\pm}$.

REMARK 3.10

(i) It's also possible to obtain absolute continuity of $\sigma_{\pm}$ using the m functions of exercise 1.8 and following the lines of III.1.2.

(ii) The proof given here excludes the case of random B_n. In this case the operator H restricted to a box has off-diagonal disorder and lemma 3.8 can't be used.

(iii) It has been announced that all the Lyapunov exponents are distinct under the hypothesis of proposition 3.1 [23]. This result seems natural but we don't know any proof ....

IV.4 The Laplace transform on $SP(\ell,\mathbb{R})$

Most of the results established in III.2 extend to the case of a locally compact metric group G acting on a metric compact G-space B (see A.I.3). In the sequel, a cocycle on (G,B) means a multiplicative cocycle (that is the exponential of an additive cocycle). We suppose that there exists a continuous map s from B to G and a fixed b_o in B such that $s(b) \cdot b_o = b$ for all b in B, thus a sequence b_n converges to b in B iff there exists a sequence g_n in G such that g_n converges to the identity of G and $b_n = g_n \cdot b$. Let μ be a probability measure on G, ρ a multiplicative cocycle, we define an operator on C(B) the Banach space of complex continuous functions on B by :

$$T_\rho f(b) = \int \rho(g,b)\ f(g\cdot b)\ d\mu(g)\ , \quad f \in C(B).$$

We say that ρ is μ integrable if $\sup_{b\in B} \rho(g,b) \in L_1(\mu)$. In this case T_ρ is a bounded operator on C(B).

The next propositions are identical to III.2.2, III.2.6 and the proofs are the same.

PROPOSITION 4.1 *Assume that*

(i) There exists a strictly positive t *such that* ρ^t *is* μ *integrable.*

(ii) μ *has a unique invariant probability measure* ν *on* B.

(iii) $\gamma = -\iint \text{Log}\,\rho(g,b)\,d\mu(g)\,d\nu(b) > 0.$

Then there exists a strictly positive number t_o *such that the spectral radius of the operator* T *associated to* ρ^t *is strictly less than* 1 *for* $t \in]0,t_o]$.

PROPOSITION 4.2 *Assume that*

(i) ρ *is* μ *integrable.*

(ii) μ *has a density with respect to an Haar measure on* G. *Then* T_ρ *is a compact operator on* C(B).

Suppose that there exists a probability measure m on B such that $g^{-1}m$ has the same null sets as m and that $r(g,b) = \frac{dg^{-1}m}{dm}(b)$ is a continuous function of (g,b). Then r is a multiplicative cocycle called a Radon Nikodym cocycle (R.N. cocycle) and we can prove exactly in the same way as in III.2.5 and III.2.8. :

PROPOSITION 4.3 *Assume that* ρ *is* $\check{\mu}$ *integrable and* r *is a* R.N. *cocycle such that* $r\rho^{-1}$ *is* μ *integrable. If* μ *has a density on* G *with respect to an Haar measure, then for* $k \in \mathbb{N}$ *and* $\lambda \in \mathbb{C}\setminus\{0\}$:

(i) $(\text{Ker}(\check{T}_\rho - \lambda I)^k)m = \text{Ker}(T^*_{r\rho^{-1}} - \lambda I)^k$

(ii) $\dim(\text{Ker}(\check{T}_\rho - \lambda I)^k) = \dim(\text{Ker}(T_{r\rho^{-1}} - \lambda I)^k)$

COROLLARY 4.4 *Together with the hypothesis of* 4.3 *assume that the sequence* $\|\check{T}^n_\rho\|$ *is bounded. Then the sequence* $\|T^n_{r\rho^{-1}}\|$ *has the same property.*

Proof : It is readily seen from the Jordan form of a finite dimensional matrix that for a compact operator T the sequence $\|T^n\|$ is bounded iff the spectral radius of T is no greater than one and for each eigenvalue λ of modulus 1 one have $\text{Ker}(T-\lambda I) = \text{Ker}(T-\lambda I)^2$. Thus we have only to use (ii) of 4.3.

We now apply these results to the case of the action of $G = SP(\ell, \mathbb{R})$ on some boundaries B. For $k \in (1, \ldots, \ell)$ a non zero k-vector $x = u_1 \wedge u_2 \wedge \ldots \wedge u_k$ of $\Lambda^k(\mathbb{R}^{2\ell})$ is said Lagrangian if $u_i' J u_j = 0$ where $i,j \in (1, \ldots, k)$ and J is the element $\begin{pmatrix} 0 & I \\ -I & 0 \end{pmatrix}$ of G. If $x \in \Lambda^k(\mathbb{R}^{2\ell})$ we denote by $\bar{x}$ its image in the projective space $P(\Lambda^k(\mathbb{R}^{2\ell}))$ and the Lagrangian manifold L_k is the compact part of the latter projective space associated to Lagrangian k-vectors. We say that $\bar{x} \in L_k$ is contained in $\bar{y} \in L_{k+1}$ if there exists $z \in \mathbb{R}^{2\ell}$ such that $\bar{x} = \overline{y \wedge z}$. As seen in A.IV, G acts on L_k by $g \cdot \overline{u_1 \wedge u_2 \wedge \ldots \wedge u_k} = \overline{g u_1 \wedge g u_2 \ldots \wedge g u_k}$ and if $(e_1, \ldots, e_{2\ell})$ is the usual basis of $\mathbb{R}^{2\ell}$, $L_k = G \cdot \overline{e_1 \wedge \ldots \wedge e_k}$.

We only consider two boundaries of G, namely $X_1 = L_{\ell-1}$ and $X_{1,2}$ the partial "flag-manifold" defined by :

$$X_{1,2} = \{(\overline{x_1}, \overline{x_2}) \ / \ \overline{x_1} \in L_{\ell-1} \ , \ \overline{x_2} \in L_\ell \ , \ \overline{x_1} \subset \overline{x_2}\}$$

We define on (G, L_k) the multiplicative cocycle $\rho_k(g, \bar{x}) = \dfrac{\|x\|}{\|\Lambda^k g x\|}$ where the norm is taken in $\Lambda^k(\mathbb{R}^{2\ell})$. Actually if we take $x = u_1 \wedge \ldots \wedge u_k$ and if we denote by M the $2\ell \times k$ matrix associated to the coordinates of $u_1, \ldots, u_k$ in the usual basis of $\mathbb{R}^{2\ell}$ then $\|x\|^2 = \det(M'M)$.

There exist on X_1 and $X_{1,2}$ two probability measures m_1 and $m_{1,2}$ with the invariance property $k m_1 = m_1$, $k m_{1,2} = m_{1,2}$ for k in the subgroup $K = SO(2\ell) \cap G$ (K is a maximal compact subgroup of G). Furthermore the associated Radon Nikodym cocycles can be computed, using a method explained by H. Furstenberg and Tzkoni in [22]. We give below a sketch of the proof (see exercise 4.7, 4.8 or [41] for a detailed computation).

PROPOSITION 4.5

$$r_1(g, \bar{x}_1) = \frac{dg^{-1} m_1}{dm_1} (\bar{x}_1) = (\rho_{\ell-1}(g, \bar{x}_1))^{\ell+2}$$

$$r_{1,2}(g, (\bar{x}_1, \bar{x}_2)) = \frac{dg^{-1} m_{1,2}}{dm_{1,2}} (\bar{x}_1, \bar{x}_2) = (\rho_{\ell-1}(g, \bar{x}_1))^{\ell} (\rho_\ell(g, \bar{x}_2))^2$$

Proof : The cocycles r_1 and $r_{1,2}$ are K invariant (that is $r_1(k, \bar{x}_1) = r_{1,2}(k, (\bar{x}_1 \bar{x}_2)) = 1 \ \forall k \in K$) and thus they must be (see [22]) of the form $r_1 = (\rho_{\ell-1})^\alpha$, $r_2 = (\rho_{\ell-1})^\beta (\rho_\ell)^\gamma$ where α, β, γ are three real numbers we ought to compute. In this purpose we choose a point b_o in B ($B = X_1$ or $B = X_{1,2}$) and g_o in G with $g_o \cdot b_o = b_o$. Then if ℓ is a positive measure on B with the same null sets as m, we can write :

$$\frac{dg^{-1}m}{dm} = \frac{dg^{-1}m}{dg^{-1}\ell} \cdot \frac{dg^{-1}\ell}{d\ell} \cdot \frac{d\ell}{dm}$$

Constructing a parametrical representation of B and choosing ℓ as the image of the Lebesgue measure on the parameters, the computation of $\frac{dg_o^{-1}\ell}{d\ell}(b_o)$ reduces to a Jacobian computation. This is enough to obtain α, and we have to use two pairs (g_o, b_o) in order to obtain γ and β.

The interest of this computation lies in the fact that in our case the important cocycle is $\theta = \rho_{\ell-1}(\rho_\ell)^{-1}$ and from 4.5 we obtain $\theta^2 = r_{1,2}(r_1)^{-1}$. The following theorem summarizes the essential results needed in the following section. We denote by T_t the operator associated to θ^t on $C(X_{1,2})$ and by $r(t)$ the spectral radius of T_t.

THEOREM 4.6 *Assume that* μ *has a density with respect to the Haar measure on* $SP(\ell,\mathbb{R})$ *and that* $\int \|g\|^{\ell(\ell+1)} d\mu(g) < +\infty$. *Then* $r(0) = r(2) = 1$, $r(1) < 1$, *and the sequence* $\|T_2^n\|$ *is bounded. Furthermore assume that there exists a map* $\lambda \to \mu_\lambda$ *such that* $\lambda \to T_{t,\lambda}$ *is continuous from* $\mathbb{R}$ *to the space of bounded operators on* $C(X_{1,2})$ *for* $t \in [0,2]$. *Then for a compact subset* J *of* $\mathbb{R}$ *there exist positive constants* A_J, B_J, C_J *with* $A_J < 1$ *and*

(i) $\|T^n_{1,\lambda}\| \leq B_J A_J^n \quad , n \in \mathbb{N} \ , \ \lambda \in J$

(ii) $\|T^n_{2,\lambda}\| \leq C_J \quad , n \in \mathbb{N} \ , \ \lambda \in J$

Moreover (i) and (ii) are true if we only suppose that there exists an integer n_o *such that* $\mu_\lambda^{n_o}$ *has a density with respect to the Haar measure on* $SP(\ell,\mathbb{R})$.

Proof : From corollary 4.4 we know that the sequence $\|\check{T}^n_{r_1}\|$ is bounded. We may consider that $\check{T}_{r_1}$ is an operator on $C(X_{1,2})$ and it is readily seen that the norm of $\check{T}^n_{r_1}$ is the same on $C(X_1)$ and $C(X_{1,2})$. Thus applying again corollary 4.4 we obtain that the sequence $\|T_2^n\|$ is bounded (taking in account that $\theta^2 = r_{1,2}(r_1)^{-1}$). From 4.3 (ii), the spectral

radius of T_2 is equal to the spectral radius of $\check{T}_{r_1}$ acting on $C(X_{1,2})$ and hence to the spectral radius of $\check{T}_{r_1}$ on $C(X_1)$ thank's to the above remark. Applying again 4.3 (ii) this spectral radius is equal to 1 and thus r(o) = r(2) = 1. We know from A.IV. 4.2. that there exists a unique μ invariant probability measure $\nu_{1,2}$ on $X_{1,2}$ and taking in account that $\theta = \rho_{\ell-1}(\rho_\ell)^{-1}$ we obtain $\gamma_\ell = -\iint \text{Log}\,\theta \, d\nu_{1,2}\, d\mu$. Looking at A.IV we see that $\gamma_\ell > 0$.
Proposition 4.1 implies that r(t) is strictly negative for $t \in]0,2[$ since we know that r(0) = r(2) = 1, that Log r(t) is a convex function on $[0,2]$ and that this function is strictly negative at some point of $]0,2[$.

If J is a compact set, conclusion (i) follows from lemma III.3.2. To obtain (ii) we write $T^n_{2,\lambda} = P^n_\lambda + Q^n_\lambda$, where P_λ is the projection operator on the sum of the eigenspaces associated to the eigenvalues of $T_{2,\lambda}$ of modulus one and Q_λ is an operator of spectral radius strictly less than one. From continuous perturbation theory (see Kato [31]) we know that $\lambda \to P_\lambda$ and $\lambda \to Q_\lambda$ are continuous. Looking at the proof of corollary 4.4 we obtain that P_λ is a diagonalizable matrix, thus the sequence $\|P^n_\lambda\|$ is uniformly bounded on a compact set. Result (ii) now follows from a new application of lemma III.3.2 to the operator Q_λ. The last assertion of the theorem is obvious.

<u>Exercise 4.7</u> To a Lagrangian element $x = u_1 \wedge u_2 \wedge \ldots \wedge u_{\ell-1}$ we associate the $2\ell \times \ell-1$ matrix of coordinates of $(u_1, \ldots, u_{\ell-1})$ in the usual basis of $\mathbb{R}^{2\ell}$ and this matrix has the form $\begin{bmatrix} a' \\ c \\ b' \\ d \end{bmatrix}$ where a and b are in $\mathbb{R}^{\ell-1}$, c and d are square matrices of order $\ell-1$.

(i) Prove that the mapping Π defined by :

$\Pi(\bar{x}) = (a'd^{-1}, b'd^{-1}, cd^{-1} - d'^{-1}ab'd^{-1})$

is one to one from the elements of X_1 with d invertible, onto the product $\mathbb{R}^{\ell-1} \times \mathbb{R}^{\ell-1} \times \mathbb{R}^{\frac{\ell(\ell-1)}{2}}$ (remark that the last coordinate of Π is a symmetric matrix of order $\ell-1$).

(ii) Let $b_o = \overline{e_{\ell+2} \wedge \ldots \wedge e_{2\ell}}$ and $g_o = \begin{pmatrix} \Delta & 0 \\ 0 & \Delta^{-1} \end{pmatrix}$ where Δ is a diagonal matrix, diag $\Delta = (1, \ldots, 1, \alpha)$ with $\alpha > 0$. Prove that $g_o \cdot b_o = b_o$

(iii) Π is well defined in a neighborhood U of b_o and for $\bar{x} \in U$ we set

$\Pi(\bar{x}) = (r',s',N)$. Prove that $\Pi(g_o \cdot \bar{x}) = (r'\delta,\ s'\delta,\ \delta N\delta)$ where δ is the diagonal matrix of order $\ell-1$ defined by $\operatorname{diag}\delta = (1,\ 1,\ \ldots,\ 1,\ \alpha)$.

(iv) Using the proof of 4.5, prove that $r_1 = (\rho_{\ell-1})^{\ell+2}$.

Exercise 4.8 We associate to a pair of Lagrangian elements $x_1 = u_1 \wedge \ldots \wedge u_{\ell-1}$, $x_2 = x_1 \wedge u_\ell$ the matrix of coordinates of $(u_1, \ldots, u_\ell)$ in the usual basis of $\mathbb{R}^{2\ell}$, of the form $\begin{pmatrix} a \\ b \end{pmatrix}$ where a and b are square matrices of order ℓ. When b is invertible we set $t = (b^{-1})_{1,1}$

(i) Prove that the mapping $\tilde{\Pi}$ defined by : $\tilde{\Pi}(\bar{x}_1,\bar{x}_2) = (\Pi(\bar{x}_1),t)$ where Π is defined in 4.7 is one to one from the elements of $X_{1,2}$ with b invertible onto the product $\mathbb{R}^{\ell-1} \times \mathbb{R}^{\ell-1} \times \mathbb{R}^{\frac{\ell(\ell-1)}{2}} \times \mathbb{R}$.

(ii) Let $b_o = (\overline{e_{\ell+1} \wedge \ldots \wedge e_{2\ell}},\ \overline{e_\ell \wedge \ldots \wedge e_{2\ell}})$ and $g_o = \begin{pmatrix} \Delta & 0 \\ 0 & \Delta^{-1} \end{pmatrix}$ where $\Delta = \operatorname{diag}(\beta,\ 1,\ \ldots,\ 1,\ \alpha)$ with α and $\beta > 0$. Prove that $g_o \cdot b_o = b_o$.

(iii) $\tilde{\Pi}$ is well defined in a neighborhood U of b_o, and for $(\bar{x}_1,\bar{x}_2) \in U$ we set $\tilde{\Pi}(\bar{x}_1,\bar{x}_2) = (r',s',N,t)$. Prove that $\tilde{\Pi}(g_o \cdot \bar{x}_1,\ g_o \cdot \bar{x}_2) = (\beta r'\delta,\ \frac{1}{\beta} s'\delta,\ \delta N\delta,\ \beta^2 t)$ where δ is the diagonal matrix of order $\ell-1$ with $\delta = \operatorname{diag}(1,\ \ldots,\ 1,\alpha)$

(iv) Prove, using 4.5, that $r_{1,2} = (\rho_{\ell-1})^{\ell}(\rho_\ell)^2$

IV.5 The pure point spectrum, second proof

If we want to apply the results of the preceding section together with criterion 1.6, we have to remark that the entries of the matrix of density of ${}^\wedge\tilde{\sigma}$ can be constructed from the cocycle θ.

For a box $\Lambda = [M,N]$, $n \in \Lambda$, we denote by ${}^\wedge R_n$ the matrix of the form $\begin{bmatrix} {}^\wedge\Delta_n^{-1/2} & 0 \\ 0 & {}^\wedge\Delta_n^{1/2} \end{bmatrix}$ where ${}^\wedge\Delta_n = {}^\wedge P'_{n+1}\ {}^\wedge P_{n+1} + {}^\wedge P'_n {}^\wedge P_n$. If m,n are such that $M \leq m \leq n \leq N$ we denote by Z_3, Z_2, Z_1 the products $Z_3 = Y_m Y_{m-2}, \ldots, Y_M$ $Z_2 = Y_n Y_{n-1} \ldots Y_{m+1}$, $Z_1 = Y_N Y_{N-1} \ldots Y_{n+1}$. Furthermore if $e_1, \ldots, e_{2\ell}$ is the usual basis of $\mathbb{R}^{2\ell}$ we set

$$u_i = e_1 \wedge \ldots \wedge e_{i-1} \wedge e_{i+1} \wedge \ldots \wedge e_\ell$$

$u = e_1 \wedge e_2 \wedge \ldots \wedge e_\ell$

PROPOSITION 5.1 *With the above notation we obtain*

$$\sum_{i,j} \left[(\wedge P_m (\wedge\Delta)^{-1} \wedge P'_n)_{i,j} \right] \leq \ell \sum_{i,j} \theta(Z_1, Z_2 Z_3 \wedge R_n \cdot (\bar{u}_i, \bar{u}))$$

$$\theta(Z_1, Z_2 Z_3 \wedge R_m \cdot (\bar{u}_j, \bar{u}))\, \theta(Z_2, Z_3 \wedge R_m \cdot (\bar{u}_j, \bar{u}))]$$

Proof : If A, B, C are square matrices of order ℓ with $C = AB$ then $\sum_{i,j} |C_{ij}| \leq \ell (\mathrm{tr}\, A'A \cdot \mathrm{tr}\, B'B)^{1/2}$.
Choosing $A = \wedge P_m (\wedge\Delta)^{-1/2}$, $B = (\wedge\Delta)^{-1/2} \wedge P'_n$ we obtain that the left member of the inequality in 5.1 is less than

$$\ell (\mathrm{tr}\, \wedge\Delta_n (\wedge\Delta)^{-1}\ \mathrm{tr}\, \wedge\Delta_m (\wedge\Delta)^{-1})^{1/2}$$

We now remark that

$$\mathrm{tr}(\wedge\Delta_n (\wedge\Delta)^{-1}) = \sum_{i=1}^{\ell} \theta^2 (Z_1 Z_2 Z_3 \wedge R_n , (\bar{u}_i, \bar{u}))$$

The same is true for the index m and taking in account the cocycle property for θ^2 and $\theta(Z_2 Z_3 \wedge R_n, (\bar{u}_i, \bar{u})) = \theta(Z_3 \wedge R_m, (\bar{u}_j, \bar{u})) = 1$ we only have to use the inequality $(\sum_i a_i^2)^{1/2} (\sum_j b_j^2)^{1/2} \leq \sum_{i,j} |a_i|\, |b_j|$ to obtain the announced result.

COROLLARY 5.2 *Let* $T_{t,\lambda}$ *be the operator introduced in theorem* 4.6 *and assume that the sequence* (A_n, B_n) *is an independent process. Then for a box* $\Lambda = [M,N]$, $M \leq m \leq n \leq N$ *and a Borel subset* I *of* $\mathbb{R}$ *we have :*

$$\mathbb{E} \sum_{i,j} |\wedge \tilde{\sigma}_{(i,m),(j,n)}|(I) \leq \frac{\ell^3}{\pi} \int_I \| T_{1,\lambda}^{n-m} \|\, \| T_{2,\lambda}^{N-n} \| \, d\lambda$$

Proof : First integrating with respect to Z_1 in 5.1, proposition 1.5 and the Schwarz inequality imply that the above expectation is less than

$$\frac{\ell^2}{\pi} \int_I \| T_{2,\lambda}^{N-n} \| \sum_j \theta(Z_2, Z_3 \wedge R_m \cdot (\bar{u}_j, \bar{u}))\, d\lambda$$

A new integration with respect to Z_2 gives the result.

If the independent random process (A_n, B_n) has the structure described at the beginning of the section 3, it's now clear that the following theorem holds :

THEOREM 5.3 *Let* H *be the Schrödinger operator associated to an independent identically distributed process* (A_n, B_n) *with the structure described in* V.3. *Let* I *be an open subset of* $\mathbb{R}$ *with a compact closure* J *such that for* $\lambda \in J$ *the law of* $a_i - \lambda b_i$ $(i = 1, \ldots, \ell)$ *has a density* $\Psi_{i,\lambda}$ *with respect to the Lebesgue measure on* $\mathbb{R}$ *and* $\lambda \to \Psi_{i,\lambda}$ *is continuous from* J *to* $L_1(\mathbb{R})$. *If we assume that for* $i = 1, \ldots, \ell$. $\mathbb{E}(|a_i|^{\ell(\ell+1)} + b_i^{\ell(\ell+1)}) < +\infty$, *then for almost all* ω *the spectrum of* $H(\omega)$ *is pure point on* I *(if any) and the eigenfunctions fall off exponentially fast.*

Proof : We only have to collect the results of 4.6, 3.6(ii) and 5.2. As in the one dimensional case the hypothesis about $\Psi_{i,\lambda}$ is easily checked in general.

APPENDIX

The spectral theory of random Schrödinger operators still gives rise to a lot of papers and some questions left without solution in the preceding chapters have received partial answers during the redaction of this book.

1 The one dimensional case

F. Ledrappier has given a simpler proof of proposition II.3.3 avoiding utilization of the Thouless formula. This result now appears as a direct application of a general fact in ergodic theory (F. Ledrappier "Positivity of the exponent for stationnary sequences of matrices" preprint).

S. Kotani has pointed out that in corollary III.1.2, it is possible to handle the Helmotz case assuming now that there exists two consecutive indexes $n = n_o, n_o+1$ such that :

(i) b_n is bounded

(ii) for each $\lambda \in A$, the law of $a_n - \lambda b_n$ has a bounded density $\phi_{n,\lambda}$ such that $\sup_{\lambda \in A} \|\phi_{n,\lambda}\| < +\infty$.

The proof is the same as in III.1.2 and Corollary III.1.3 is obtained under the same hypothesis. (S. Kotani, private communication).

In [61], B. Simon and M. Taylor prove that for the classical independant Schrödinger operator, for any $\alpha > 0$ there exists a Bernouilli distribution of the potential such that the distribution function $K(\lambda)$ of the distribution of states is not Hölder of order α. This shows that the result (ii) of I.6.9 cannot be improved in general.

2. The strip

In their paper "Anderson localization for one and quasi one dimensional systems" (preprint) F. Delyon, Y. Levy, B. Souillard precise the theorem IV.3.9, proving that the eigenspaces of $H(\omega)$ are $\mathbb{P}$ almost surely one-dimensional. (See also the discussion following proposition IV.2.5).

3. The multidimensional case

At the beginning of 1985 the first proofs of the Anderson's conjecture (see the introduction) appear almost simultaneously in three preprints :

- J. Frölich, F. Martinelli, E. Scopolla and T. Spencer "Anderson localization for large disorder or low energy"
- F. Delyon, Y. Levy, B. Souillard "Anderson localization for multidimensional systems at large disorder or low energy"
- B. Simon, M. Taylor, T. Wolff "Some rigorous results for the Anderson model"

In these three papers, the localization property is proved in any dimension for the classical independent Schrödinger operator, assuming the disorder "large enough" and with a bounded density. The Kotani's criterion and Frölich and Spencer estimates (replacing the positivity of Lyapunov exponents in this context) are the essential tools.

REFERENCES

[1] ANDERSON P.W. (1958). Absence of diffusion in certain random lattices. Physical Review 109, 1492-1505.

[2] ATKINSON F.V. (1964). Discrete and continuous boundary problems. Mathematics in science and engineering Vol 8, Academic Press, New-York.

[3] AVRON J., SIMON B. (1983). Almost periodic Schrödinger operators II. The integrated density of states. Duke. Math. J. 50, 369-391.

[4] BEREZANSKII Ju. M. (1968). Expansions in eigenfunctions of self adjoint operators. Transl. of Math. Mono. 17, A.M.S. Providence.

[5] BORLAND R.E. (1963). Proc. R. Soc. London A 274, 529.

[6] BRUNEL A., REVUZ D. (1974). Quelques applications probabilistes de la quasi-compacité. Ann. Inst. Henri Poincaré, Vol X, 3, 301-337.

[7] CARMONA R. (1982). Exponential localization in one dimensional disordered systems. Duke Math. J. 49, 191-213.

[8] CARMONA R. (1983). One dimensional Schrödinger operators with random or deterministic potentials : new spectral types. J. Functionnal Anal 51, 229-258.

[9] CARMONA R. (1985). Ecole d'été de probabilités de Saint-Flour. To appear in Springer Lecture Notes Series.

[10] CASHER A., LEBOWITZ J.L. (1971). J. Math. Phys. N.Y. 12 1701.

[11] CRAIG W., SIMON B. (1983). Subharmonicity of the Lyapunov index. Duke Math. J. 50, 551-559.

[12] CRAIG W., SIMON B. (1983). Log Hölder continuity of the integrated density of states for stochastic Jacobi matrices. Comm. Math. Phys. 90, 207-218.

[13] DELYON F., KUNZ H., SOUILLARD B. (1983). One dimensional wave equations in random media. J. Phys. A16, 25-42.

[14] DELYON F., LEVY Y., SOUILLARD B. (1985). Anderson localization for multi-dimensional systems at large disorder or large energy. To appear in Comm. Math. Phys.

[15] DELYON F., SOUILLARD B. (1983). The rotation number for finite difference operators and its properties. Comm. Math. Phys. 89, 415-427.

[16] DERRIDA B., HILHORST H.J., (1983). Singular behaviour of certain infinite products of random 2×2 matrices. J. Phys. A. Math. Gen. 16, 2641-2654.

[17] DERRIENNIC Y., GUIVARCH Y. (1974). Théorème de renouvellement pour les groupes non moyennables. Com. Rend. Acad. Sc. PARIS 277, 613-615.

[18] FIGOTIN A.L., PASTUR L.A. (1984). An exactly solvable model of a multidimensional incommensurate structure. Comm. Math. Phys. 95, 401-425.

[19] FURSTENBERG H. (1963). Non commuting random products. Trans. Amer. Soc. 108, 377-428.

[20] FURSTENBERG H., KESTEN H. (1960). Products of random matrices. Ann. Math. Stat. 31, 457-469.

[21] FURSTENBERG H., KIFER Y. (1983). Random matrix products and measures on projective spaces. Israël J. of Math. 46, 12-32.

[22] FURSTENBERG H., TZKONI I. (1971). Spherical functions and integral geometry. Israël J. of Math 10, 327-338.

[23] GOLSHEID I. Ja. (1980). The structure of the spectrum of the Schrödinger random difference operator. Soviet Math Dokl 22, 670-675.

[24] GOLSHEID I. Ja. MOLCANOV S.A., PASTUR L.A. (1977). A pure point spectrum of the one dimensional Schrödinger operator. Funct. Anal. Appl. 11, 1-10.

[25] HENNION H. (1984). Loi des grands nombres et perturbations pour des produits réductibles de matrices aléatoires indépendantes. Wahrs. Verw. Gebiete 67, 265-278.

[26] HERMAN M. (1981). Ecole Polytechnique, preprint.

[27] HERMAN M. (1983). Une méthode pour minorer les exposants de Lyapounov et quelques exemples montrant le caractère local d'un théorème d'Arnold et de Moser sur le tore de dimension 2. Comment. Math. Helvetici 58 , 453-502.

[28] HUA L.K. (1963). Harmonic analysis of several complex variables in the classical domains. A.M.S. Providence.

[29] ISHII K. (1973). Localization of eigenstates and transport phenomena in the one dimensional disordered system. Suppl. Prog. Theor. Phys. 53, 77-138.

[30] JHONSTON R., KUNZ H. (1983). The conductance of a disordered wire. J.Phys. C : Solid State Phys. 16, 3895-3912.

[31] KATO T. (1966). Perturbation theory for linear operators. Springer Verlag, New-York.

[32] KATZNELSON Y. (1976). An introduction to harmonic analysis. New-York : Dover.

[33] KIRSH W., KOTANI S., SIMON B. (1985). Absence of absolutely continuous spectrum for some one dimensional random but deterministic Schrödinger operator. To appear in Ann. Inst. Henri Poincaré A.

[34] KIRSH W., MARTINELLI F. (1982). On the density of states of Schrödinger operators with a random potential. J. Phys. A 15, 2139-2156.

[35] KOTANI S. (1983). Lyapunov indices determine absolute continuous spectra of stationary one dimensional Schrödinger operators. Proc. Kyoto Stoch. Conf.

[36] KOTANI S. (1984). Lyapunov exponents and spectra for one dimensional random Schrödinger operators. Proceedings of the A.M.S. meeting on "Random matrices". Brunswick.

[37] KUNZ H., SOUILLARD B. (1980). Sur le spectre des opérateurs aux différences finies aléatoires. Comm. Math. Phys. 78, 201-246.

[38] LACROIX J. (1982). Problèmes probabilistes liés à l'étude des opérateurs aux différences aléatoires. Ann. Inst. Elie Cartan 7, Marches aléatoires et processus stochastiques sur les groupes de Lie, 80-95.

[39] LACROIX J. (1983). Singularité du spectre de l'opérateur de Schrödinger aléatoire dans un ruban ou un demi-ruban. Ann. Inst. H. Poincaré A 38, 385-399.

[40] LACROIX J. (1984). Localisation pour l'opérateur de Schrödinger aléatoire dans un ruban. Ann. Inst. Henri Poincaré A 40, 97-116.

[41] LACROIX J. (1984). The Schrödinger operator in a strip. Probability measures on groups VII, Proceedings Oberwolfach, Springer lecture notes series 1064, 280-297.

[42] LEDRAPPIER F. (1984). Ecole d'été de probabilités de Saint-Flour XII. Springer lecture notes séries 1097.

[43] LE PAGE E. (1984). Répartition d'état d'un opérateur de Schrödinger aléatoire. Probability measures on groups VII, Proceedings Oberwolfach, Springer lectures notes series 1064, 309-367.

[44] MATSUDA, ISHII K. (1970). Localization of normal modes and energy transport in the disordered harmonic chain. Supp. Prog. Théor. Phys. 45, 56-89.

[45] MOLCANOV S.A. (1978). The structure of eigenfunctions of one dimensional unordered structures. Math. U.S.S.R Izvestija 12, 69-101.

[46] MOTT N. TWOSE W.D. (1961). The theory of impurity conduction. Adv. in Physics 10, 107-155.

[47] NERI U. (1971). Singular Integrals VII. Springer Lectures Notes in mathematics series 200.

[48] O'CONNOR (1975). Disordered harmonic chain. Comm. Math. Phys. 45, 63-77.

[49] OSSELEDEC V. (1968). A multiplicative ergodic theorem. Lyapunov characteristic numbers for dynamical systems". Trans. Moscow Math. Soc. 19, 197-231.

[50] PASTUR L.A. (1980). Spectral properties of disordered systems in the one body approximation. Comm. Math. Phys. 75, 167-196.

[51] PICHARD J.L. (1984). Contribution à une théorie quantique des phénomènes de transport ... . Thesis, Université de Paris-Sud Orsay.

[52] ROYER G. (1982). Etude des opérateurs de Schrödinger à potentiel aléatoire en dimension un. Bull. Soc. Math. France 110, 27-48.

[53] RUELLE D. (1969). A remark on bound states in potential scattering theory. Nuovo Cimento A 61, 655.

[54] RUELLE D. (1978). Analyticity properties of the characteristic exponents of random matrix products. Adv. Math. 32, 68.

[55] SARNAK P. (1982). Spectral behavior of quasi periodic potentials. Comm. Math. Phys. 84, 377-401.

[56] SCHNEIDER W. (1984). Rigorous scaling laws for Dyson measures. To appear in Bibos symposium on stochastic processes, Springer lecture notes series.

[57] SIMON B. (1982). Almost periodic operators : a review. Adv. Appl. Math 3, 463-490.

[58] SIMON B. (1983). Kotani theory for one dimensional stochastic Jacobi matrices. Comm. Math. Phys. 89 227.

[59] SIMON B. (1985). Almost periodic Schrödinger operators IV. The Maryland model. To appear in Ann. Phys.

[60] SIMON B. (1983). Equality of the density of states in a wide class of tight - binding Lorentzian random models. Phys. Rev. B 27, 3859-3860.

[61] SIMON B. TAYLOR M. (1985). Harmonic analysis on SL(2,ℝ) and smoothness of the density of states in the one dimensional Anderson model. To appear in Comm. Math. Phys.

[62] THOULESS D. (1972). A relation between the density of states and range of localization for one dimensional random systems. J. Phys. C 5, 77-81.

[63] VERHEGGEN T. (1979). Transmission coefficient and heat conduction

of a harmonic chain with random masses ... Comm. Math. Phys. 68, 69-82.

[64] YOSHIOKA Y. (1973). On the singularity of the spectral measures of a semi-infinite random system. Proc. Japan Acad. 49, 665-668.

Already Published in Progress in Probability and Statistics

PPS 1 **Seminar on Stochastic Processes, 1981**
E. Çinlar, K. L. Chung, R. K. Getoor, editors
ISBN 3-7643-3072-4, 248 pages, hardcover

PPS 2 **Percolation Theory for Mathematicians**
Harry Kesten
ISBN 3-7643-3107-0, 432 pages, hardcover

PPS 3 **Branching Processes**
S. Asmussen, H. Hering
ISBN 3-7643-3122-4, 472 pages, hardcover

PPS 4 **Introduction to Stochastic Integration**
K. L. Chung, R. J. Williams
ISBN 0-8176-3117-8
ISBN 3-7643-3117-8, 204 pages, hardcover

PPS 5 **Seminar on Stochastic Processes, 1982**
E. Çinlar, K. L. Chung, R. K. Getoor, editors
ISBN 0-8176-3131-3
ISBN 3-7643-3131-3, 302 pages, hardcover

PPS 6 **Least Absolute Deviation**
Peter Bloomfield, W. L. Steiger
ISBN 0-8176-3157-7
ISBN 3-7643-3157-7

PPS 7 **Seminar on Stochastic Processes, 1983**
E. Çinlar, K. L. Chung, R. K. Getoor, editors
ISBN 0-8176-3293-X
ISBN 3-7643-3293-X, 290 pages, hardcover

GPSR Compliance

The European Union's (EU) General Product Safety Regulation (GPSR) is a set of rules that requires consumer products to be safe and our obligations to ensure this.

If you have any concerns about our products, you can contact us on ProductSafety@springernature.com

In case Publisher is established outside the EU, the EU authorized representative is:

Springer Nature Customer Service Center GmbH
Europaplatz 3
69115 Heidelberg, Germany

Batch number: 10577896

Printed by Printforce, the Netherlands